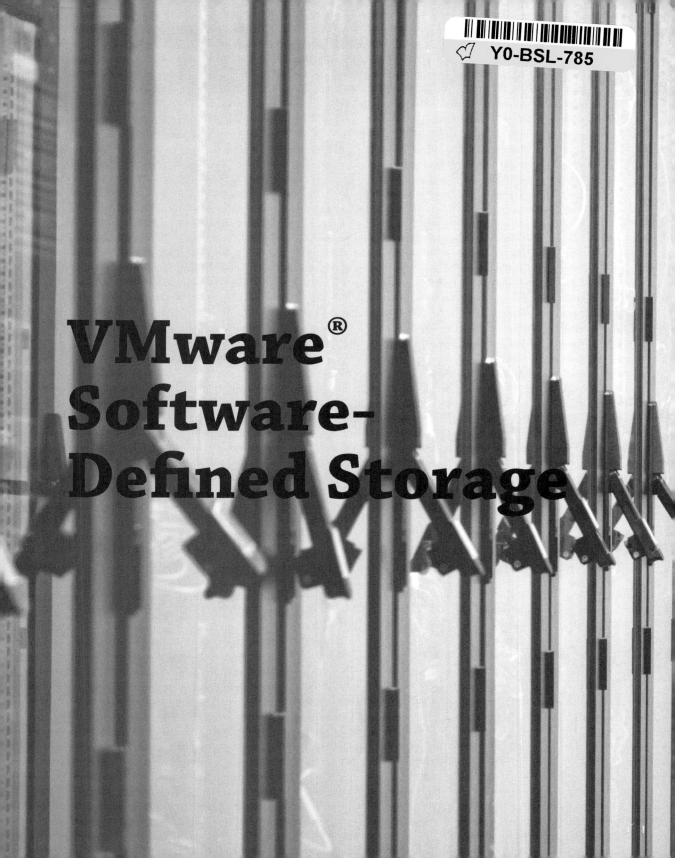

VMware® Software-Defined Storage

VMware® Software-Defined Storage

A Design Guide to the Policy-Driven, Software-Defined Storage Era

Martin Hosken, VCDX

A Wiley Brand

Executive Editor: Jody Lefevere
Development Editor: David Clark
Technical Editor: Ray Heffer
Production Editor: Barath Kumar Rajasekaran
Copy Editor: Sharon Wilkey
Editorial Manager: Mary Beth Wakefield
Production Manager: Kathleen Wisor
Proofreader: Nancy Bell
Indexer: Nancy Guenther
Project Coordinator, Cover: Brent Savage
Cover Designer: Wiley
Cover Image: ©Mikhail hoboton Popov/Shutterstock

Copyright © 2016 by John Wiley & Sons, Inc., Indianapolis, Indiana

Published simultaneously in Canada

ISBN: 978-1-119-29277-7
ISBN: 978-1-119-29279-1 (ebk.)
ISBN: 978-1-119-29278-4 (ebk.)

Manufactured in the United States of America

No part of this publication may be reproduced, stored in a retrieval system, or transmitted in any form or by any means, electronic, mechanical, photocopying, recording, scanning, or otherwise, except as permitted under Sections 107 or 108 of the 1976 United States Copyright Act, without either the prior written permission of the Publisher, or authorization through payment of the appropriate per-copy fee to the Copyright Clearance Center, 222 Rosewood Drive, Danvers, MA 01923, (978) 750-8400, fax (978) 646-8600. Requests to the Publisher for permission should be addressed to the Permissions Department, John Wiley & Sons, Inc., 111 River Street, Hoboken, NJ 07030, (201) 748-6011, fax (201) 748-6008, or online at http://www.wiley.com/go/permissions.

Limit of Liability/Disclaimer of Warranty: The publisher and the author make no representations or warranties with respect to the accuracy or completeness of the contents of this work and specifically disclaim all warranties, including without limitation warranties of fitness for a particular purpose. No warranty may be created or extended by sales or promotional materials. The advice and strategies contained herein may not be suitable for every situation. This work is sold with the understanding that the publisher is not engaged in rendering legal, accounting, or other professional services. If professional assistance is required, the services of a competent professional person should be sought. Neither the publisher nor the author shall be liable for damages arising herefrom. The fact that an organization or website is referred to in this work as a citation and/or a potential source of further information does not mean that the author or the publisher endorses the information the organization or website may provide or recommendations it may make. Further, readers should be aware that Internet websites listed in this work may have changed or disappeared between when this work was written and when it is read.

For general information on our other products and services or to obtain technical support, please contact our Customer Care Department within the U.S. at (877) 762-2974, outside the U.S. at (317) 572-3993 or fax (317) 572-4002.

Wiley publishes in a variety of print and electronic formats and by print-on-demand. Some material included with standard print versions of this book may not be included in e-books or in print-on-demand. If this book refers to media such as a CD or DVD that is not included in the version you purchased, you may download this material at http://booksupport.wiley.com. For more information about Wiley products, visit www.wiley.com.

Library of Congress Control Number: 2016944021

TRADEMARKS: Wiley, the Wiley logo, and the Sybex logo are trademarks or registered trademarks of John Wiley & Sons, Inc. and/or its affiliates, in the United States and other countries, and may not be used without written permission. VMware is a registered trademark of VMware, Inc. All other trademarks are the property of their respective owners. John Wiley & Sons, Inc. is not associated with any product or vendor mentioned in this book.

10 9 8 7 6 5 4 3 2 1

About the Author

Martin Hosken is employed as a global cloud architect within the VMware Global Cloud Practice, which is part of its Cloud Provider Software Business Unit.

He has extensive experience architecting and consulting with international customers and designing the transition of organizations' legacy infrastructure onto VMware cloud-based platforms. His broad and deep knowledge of physical and virtualized services, platforms, and cloud infrastructure solutions is based on involvement and leadership in the global architecture, design, development, and implementation of large-scale, complex, multitechnology projects for enterprises and cloud service providers. He is a specialist in designing, implementing, and integrating best-of-breed, fully redundant Cisco, EMC, IBM, HP, Dell, and VMware systems into enterprise environments and cloud service providers' infrastructure.

In addition, Martin is a double VMware Certified Design Expert (VCDX #117) in Data Center Virtualization and Cloud Management and Automation. (See the Official VCDX directory available at http://vcdx.vmware.com.) Martin also holds a range of industry certifications from other vendors such as EMC, Cisco, and Microsoft, including MCITP and MCSE in Windows Server and Messaging.

He has been awarded the annual *VMware vExpert* title for a number of years for his significant contribution to the community of VMware users. (See the VMware Community vExpert Directory available at https://communities.vmware.com/vexpert.jspa.) This title is awarded to individuals for their commitment to the sharing of knowledge and their passion for VMware technology beyond their job requirements. Martin is also a part of the CTO Ambassador Program, and as such is responsible for connecting the R&D team at VMware with customers, partners, and field employees.

Follow Martin on Twitter: @hoskenm.

About the Technical Reviewer

Ray Heffer is employed as a global cloud architect for VMware's Cloud Provider Software Business Unit. He is also a double VCDX #122 (Desktop and Datacenter). In his previous roles with End User Computing (EUC), Technical Marketing, and Professional Services at VMware, he has led many large-scale platform designs for service providers, manufacturing, and government organizations.

Since 1997 Ray has specialized in administering, designing, and implementing solutions ranging from Microsoft Exchange, Linux, Citrix, and VMware. He deployed his first VMware environment in 2004 while working at a hosting company in the United Kingdom.

Ray is also a regular presenter at VMworld and VMUG events, covering topics such as Linux desktops and VMware Horizon design best practices.

Contents at a Glance

Foreword by Duncan Epping . . . . *xvii*

Introduction . . . . *xix*

    Chapter 1 • Software-Defined Storage Design 1

    Chapter 2 • Classic Storage Models and Constructs 21

    Chapter 3 • Fabric Connectivity and Storage I/O Architecture 101

    Chapter 4 • Policy-Driven Storage Design with Virtual SAN 187

    Chapter 5 • Virtual SAN Stretched Cluster Design 315

    Chapter 6 • Designing for Web-Scale Virtual SAN Platforms 367

    Chapter 7 • Virtual SAN Use Case Library 381

    Chapter 8 • Policy-Driven Storage Design with Virtual Volumes 429

    Chapter 9 • Delivering a Storage-as-a-Service Design 453

    Chapter 10 • Monitoring and Storage Operations Design 473

Index 509

Contents

Foreword by Duncan Epping .... *xvii*

Introduction .... *xix*

Chapter 1 • Software-Defined Storage Design **1**
Software-Defined Compute 2
Software-Defined Networking 2
Software-Defined Storage 3
Designing VMware Storage Environments 4
    Technical Assessment and Requirements Gathering 5
    Establishing Storage Design Factors 6
The Economics of Storage 10
    Calculating the Total Cost of Ownership for Storage Resources ... 11
    Information Lifecycle Management 13
Implementing a Software-Defined Storage Strategy 15
Software-Defined Storage Summary 16
    Hyper-Converged Infrastructure and Virtual SAN 18
    Virtual Volumes 18
    Classic and Next-Generation Storage Models 19

Chapter 2 • Classic Storage Models and Constructs **21**
Classic Storage Concepts 21
    RAID Sets 25
    Virtual Provisioning 44
    Storage Tiering 49
    Storage Scalability Design 54
    Storage Management Tools 57
    Multitenanted Storage Design 58
    Quality of Service 59
    Data Deduplication and Data Compression 60
    Storage Device Security 61
    Hardware High Availability 61
    Storage Array–Based Disaster Recovery and Backups 62
    Storage Array Snapshots and Clones in a Classic Storage Environment 63
    vSphere Metro Storage Cluster 65
    All-Flash Disk Arrays 65
vSphere Storage Technologies 67
    Virtual Disks 68
    Virtual Machine Storage Controllers (vSCSI Adapters) 71
    Datastore 73

Raw Device Mapping... 79
When to Use RDMs over VMFS or NFS?............................... 81
Storage vMotion and Enhanced vMotion Operations.................... 81
Datastore Clusters.. 82
Storage Distributed Resource Scheduler............................... 83
Storage I/O Control... 85
Classic Storage Model—vStorage APIs for Array Integration............. 89
Classic Storage Model—VASA 1.0.................................... 90
VADP and VAMP.. 91
Boot from SAN... 92
Classic Storage Model—vSphere Storage Policies....................... 94
Tiered Storage Design Models in vSphere.............................. 95
Sub-LUN System Access... 98

Chapter 3 • Fabric Connectivity and Storage I/O Architecture........ 101
Fibre Channel SAN.. 102
   Fibre Channel Protocol... 102
   Fibre Channel Topologies... 115
   Switch-Based Fabric Architecture.................................. 117
   Security and Traffic-Isolation Features............................. 125
   N_Port Virtualization and N_Port ID Virtualization.................. 131
   Boot from SAN.. 132
   Fibre Channel Summary.. 132
iSCSI Storage Transport Protocol...................................... 135
   iSCSI Protocol Components....................................... 135
   iSCSI Traffic Isolation.. 137
   Jumbo Frames... 138
   iSCSI Device-Naming Standards................................... 138
   CHAP Security.. 139
   iSCSI Network Adapters.. 140
   Virtual Switch Design.. 143
   iSCSI Boot from SAN... 148
   iSCSI Protocol Summary.. 148
NFS Storage Transport Protocol....................................... 149
   Comparing NAS and SAN.. 149
   NFS Components.. 149
   NAS Implementation... 152
   Single Virtual Switch / Single Network Design...................... 157
   Single Virtual Switch / Multiple Network Design.................... 159
   vSphere 6 NFS Version 4.1 Limitations............................. 161
   NFS Protocol Summary... 161
Fibre Channel over Ethernet Protocol.................................. 161
   Fibre Channel over Ethernet Protocol.............................. 163
   Fibre Channel over Ethernet Physical Components................... 165
   Fibre Channel over Ethernet Infrastructure......................... 167

Fibre Channel over Ethernet Design Options 167
　　　Fibre Channel over Ethernet Protocol Summary 170
　Multipathing Module 170
　　　Pluggable Storage Architecture 174
　　　iSCSI Multipathing. 177
　　　NAS Multipathing 178
　Direct-Attached Storage 180
　Evaluating Switch Design Characteristics. 182
　Fabric Connectivity and Storage I/O
　　Architecture Summary. 184

Chapter 4 • Policy-Driven Storage Design with Virtual SAN. 187
　　　Challenges with Legacy Storage. 187
　　　Policy-Driven Storage Overview 190
　　　VMware Object Storage Overview. 191
　Virtual SAN Overview 192
　Virtual SAN Architecture. 194
　　　Virtual SAN Disk Groups 194
　　　Comparing Virtual SAN Hybrid and All-Flash Models 200
　　　All-Flash Deduplication and Compression 202
　　　Data Locality and Caching Algorithms 205
　　　Virtual SAN Destaging Mechanism 206
　　　Virtual SAN Distributed Datastore 206
　　　Objects, Components, and Witnesses 207
　　　On-Disk Formats. 212
　　　Swap Efficiency / Sparse Swap 214
　　　Software Checksum 215
　Virtual SAN Design Requirements. 216
　　　Host Form Factor. 216
　　　Host Boot Architecture 217
　　　Virtual SAN Hardware Requirements 222
　Virtual SAN Network Fabric Design 236
　　　vSphere Network Requirements 236
　　　Physical Network Requirements 240
　Virtual SAN Storage Policy Design. 250
　　　Storage Policy–Based Management Framework 250
　　　Virtual SAN Rules 251
　　　Virtual SAN Rule Sets 253
　　　Default Storage Policy 267
　　　Application Assessment and Storage-Policy Design. 268
　Virtual SAN Datastore Design and Sizing. 271
　　　Hosts per Cluster 273
　　　Storage Capabilities 275
　　　Configuring Multiple Disk Groups 276
　　　Endurance Flash Sizing. 278

Objects, Components, and Witness Sizing 279
Datastore Capacity Disk Sizing 281
Capacity Disk Size 282
Designing for Availability 287
Designing for Hardware Component Failure 289
Host Cluster Design and Planning for Host Failure. 292
Quorum Logic Design and vSphere High Availability 302
Fault Domains. 302
Virtual SAN Internal Component Technologies. 308
Reliable Datagram Transport 308
Cluster Monitoring, Membership, and Directory Services 308
Cluster-Level Object Manager 310
Distributed Object Manager 310
Local Log-Structured Object Manager 310
Object Storage File System 311
Storage Policy–Based Management 312
Virtual SAN Integration and Interoperability. 312

Chapter 5 • Virtual SAN Stretched Cluster Design 315
Stretched Cluster Use Cases. 317
Fault Domain Architecture. 318
Witness Appliance 318
Network Design Requirements 320
Distance and Latency Considerations. 322
Bandwidth Requirements Calculations 325
Stretched Cluster Deployment Scenarios. 327
Default Gateway and Static Routes 327
Stretched Cluster Storage Policy Design 327
Preferred and Nonpreferred Site Concepts 329
Stretched Cluster Read/Write Locality 329
Distributed Resource Scheduler Configurations 332
High Availability Configuration 335
Stretched Cluster WAN Interconnect Design 339
Evaluating WAN Platforms for Stretched Clusters. 339
Deploying Stretched VLANs 347
WAN Interconnect High Availability 353
Secure Communication 353
Data Center Interconnect Design Considerations Summary 354
Stretched Cluster Solution Architecture Example 356
Cisco vPC over DWDM and Dark Fiber 358
OTV over DWDM and Dark Fiber 360
Cisco LISP Configuration Overview 363
Stretched Cluster Failure Scenarios 363
Stretched Cluster Interoperability. 365
Support Limitations 365

Chapter 6 • Designing for Web-Scale Virtual SAN Platforms 367

Scale-up Architecture 368
Scale-out Architecture. 370
Designing vSphere Host Clusters for Web-Scale 372
Building-Block Clusters and Scale-out Web-Scale Architecture 372
Scalability and Designing Physical Resources
   for Web-Scale 373
Leaf-Spine Web-Scale Architecture. 377

Chapter 7 • Virtual SAN Use Case Library 381

Use Cases Overview 383
   Two-Node Remote Office / Branch Office Design 386
   Horizon and Virtual Desktop Infrastructure. 392
   Virtual SAN File Services 395
Solution Architecture Example: Building a
   Cloud Management Platform with Virtual SAN 395
   Introduction and Conceptual Design 395
   Customer Design Requirements and Constraints. 398
   Cluster Configuration 404
   Network-Layer Design. 408
   Storage-Layer Design. 412
   Cloud Management Platform Security Design 423

Chapter 8 • Policy-Driven Storage Design with Virtual Volumes 429

Introduction to Virtual Volumes Technology 430
   Virtual Volumes Component Technology Architecture. 434
   Virtual Volumes Object Architecture 434
Management Plane 436
   VASA 2.0 Specification 436
   VASA Provider 436
Data Plane 437
   Storage Container 437
   Protocol Endpoints 440
   Binding Operations 442
Storage Policy–Based Management with Virtual Volumes. 444
   Published Capabilities 446
   Storage Capabilities 448
   Storage Capabilities Summary 449
Benefits of Designing for Virtual Volumes 449
   Enhanced Performance 450
   Greater Application Control 450
   Operational Simplification 450
   Reduced Wasted Capacity. 450
Virtual Volumes Key Design Requirements 450
vSphere Storage Feature Interoperability. 451
VAAI and Virtual Volumes. 451
Virtual Volumes Summary. 451

Chapter 9 • Delivering a Storage-as-a-Service Design 453
STaaS Service Definition 457
Cloud Platforms Overview 458
Cloud Management Platform Architectural Overview 461
    vRealize Automation Cloud Management Platform 461
    vRealize Orchestrator 465
The Combined Solution Stack 468
Workflow Examples 468
Summary 472

Chapter 10 • Monitoring and Storage Operations Design 473
Storage Monitoring 473
    Monitoring Component Health 474
    Monitoring Capacity 474
    Monitoring Storage Performance 475
    Monitoring Security 476
Storage Component Monitoring 477
    Monitoring Storage on Host Servers 477
    Monitoring the Storage Fabric 477
    Monitoring a Storage Array System 480
Storage Monitoring Challenges 481
Common Storage Management and Monitoring Standards 483
Virtual SAN Monitoring and Operational Tools 486
vRealize Operations Manager 492
    Management Pack for Storage Devices 492
    Storage Partner Solutions 494
vRealize Log Insight 497
Log Insight Syslog Design 498
End-to-End Monitoring Solution Summary 499
Storage Capacity Management and Planning 499
    Management Strategy Design 502
    Process and Approach 503
    Capacity Management for Virtual SAN 505
Summary 505

Index 509

Foreword by Duncan Epping

I had just completed the final chapter of the Virtual SAN book I was working on when Martin reached out and asked if I wanted to write a foreword for his book. You can imagine I was surprised to find out that there was another person writing a book on software-defined storage, and pleasantly surprised to find out that VSAN is one of the major topics in this book. Not just surprised, but also very pleased. The world is changing rapidly, and administrators and architects need guidance along this journey, the journey toward a software-defined data center.

When talking to customers and partners on the subject of the software-defined data center, a couple of concerns typically arise. Two parts of the data center have always been historically challenging and/or problematic—namely, networking and storage. Networking problems and concerns (and those related to security, for that matter) have been largely addressed with VMware NSX, which allows virtualization and networking administrators to work closely together on providing a flexible yet very secure foundation for the workloads they manage. This is done by adding an abstraction layer on top of the physical environment and moving specific services closer to the workloads (for instance, firewalling and routing), where they belong.

Over 30 years ago, RAID was invented, which allowed you to create logical devices formed out of multiple hard disk drives. This allowed for more capacity, higher availability, and of course, depending on the type of RAID used, better performance. It is fair to say, however, that the RAID construct was created as a result of the many constraints at the time. Over time, all of these constraints have been lifted, and the hardware evolution started the (software-defined) storage revolution. SSDs, PCIe-based flash, NVMe, 10GbE, 25GbE (and higher), RDMA, 12 Gbps SAS, and many other technologies allowed storage vendors to innovate again and to make life simpler. No longer do we need to wide-stripe across many disks to meet performance expectations, as that single SSD device can now easily serve 50,000 IOPS. And although some of the abstraction layers, such as traditional RAID or disk groups, may have been removed, most storage systems today are not what I would consider admin/user friendly.

There are different protocols (iSCSI, FCoE, NFS, FC), different storage systems (spindles, hybrid, all flash), and many different data services and capabilities these systems provide. As a result, we cannot simply place an abstraction layer on top as we have done for networking with NSX. We still need to abstract the resources in some shape or form and most definitely present them in a different, simpler manner. Preferably, we leverage a common framework across the different types of solutions, whether that is a hyper-converged software solution like Virtual SAN or a more traditional iSCSI-based storage system with a combination of flash and spindles.

Storage policy–based management is this framework. If there is anything you need to take away from this book, then it is where your journey to software-defined storage should start, and that is the SPBM framework that comes as part of vSphere. SPBM is that abstraction layer that allows you to consume storage resources across many different types of storage (with different protocols) in a simple and uniform way by allowing you to create policies that are passed down to the respective storage system through the VMware APIs for Storage Awareness.

In order to be able to create an infrastructure that caters to the needs of your customers (application owners/users), it is essential that you, the administrator or architect, have a good understanding of all the capabilities of the different storage platforms, the requirements of the application, and how architectural decisions can impact availability, recoverability, and performance of your workloads.

But before you even get there, this book will provide you with a good foundational understanding of storage concepts including thin LUNs, protocols, RAID, and much more. This will be quickly followed by the software-defined storage options available in a VMware-based infrastructure, with a big focus on Virtual Volumes and Virtual SAN.

Many have written on the subject of software-defined storage, but not many are as qualified as Martin. Martin is one of the few folks who have managed to accrue two VCDX certifications, and as a global cloud architect has a wealth of experience in this field. He is going to take you on a journey through the world of software-defined storage in a VMware-based infrastructure and teach you the art of architecture along the way.

I hope you will enjoy reading this book as much as I have.

Duncan Epping
Chief Technologist, Storage and Availability, VMware

Introduction

Storage is typically the most important element of any virtual data center. It is the key component in system performance, availability, scalability, and manageability. It has also traditionally been the most expensive component from a capital and operational cost perspective.

The storage infrastructure must meet not only today's requirements, but also the business needs for years to come, because of the capital expenditure costs historically associated with the hardware. Storage and vSphere architects must therefore make the most informed choices possible, designing solutions that take into account multiple complex and contradictory business requirements, technical goals, forecasted data growth, constraints, and of course, budget.

In order for you to be confident about undertaking a vSphere storage design that can meet the needs of a whole range of business and organization types, you must understand the capabilities of the platform. Designing a solution that can meet the requirements and constraints set out by the customer requires calling on your experience and knowledge, as well as keeping up with advances in the IT industry. A successful design entails collecting information, correlating it into a solid design approach, and understanding the design trade-offs and design decisions.

The primary content of this book addresses various aspects of the VMware vSphere software-defined storage model, which includes separate components. Before you continue reading, you should ensure that you are already well acquainted with the core vSphere products, such as VMware vCenter Server and ESXi, the type 1 hypervisor on which the infrastructure's virtual machines and guest operating systems reside.

It is also assumed that you have a good understanding of shared storage technologies and networking, along with the wider infrastructure required to support the virtual environment, such as physical switches, firewalls, server hardware, array hardware, and the protocols associated with this type of equipment, which include, but are not limited to, Fibre Channel, iSCSI, NFS, Ethernet, and FCoE.

Who Should Read This Book?

This book will be most useful to infrastructure architects and consultants involved in designing new vSphere environments, and administrators charged with maintaining existing vSphere deployments who want to further optimize their infrastructure or gain additional knowledge about storage design. In addition, this book will be helpful for anyone with a VCA, VCP, or a good foundational knowledge who wants an in-depth understanding of the design process for new vSphere storage architectures. Prospective VCAP, VCIX, or VCDX candidates who already have a range of vSphere expertise but are searching for that extra bit of detailed knowledge will also benefit.

What Is Covered in This Book?

VMware-based storage infrastructure has changed a lot in recent years, with new technologies and new storage vendors stepping all over the established industry giants, such as EMC, IBM, and NetApp. However, life-cycle management of the storage platform remains an ongoing challenge for enterprise IT organizations and service providers, with hardware renewals occurring on an ongoing basis for many of VMware's global customer base.

This book aims to help vSphere architects, storage architects, and administrators alike understand and design for this new generation of VMware-focused software-defined storage, and to drive efficiency through simple, less complex technologies that do not require large numbers of highly trained storage administrators to maintain.

In addition, this book aims to help you understand the design factors associated with these new vSphere storage options. You will see how VMware is addressing these data-center challenges through its software-defined storage offerings, Virtual SAN and Virtual Volumes, as well as developing cloud automation approaches to these next-generation storage solutions to further simplify operations.

This book offers you deep knowledge and understanding of these new storage solutions by

- Providing unique insight into Virtual SAN and Virtual Volumes storage technologies and design

- Providing a detailed knowledge transfer of these technologies and an understanding of the design factors associated with the architecture of this next generation of VMware-based storage platform

- Providing guidance over delivering storage as a service (STaaS) and enabling enterprise IT organizations and service providers to deploy and maintain storage resources via a fully automated cloud platform

- Providing detailed and unique guidance in the design and implementation of a stretched Virtual SAN architecture, including an example solution

- Providing a detailed knowledge transfer of legacy storage and protocol concepts, in order to help provide context to the VMware software-defined storage model

Finally, in writing this book, I hope to help you understand all of the design factors associated with these new vSphere storage options, and to provide a complete guide for solution architects and operational teams to maximize quality storage design for this new generation of technologies.

The following provides a brief summary of the content in each of the 10 chapters:

Chapter 1: Software-Defined Storage Design   This chapter provides an overview of where vSphere storage technology is today, and how we've reached this point. This chapter also introduces software-defined storage, the economics of storage resources, and enabling storage as a service.

Chapter 2: Classic Storage Models and Constructs   This chapter covers the legacy and classic storage technologies that have been used in the VMware infrastructure for the last decade. This chapter provides the background required for you to understand the focus of this book, VMware vSphere's next-generation storage technology design.

Chapter 3: Fabric Connectivity and Storage I/O Architecture   This chapter presents storage connectivity and fabric architecture, which is relevant for legacy storage technologies as well as next-generation solutions including Virtual Volumes.

Chapter 4: Policy-Driven Storage Design with Virtual SAN   This chapter addresses all of the design considerations associated with VMware's Virtual SAN storage technology. The chapter provides detailed coverage of Virtual SAN functionality, design factors, and architectural considerations.

Chapter 5: Virtual SAN Stretched Cluster Design   This chapter focuses on one type of Virtual SAN solution, stretched cluster design. This type of solution has specific design and implementation considerations that are addressed in depth. This chapter also provides an example Virtual SAN stretched architecture design as a reference.

Chapter 6: Designing for Web-Scale Virtual SAN Platforms   This chapter addresses specific considerations associated with large-scale deployments of Virtual SAN hyper-converged infrastructure, commonly referred to as *web-scale*.

Chapter 7 Virtual SAN Use Case Library   This chapter provides an overview of Virtual SAN use cases. It also provides a detailed solution architecture for a cloud management platform that you can use as a reference.

Chapter 8: Policy-Driven Storage Design with Virtual Volumes   This chapter provides detailed coverage of VMware's Virtual Volumes technology and its associated policy-driven storage concepts. This chapter also provides a low-level knowledge transfer, as well as addressing in detail the design factors and architectural concepts associated with implementing Virtual Volumes.

Chapter 9: Delivering a Storage-as-a-Service Design   This chapter explains how IT organizations and service providers can design and deliver storage as a service in a cloud-enabled data center by using VMware's cloud management platform technologies.

Chapter 10: Monitoring and Storage Operations Design   To ensure that a storage design can deliver an operationally efficient storage platform end to end, this final chapter covers storage monitoring and alerting design in the software-defined storage data center.

Chapter 1

Software-Defined Storage Design

VMware is the global leader in providing virtualization solutions. The VMware ESXi software provides a hypervisor platform that abstracts CPU, memory, and storage resources to run multiple virtual machines concurrently on the same physical server.

To successfully design a virtual infrastructure, other products are required in addition to the hypervisor, in order to manage, monitor, automate, and secure the environment. Fortunately, VMware also provides many of the products required to design an end-to-end solution, and to develop an infrastructure that is software driven, as opposed to hardware driven. This is commonly described as the *software-defined data center* (SDDC), illustrated in Figure 1.1.

FIGURE 1.1
Software-defined data center conceptual model

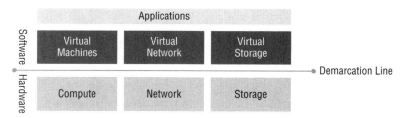

The SDDC is not a single product sold by VMware or anyone else. It is an approach whereby management and orchestration tools are configured to manage, monitor, and operationalize the entire infrastructure. This might include products such as vSphere, NSX, vRealize Automation, vRealize Operations Manager, and Virtual SAN from VMware, but it could also include solutions such as VMware Integrated OpenStack, CloudStack, or any custom cloud-management solution that can deliver the required platform management and orchestration capabilities.

The primary aim of the SDDC is to decouple the infrastructure from its underlying hardware, in order to allow software to take advantage of the physical network, server, and storage. This makes the SDDC location-independent, and as such, it may be housed in a single physical data center, span multiple private data centers, or even extend into hybrid and public cloud facilities.

From the end user's perspective, applications that are delivered from an SDDC are consumed in exactly the same way as they otherwise would be—through mobile, desktop, and virtual desktop interfaces—from anywhere, any time, with any device.

However, with the SDDC infrastructure decoupled from the physical hardware, the operational model of a virtual machine—with on-demand provisioning, isolation, mobility, speed, and agility—can be replicated for the entire data-center environment (including networking and storage), with complete visibility, security, and scale.

The overall aim is that an SDDC can be achieved with the customer's existing physical infrastructure, and also provide the flexibility for added capacity and new deployments.

Software-Defined Compute

In this book, *software-defined compute* refers to the compute virtualization of the *x*86 architecture. What is *virtualization*? If you don't know the answer to this question, you're probably reading the wrong book, but in any case, let's make sure we're on the same page.

In the IT industry, the term *virtualization* can refer to various technologies. However, from a VMware perspective, virtualization is the technique used for abstracting the physical hardware away from the operating system. This technique allows multiple guest operating systems (logical servers or desktops) to run concurrently on a single physical server. This allows these logical servers to become a portable virtual compute resource, called *virtual machines*. Each virtual machine runs its own guest operating system and applications in an isolated manner.

Compute virtualization is achieved by a hypervisor layer, which exists between the hardware of the physical server and the virtual machines. The hypervisor is used to provide hardware resources, such as CPU, memory, and network to all the virtual machines running on that physical host. A physical server can run numerous virtual machines, depending on the hardware resources available.

Although a virtual machine is a logical entity, to its operating system and end users, it seems like a physical host with its own CPU, memory, network controller, and disks. However, all virtual machines running on a host share the same underlying physical hardware, but each taking its own share in an isolated manner. From the hypervisor's perspective, each virtual machine is simply a discrete set of files, which include a configuration file, virtual disk files, log files, and so on.

It is VMware's ESXi software that provides the hypervisor platform, which is designed from the ground up to run multiple virtual machines concurrently, on the same physical server hardware.

Software-Defined Networking

Traditional physical network architectures can no longer scale sufficiently to meet the requirements of large enterprises and cloud service providers. This has come about as the daily operational management of networks is typically the most time-consuming aspect in the process of provisioning new virtual workloads. *Software-defined networking* helps to overcome this problem by providing networking to virtual environments, which allows network administrators to manage network services through an abstracted higher-level functionality.

As with all of the components that make up the SDDC model, the primary aim is to provide a simplified and more efficient mechanism to operationalize the virtual data-center platform. Through the use of software-defined networking, the majority of the time spent provisioning and configuring individual network components in the infrastructure can be performed programmatically, in a virtualized network environment. This approach allows network administrators to get around this inflexibility of having to pre-provision and configure physical networks, which has proved to be a major constraint to the development of cloud platforms.

In a software-defined networking architecture, the control and data planes are decoupled from one another, and the underlying physical network infrastructure is abstracted from the applications. As a result, enterprises and cloud service providers obtain unprecedented programmability, automation, and network control. This enables them to build highly scalable, flexible networks with cloud agility, which can easily adapt to changing business needs by

- Providing centralized management and control of networking devices from multiple vendors.
- Improving automation and management agility by employing common application program interfaces (APIs) to abstract the underlying networking from the orchestration and provisioning processes, without the need to configure individual devices.
- Increasing network reliability and security as a result of centralized and automated management of the network devices, which provides this unified security policy enforcement model, which in turn reduces configuration errors.
- Providing more-granular network control, with the ability to apply a wide range of policies at the session, user, device, or application level.

NSX is VMware's software-defined networking platform, which enables this approach to be taken through an integrated stack of technologies. These include the NSX Controller, NSX vSwitch, NSX API, vCenter Server, and NSX Manager. By using these components, NSX can create layer 2 logical switches, which are associated with logical routers, both north/south and east/west firewalling, load balancers, security policies, VPNs, and much more.

Software-Defined Storage

Where the data lives! That is the description used by the marketing department of a large financial services organization that I worked at several years ago. The marketing team regularly used this term in an endearing way when trying to describe the business-critical storage systems that maintained customer data, its availability, performance level, and compliance status.

Since then, we have seen a monumental shift in the technologies available to vSphere for virtual machine and application storage, with more and more storage vendors trying to catch up, and for some, steam ahead. The way modern data centers operate to store data has been changing, and this is set to continue over the coming years with the continuing shift toward the next-generation data center, and what is commonly described as *software-defined storage*.

VMware has undoubtedly brought about massive change to enterprise IT organizations and service-provider data centers across the world, and has also significantly improved the operational management and fundamental economics of running IT infrastructure. However, as application workloads have become more demanding, storage devices have failed to keep up with IT organizations' requirements for far more flexibility from their storage solutions, with greater scalability, performance, and availability. These design challenges have become an everyday conversation for operational teams and IT managers.

The primary challenge is that many of the most common storage systems we see in data centers all over the world are based on outdated technology, are complex to manage, and are highly proprietary. This ties organizations into long-term support deals with hardware vendors.

This approach is not how the biggest cloud providers have become so successful at scaling their storage operations. The likes of Amazon, Microsoft, and Google have scaled their cloud storage platforms by trading their traditional storage systems for low-cost commodity hardware, and employed the use of powerful software around it to achieve their goals, such as availability, data protection, operational simplification, and performance. With this approach, and through the economies of scale, these large public cloud providers have achieved their supremacy at a significantly lower cost than deploying traditional monolithic centralized storage systems. This methodology, known as web-scale, is addressed further in Chapter 6, "Designing for Web-Scale Virtual SAN Platforms (10,000 VMS+)."

The aim of this book is to help you understand the new vSphere storage options, and how VMware is addressing these data-center challenges through its software-defined storage offerings, Virtual SAN and Virtual Volumes. The primary aim of these two next-generation storage solutions is to drive efficiency through simple, less complex technologies that do not require large numbers of highly trained storage administrators to maintain. It is these software-defined data-center concepts that are going to completely transform all aspects of vSphere data-center storage, allowing these hypervisor-driven concepts to bind together the compute, networking, and software-defined storage layers.

The goal of software-defined storage is to separate the physical storage hardware from the logic that determines *where the data lives*, and what storage services are applied to the virtual machines and data during read and write operations.

As a result of VMware's next-generation storage offerings, a storage layer can be achieved that is more flexible and that can easily be adjusted based on changing application requirements. In addition, the aim is to move away from complex proprietary vendor systems, to a virtual data center made up of a coherent data fabric that provides full visibility of each virtual machine through a single management toolset, the so-called *single pane of glass*. These features, along with lowered costs, automation, and application-centric services, are the primary drivers for enterprise IT organizations and cloud service providers to begin to rethink their entire storage architectural approach.

The next point to address is what software-defined storage isn't, as it can sometimes be hard to wade through all the marketing hype typically generated by storage vendors. Just because a hardware vendor sells or bundles management software with their products, doesn't make it a software-defined solution. Likewise, a data center full of different storage systems from a multitude of vendors, managed by a single common software platform, does not equate to a software-defined storage solution. As each of the underlining storage systems still has its legacy constructs, such as disk pools and LUNs, this is referred to as a *federated storage solution* and not software-defined. These two approaches are sometimes confused by storage vendors, as understandably, manufacturers always want to use the latest buzzwords in their marketing material.

Despite everything that has been said up until now, software-defined storage isn't just about software. At some point, you have to consider the underlying disk system that provides the storage capacity and performance. If you go out and purchase a lot of preused 5,400 RPM hard drives from eBay, you can't then expect solid-state flash-like performance just because you've put a smart layer of software on top of it.

Designing VMware Storage Environments

Gathering requirements and documenting driving factors is a key objective for you, the architect. Understanding the customer's business objectives, challenges, and requirements should always be the first task you undertake, before any design can be produced. From this activity, you can translate the outcomes into design factors, requirements, constraints, risks, and assumptions, which are all critical to the success of the vSphere storage design.

Architects use many approaches and methodologies to provide customers with a meaningful design that meets their current and future needs. Figure 1.2 illustrates one such method, which provides an elastic sequence of activities that can typically fulfill all stages of the design process. However, many organizations have their own approach, which may dictate this process and mandate specific deliverables and project methodologies.

FIGURE 1.2
Example of a design sequence methodology

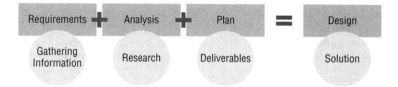

Technical Assessment and Requirements Gathering

The first step toward any design engagement is discovery, and the process of gathering the requirements for the environment in which the vSphere-based storage will be deployed. Many practices are available for gathering requirements, with each having value in different customer scenarios. As the architect, you must use the best technique to gain a complete picture from various stakeholders. This may include one-to-one meetings with IT organizational leaders and sponsors, facilitated sessions or workshops with the team responsible for managing the storage operations, and review of existing documents. Table 1.1 lists key questions that you need to ask stakeholder and operational teams.

TABLE 1.1: Requirements gathering

ARCHITECT QUESTION	ARCHITECTURAL OBJECTIVE
What will it be used for?	Focus on applications and systems
Who will be using it?	Users and stakeholders
What is the purpose?	Objectives and goals
What will it do? When? How?	Help create a scenario
What if something goes wrong with it?	Availability and recoverability
What quality? How fast? How reliable? How secure? How many?	Scaling, security, and performance

After all design factors and business drivers have been reviewed and analyzed, it is essential to take into account the integration of all components into the design, before beginning the qualification effort needed to sort through the available products and determine which solution will meet the customer's objectives. The integration of all components within a design can take place only if factors such as data architecture, business drivers, application architecture, and technologies are put together.

The overall aim of all the questions is to quantify the objectives and business goals. For instance, these objectives and goals might include the following:

Performance User numbers and application demands: Does the organization wish to implement a storage environment capable of handling an increase in user numbers and application storage demands, without sacrificing end-user experience?

Total Cost of Ownership Does the organization wish to provide separate business units with a storage environment that provides significant cost relief?

Scalability Does the organization wish to ensure capability and sustainability of the storage infrastructure for business continuity and future growth?

Management Does the organization wish to provide a solution that simplifies the management of storage resources, and therefore requires improved tools to support this new approach?

Business Continuity and Disaster Recovery Does the organization wish to provide a solution that can facilitate high levels of availability, disaster avoidance, and quick and reliable recovery from incidents?

In addition to focusing on these goals, you need to collect information relating to the existing infrastructure and any new technical requirements that might exist. These technical requirements will come about as a result of the business objectives and the current state analysis of the environment. However, these are likely to include the following:

- Application classification
- Physical and virtual network constraints
- Host server options
- Virtual machines and workload deployment methodology
- Network-attached storage (NAS) systems
- Storage area network (SAN) systems

Understanding the customer's business goals is critical, but what makes it such a challenge is that no two projects are ever the same. Whether it is different hardware, operating systems, maintenance levels, physical or virtual servers, or number of volumes, the new design must be validated for each component within each customer's specific infrastructure. In addition, just as every environment is different, no two workloads are the same either. For instance, peak times can vary from site to site and from customer to customer. These individual differentiators must be validated one by one, in order to determine the configuration required to meet the customer's design objectives.

Establishing Storage Design Factors

Establishing storage design factors is key to any architecture. However, as previously stated, the elements will vary from one engagement to another. Nevertheless, and this is important, the design should focus on the business drivers and design factors, and not the product features or latest technology specification from the customer's preferred storage hardware vendor.

A customer-preferred storage device could well be the best product ever, but may not align with the customer use cases, regardless of what they're being told by their supplier. Therefore, creating an architecture that focuses on the hardware specification and not the business goals is likely to introduce significant risks and ultimately fail as a design.

Although the business drivers and design factors for each customer will be different, with all having their own priorities and goals that need to be factored into the design, you likely will see many common design qualities, illustrated in Figure 1.3, time and time again.

FIGURE 1.3
Storage architecture business drivers and design factors

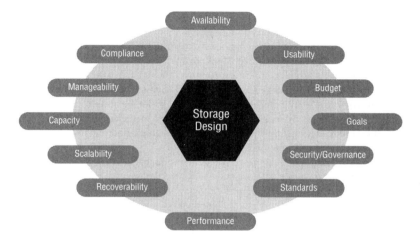

Availability

The *availability* of the storage infrastructure is typically dictated by a service-level agreement (SLA) of some sort, and is often represented as a percentage of possible uptime (such as four nines, 99.99 percent). Availability is achieved through techniques such as redundant hardware, RAID technologies, array mirroring, or eliminating single points of failure. Additionally, high levels of availability can be provided by using technologies such as storage replication, vSphere anti-affinity rules, or Virtual SAN Stretched Clusters. An available design is reliable and implements multiple mechanisms to restore services within the IT organization's agreed-upon service-level agreement.

Compliance

Compliance means conforming to a specification, policy, standard, or law. Regulatory compliance is now a part of everyday life for an information technology architect. Having a strong understanding of the requirements that the customers must comply with will help significantly in producing a design that meets the needs of the organization you're working with. Compliance goals also differ for different countries. For instance, in the United States, architects may be familiar with the Sarbanes–Oxley Act of 2002 or the Health Insurance Portability and Accountability Act of 1996 (HIPAA). In addition, global compliance standards, such as the Payment Card Industry Data Security Standard (PCI DSS), cross geographical boundaries.

Usability

Usability is the ease of use and learnability of the day-to-day operations associated with the storage platform. As the architect, one of your tasks will be to ensure that the customer's operational team or administrators are able to manage the environment after you leave and move on to the next project. This, of course, links into manageability, and you may be required to provide operational documentation, or partake in knowledge transfer and training as defined in the scope of work.

Budget

Unfortunately, few projects have unlimited budgets. Cost is always at the forefront of stakeholders' minds, and, as the architect, you will probably find that justifying costs associated with the design will often come down to you. I can assure you from personal experience that CFOs and their representatives can be scary and love to ask difficult and challenging questions. (To be fair, all they are trying to do is justify costs, so let's not be too hard on them.) Your goal is to meet the organization's business needs, while remaining within budget. If this is not possible, you must be able to explain and justify the best course of action to the organization's key stakeholders, who hold the purse strings.

The budget will depend on multiple factors. It might be too small a number, and you can think of it as a design constraint. In an ideal world, the design should focus only on system readiness, performance, and capacity, with an aim to provide a world-class solution with the future in mind, regardless of the cost. However, this is rarely the case; typically, the task of an architect is to take in all of the requirements and provide the best solution with the lowest conceivable budget. Even if, as the architect, you are not accountable for the financial aspects of the design, it's typically useful to have an understanding of budgetary constraints and to be able to demonstrate value for money, as and when required.

Manageability

For this design factor, you should keep in mind KISS: *keep it standardized and simple*. Making a design unnecessarily complex has a serious impact on the manageability of the environment. Also, a design that is unnecessarily complex can easily contribute to failure, because the operational team might not understand the design, and making a change to one component can have implications on another. Instead, your aim should be to keep the design as simple as possible, while still meeting the business goals. The objective should be to keep the design easy to deploy, easy to administer and maintain for the operational teams, and easy to update and upgrade when the time comes.

Goals

The key goals for the design will be different for each project . However, in general, a good design is not unnecessarily complex, provides detailed documentation (which includes rationales for design decisions), balances the organization's requirements with technical best practices, and involves key stakeholders and the customer's subject matter experts in every aspect of the design, delivery, testing, and hand-over of the storage platform.

Security and Governance

Needless to say, in today's world security is a key deliverable in every enterprise IT or cloud service provider project. On some of the projects involving government agencies and financial institutions that I've worked on almost every aspect of the design is governed by security considerations and requirements. This can have a significant impact on both operational considerations and budget.

Standards

An enterprise organization or cloud service provider typically has standards that must be met for every project. Hopefully, these standards include a clear methodology for identifying stakeholders, identifying the most relevant business drivers, and providing transparency and traceability for all decisions. Standards might also include a defined and repeatable approach to design, delivery, testing and verification, and hand-over to operational teams.

Performance

Like availability, performance is often governed by a service-level agreement. The design must meet the performance requirements set out by the customer. Performance is typically measured by achievable throughput, latency, I/O per second, or other defined metrics the customer deems appropriate. Storage performance is probably less understood than capacity or availability. However, in a virtualized infrastructure, not much has a greater impact on the overall performance of the environment than the storage platform.

Recoverability

Like availability and performance, recoverability is typically governed by a service-level agreement. The design should document how the infrastructure can be recovered from any kind of outage. Typically, two metrics are used to define recoverability: *recovery time objective* (RTO), which is the amount of time it takes to restore the service after the disruption began; and *recovery point objective* (RPO), which is the point in time at which data must be recovered to, after the disruption began.

Scalability

The design should be scalable—able to grow as the customer's data requirements change and the storage platform is required to expand. As part of the project, it is important to determine the business growth plans for data capacity, and any future performance requirements. This information is typically provided as a percentage of growth per year, and the design should take these factors into account. Later we address a building-block approach to storage design, but for now, it's of key importance that the customer is able to provide clear expectations on the growth of their environment, as this will almost certainly impact the design.

CAPACITY

The design's capacity requirements can typically be achieved as a business grows or shrinks. Capacity is generally predictable and can be provisioned on demand, as it is typically a relatively easy procedure to add disks and/or enclosures to most storage arrays or hosts without experiencing downtime. As a result, capacity can be managed relatively easily, but it is still an important aspect of storage design.

The Economics of Storage

At first glance, storage technologies, much like compute resource, should be priced based on a commodity hardware model; however, this is typically not the case. As illustrated in Figure 1.4, each year the cost of raw physical disk storage, on a per gigabyte basis, continues to fall, and has being doing so since the mid-1980s.

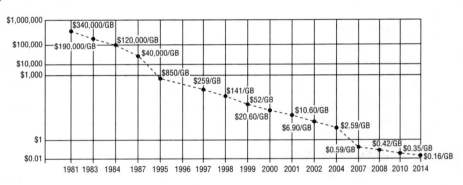

FIGURE 1.4
Hard disk drive cost per gigabyte

Alongside the falling cost of storage, as you might expect, in terms of raw disk capacity per drive, this has aligned with the falling cost per gigabyte charged by cloud service providers. This is illustrated in Figure 1.5, where the increasing capacity available on physical disks pretty much aligns with that falling cost.

Despite these falling costs in raw disk storage capacity, the chassis, the disk shelves used to create disk arrays, and the storage controllers tasked with organizing disks into large RAID (redundant array of independent disks) or JBOD (just a bunch of disks) sets, vendor prices for their technologies continue to increase year after year, regardless of this growing commoditization of the components used by them.

The reason for this is the ongoing development and sophistication of vendor software. For instance, an array made up of commoditized components, including 300 2 TB disks stacked in commodity shelves, may have a hardware cost totaling approximately $4,000. However, the end array vendor might assign a manufacturer's suggested retail price tag of $400,000. This price is based on the vendor adding their *secret source* software, enabling the commodity hardware to include features such as manageability and availability and to provide the performance aspects required by its customers, while also allowing the vendor to differentiate their product from that of their competitors. It is this aspect of storage that often adds the most significant cost component to storage technologies, regardless of the actual value added by the vendor's software, or which of those added features are actually used by their customers.

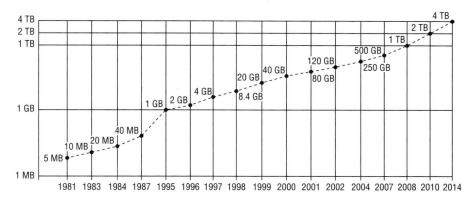

FIGURE 1.5
Hard disk drive capacity improvements

So whether you are buying or leasing, storage costs and other factors all contribute to the acquisition of storage resources, which is why IT organizations are increasingly trying to extend the useful life expectancy of their storage hardware. A decade ago, IT organizations were purchasing hardware with an expected life expectancy of three years. Today the same IT organizations are routinely acquiring hardware with the aim of achieving a five-to-seven-year useful life expectancy. One of the challenges is that most hardware and software ships with a three-year support contract and warranty, and renewing that agreement when it reaches end-of-life can sometimes cost as much as purchasing an entirely new array.

The next significant aspect of storage ownership to consider is that hardware acquisition accounts for approximately only one-fifth of the estimated annual total cost of ownership (TCO). This clearly outweighs the cost to acquire or capital expenditures (CapEx), and makes operational and management costs (OpEx) a far greater factor than many IT organizations account for in their initial design and planning cost estimations.

Calculating the Total Cost of Ownership for Storage Resources

As illustrated in Figure 1.6, the operational management, disaster recovery, and environmental costs are the real drivers behind the total cost of ownership calculations for storage devices.

FIGURE 1.6
Breakdown of total cost of ownership of storage hardware

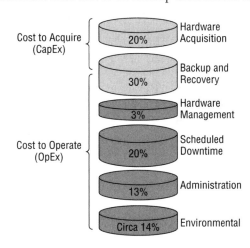

One of the factors that contributes to these operational costs is the heterogeneity of enterprise storage infrastructure. This significantly increases the challenges associated with providing a unified management approach, and as such, increases costs. Some IT organizations use this as a driver for the replacement of their heterogeneous storage platform, in favor of a more homogeneous approach. But typically, the replacement environment results from a deliberate attempt to procure the latest, best-of-breed technologies, or an attempt to facilitate storage tiering through a combination of hardware from a variety of vendors. Storage vendors often are unable to offer a varying portfolio of products for different types of workloads and data use cases. Furthermore, this problem is exacerbated by vendors who offer a wide range of products but can't typically offer a common management platform across all storage offerings. This is especially true when vendors have acquired the technology through a business acquisition.

The simplified formula in Figure 1.7 can be used to estimate the annual total cost of ownership of storage resources over the hardware's life expectancy.

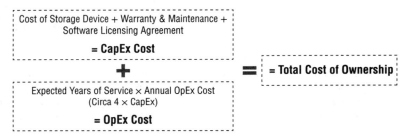

Figure 1.7
Simplified annual total cost of ownership

The next aspect of calculating storage costs relates to how efficiently storage capacity is allocated to the appropriate storage tier. *Utilization efficiency*, provides a measure of how effectively storage capacity is allocated to the correct storage type, based on factors such as frequency of access, availability requirements of the data, or required response time.

IT organizations often do not use storage capacity efficiently. They often use *tier-1* storage to host data for workloads and applications that do not require the expensive, high-performance attributes that the hardware is capable of delivering. Tiered storage is supposed to enable the placement of data based on cost-appropriate requirements for performance and capacity, as defined by the business. However, a growing movement to *flatten* storage via strategies such as those offered up by Hadoop (among others) in leveraging the *hyper-converged* storage model is, in itself, eliminating the requirement for tiered storage altogether.

IT organizations are typically charged with identifying the tiers of storage, the storage technologies employed, and the optimal percentage of business data that should represent each category of storage. Failing to do so undoubtedly leads to a significant increase in the per gigabyte cost of storage, which in turn results in an inflated total cost of ownership for the storage platform.

Figure 1.8 shows a tiered storage example. A business's IT organization uses this cost per gigabyte model to determine cost-appropriate storage for a specific workload type. For instance, if the IT organization requires a 100 TB storage estate, employing only two tiers of storage (tier 1 and tier 2 in this example), the total disk cost would be approximately $765,000. However, meeting the same storage requirements through the four tiers shown, segregated using the ratios illustrated, would cost approximately $482,250, and therefore represent a savings of $282,750, or a 37 percent reduction to the original cost.[1]

FIGURE 1.8
Storage cost per gigabyte example

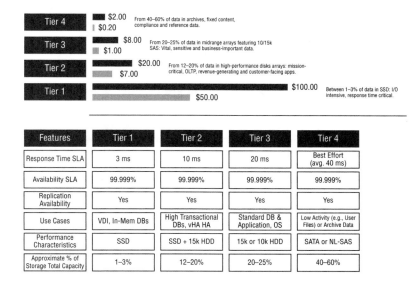

As you can see, an enterprise IT organization that fails to use this type of tiered storage strategy— which moves data across the storage estate based on its access frequency and other criteria—will suffer from poor utilization efficiency, as well as a significantly increased total cost of ownership of storage resources, within their storage platform.

Information Lifecycle Management

Information Lifecycle Management (ILM) is the primary approach used by businesses to ensure availability, capacity, and performance of data throughout its existence. When designing a storage solution for business systems, one of the key business requirements that you must understand is the ILM strategy being used by the customer for their business data.

Modern businesses and organizations must address the challenges associated with information management and its ever increasing growth, because business data and the way that it is used is playing a growing role in determining business success. For instance, companies such as Amazon and Rakuten are using their business data to gain strategic advantage over their competitors. The use of customer profiling and identifying what a customer may wish to purchase, based on their purchase history, provides a serious competitive advantage. In addition, understanding each customer's purchase habits (such as typically making all orders within the same few days each month, after payday) enables these businesses to target specific products at

---

[1] This calculation is based on 100 TB of storage, deployed at a cost of $50–$100 per gigabyte for flash (1–3 percent as tier 1), plus $7–$20 per gigabyte for fast disk (12–20 percent as tier1), plus $1 to $8 per gigabyte for capacity disk (20–25 percent as tier 3), plus $0.20–$2 per gigabyte for low-performance, high-capacity storage (40–60 percent as tier 4) totals approximately $482,250. Splitting the same capacity requirement between only tier 2 and tier 3 at the estimated cost range per gigabyte for each type of storage provides an estimated storage infrastructure cost of $765,000.

specific customers at a precise time, via a customized email based on the individual's purchase history and purchasing profile.

Another key consideration is how the value of the data changes over time. For instance, if a customer stops making purchases or closes their account, legislation might require that data to be deleted after a set period. Therefore, information that is stored may have a different value to the business, depending on its age. Understanding how an organization uses its data, and the value of its information throughout its life cycle, can be at the heart of storage design for many businesses (see Figure 1.9).

FIGURE 1.9
Information Lifecycle Management key challenges

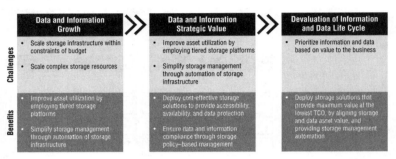

It is also important to recognize that ILM is a *strategy* adopted by a business or organization, and not a product or service. This strategy must be proactive and dynamic, in order to help plan for storage system growth, and also must reflect the value of the information to the business.

Implementing an ILM strategy throughout a large organization can take a significant period of time, but can deliver key benefits that directly address business challenges and information management and utilization. The key design considerations that relate ILM strategy to the architecture of a storage platform include the following:

- Improving utilization by employing tiered storage platforms, and providing increased visibility into all enterprise information, alongside archiving capabilities

- Providing simplified storage management tools and increasing the use of automation for daily storage operational processes

- Implementing a wide range of backup, data protection, and recovery options to balance the need for business continuity with the cost of losing data

- Simplifying compliance and regulatory requirements by providing control over data placement, and knowing what data needs to be secured and for how long

- Lowering the total cost of ownership while continuing to meet the required service levels demanded by the business, and aligning the storage management costs with the value of the data, so that storage resources are not wasted, and unnecessarily complex environments are not introduced

- Providing a tiered storage solution that ensures that low-value data is not stored at the same cost per gigabyte as high-value data

Implementing a Software-Defined Storage Strategy

As a consequence of the ever-increasing cost of enterprise business storage, as outlined previously, more IT industry attention than ever before is focused on new storage architectures and technologies designed to drive down the total cost of ownership associated with storage. This approach aims to reduce both CapEx and OpEx costs by reducing hardware to its bare commodity components, and removing *secret source* software from the controllers, in favor of placing it onto a common storage software layer provided by either the hypervisor or a software-defined storage model.

In the past, several attempts have been made to develop a common management system that can transcend storage hardware and software vendors. For example, the Storage Networking Industry Association (SNIA) developed the Storage Management Initiative Specification (SMI-S), and the World Wide Web Consortium has Representational State Transfer (REST). However, these have seen only limited adoption by the storage industry. To achieve even limited interoperability and provide a sense of single point of management and support, the only real option for large enterprise IT organizations and cloud service providers has been to deploy homogeneous storage islands from a single hardware vendor in an attempt to manage operational overhead and therefore reduce OpEx costs.

The theory behind the software-defined storage model is to facilitate management across a common plane, by breaking down the barriers to interoperability that exist with proprietary vendor storage hardware. For most IT organizations, storage from different vendors, or even different models of storage array hardware from the same vendor, create isolated storage islands. It can be difficult to interoperate, share resources, or even manage across these islands from a single pane of glass.

The software-defined storage model aims to provide OpEx cost savings by driving efficient capacity utilization and platform management in a more agile way, typically by providing automation and a common management interface for all of the storage infrastructure. Therefore, the challenge for enterprise IT organizations and cloud service providers is to find the right software-defined storage solution, one that can apply the right centralized software services to the entire infrastructure by using simple, unified operational procedures within a common user interface.

The software-defined storage model also aims to reduce CapEx costs by moving away from proprietary storage hardware, and toward technology that facilitates unified management across all components of the storage infrastructure. When considering hardware solutions to deliver a software-defined storage-based environment, IT executives may be focused on reducing the total cost of ownership of storage resources. The following list provides a buyer's guide that IT organizations can use when working with their respective storage vendors to establish core storage requirements:

- Which storage solutions can work with the applications, hypervisors, and data that we currently have and are predicting to have going forward?
- Which storage solutions can enhance application performance?
- Which storage solutions best provide the required data availability?

- Which storage solutions can be deployed, configured, and managed quickly and effectively using currently available skills?
- Which storage solutions can provide greater, and if possible, optimal, storage capacity?
- Which storage solutions can best facilitate flexibility (provide the ability to add capacity or performance in the future without impacting the applications)?
- Which storage solutions provide automation and centralized management capabilities?
- Which storage technology will meet the preceding requirements within the available budget?

The approach often taken by IT organizations is to follow the lead of a trusted storage vendor. However, a key challenge for IT decision makers is to see beyond current trends in the industry and to arrive at a strategy that will provide a solution meeting not only today's storage requirements at an acceptable level of cost, but also next year's requirements for the various lines of business, and even the next decade's. This requires a subjective and clear-headed evaluation of the options, their costs, and the alternative approaches that could deliver the required storage functionality that optimizes both CapEx and OpEx budgets.

An additional challenge, which you also shouldn't overlook, is the complication associated with educating decision makers about the intricacies of storage technologies, in order to obtain budgetary approval. Enterprise IT executives rarely question the requirement to store and retain their ever-growing volume of business data. However, explaining the differences between various storage products, and their advantages and drawbacks, often requires a transfer of technical knowledge in order for the decision makers to grasp the concepts and challenges faced by the architect, and how they relate to their storage platform design.

When finances are stretched, as they so often are, a high storage infrastructure expenditure can significantly stand out on an IT executive's annual budget spreadsheet. By examining the storage environment and calculating the total cost of ownership of storage resources, IT organizations can seek to identify new and innovative ways to address CapEx and OpEx expenditures through the software-defined storage model, without compromising application performance, capacity, availability or other data-related services.

Software-Defined Storage Summary

Just as VMware introduced $x86$ server virtualization to improve the cost metrics and utilization efficiencies of the compute platform, so too can the software-defined storage model be used to make the most efficient use of storage infrastructure, thereby reducing the total cost of ownership through storage acquisition and operational cost savings.

In the software-defined storage data center, all storage—whether it is directly attached hyper-converged Virtual SAN, or is SAN attached and leveraging Virtual Volumes–enabled arrays—can be used as part of a storage resource pool. This eliminates the requirement to *rip and replace* all of the storage infrastructure in order to adopt a fully hyper-converged unified storage model as part of a single migration project, and allows the IT organization to spread the costs associated with a full storage infrastructure refresh over a number of years.

This is only one storage strategy. Equally valid is the mixed hybrid approach of employing Virtual Volumes and Virtual SAN as a long-term design, effectively using both solutions for specific use cases and workloads, as illustrated in Figure 1.10.

SOFTWARE-DEFINED STORAGE SUMMARY | 17

FIGURE 1.10
Hybrid Virtual Volumes and Virtual SAN platform

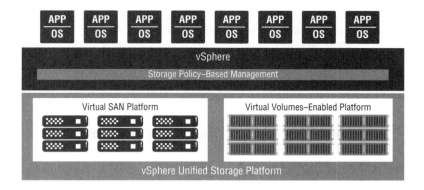

Just like the classic storage model, large enterprise customers and cloud service providers that are adopting software-defined storage typically should configure resources into pools. Each pool is composed of a different set of characteristics and services.

For instance, a Virtual SAN tier 1 pool may be optimized for performance and business-critical workloads, while a tier 0 pool may comprise all-flash disk groups and provide storage resources to specific I/O-intensive workloads. Following a similar model, high-capacity, low-cost, low-performance disks may be fashioned into a pool intended for the data that is infrequently accessed or updated. With this type of approach to storage provisioning, the software-defined storage model will continue to enable the implementation of a tiered storage strategy in order to provide improved capacity utilization and resource efficiency.

Furthermore, the implementation of a software-defined storage model allows technologies such as thin provisioning, compression, and de-duplication to be applied across an entire storage platform, rather than isolating these features behind specific hardware controllers. This helps to ensure that storage capacity can be used more efficiently, via a global storage policy.

These technologies can help slow the rate at which new capacity must be added to the infrastructure, and help ensure that where appropriate, less-expensive hardware can be deployed. In addition, centralizing this functionality through a single control plane enhances ease of administration, which in turn can also help reduce operational costs and the efforts associated with software maintenance.

The software-defined storage model is not an industry standard, and various approaches exist for the design, implementation, and function of the solution stack. Both VMware and independent software vendors (ISVs) have in recent years developed the concepts and product architecture of the software-defined storage platform for its integration into the market's leading hypervisor, to ensure that software-defined storage can operate within a robust and affordable model. These initiatives, which are the focus of much of this book, include the following:

- The introduction of the hyper-converged infrastructure product Virtual SAN, a bare-bones, hardware-agnostic model with a direct-attached storage configuration. This reduces or removes altogether the requirement for a switched fabric or LAN-attached storage infrastructure to manage, with no more proprietary storage hardware to support.

- The abstraction of advanced storage functions away from the storage vendor, and instead placed in the hypervisor software and management control plane. This approach simplifies operations, with no more proprietary software licenses and firmware levels to manage, and enables storage services to be applied to all capacity, not just specific hardware.

- ◆ The introduction of a single storage service management plane, via a unified user interface. This removes the requirement for third-party tools and specific array element managers to monitor and administer a heterogeneous storage infrastructure.

All of these attributes provide a significant improvement over the ongoing challenges associated with classic storage infrastructures, although they do not address all the problems that make proprietary storage systems expensive to own and operate.

Hyper-Converged Infrastructure and Virtual SAN

The hyper-converged infrastructure (HCI) hardware architecture model uses the hypervisor to deliver compute, networking, and shared storage from a single $x86$ server platform. This software-driven architecture enables physical storage resources to become part of commodity $x86$ servers, enabling a building-block approach with a web-scale level of scalability. Also, by adopting this commodity $x86$ server hardware approach, and combining both storage and compute hardware into a single entity, IT organizations and cloud service provider data centers can operate with agility, on a highly scalable, cost-effective, fully converged platform.

Virtual SAN is VMware's HCI platform, which enables this approach to be taken through the VMware integrated stack of technologies. Virtual SAN aggregates local storage into a unified data plane, which virtual machines can then use. Virtual SAN also uses a fully integrated policy-driven management layer, which allows virtual machines to be managed centrally, through a policy-driven storage mechanism that is integrated into the virtual machines' own settings. These policies can define reliability, redundancy, and performance characteristics that must be obeyed, independently of all other virtual machines that may reside on the same storage platform.

Virtual SAN is the foundational component of VMware's hyper-converged infrastructure solution. This model allows the convergence of compute, storage, and networking onto a single integrated layer of software that can run on any commodity $x86$ infrastructure aligned with the requirements set out on VMware's hardware compatibility list (HCL). While vSphere abstracts and aggregates compute resources into logical pools, Virtual SAN, embedded into the hypervisor's VMkernel, can pool together server-attached disk devices to create a high-performance distributed datastore.

This approach can easily meet the storage requirements of the most demanding IT organization or cloud service provider, at a lower cost than legacy monolithic SAN or NAS storage devices. Virtual SAN also allows vSphere and vSphere storage administrators to ignore concepts such as RAID sets and LUNs, and instead focus on the specific storage needs of applications. In addition, Virtual SAN can simplify capacity planning by scaling both storage and compute concurrently, allowing for the nondisruptive addition of new nodes, without the purchase of costly storage frames or disk shelves. Virtual SAN is addressed in more detail in Chapters 4–7.

Virtual Volumes

While they are not part of an HCI architecture strategy, Virtual Volumes is nevertheless an important component in VMware's software-defined storage model. Virtual Volumes uses shared storage devices in a new way, and transforms storage management by enabling full virtual machine awareness from the storage array. Based on a T10 industry standard, Virtual Volumes provides a unique level of integration between vSphere and third-party vendors' storage hardware, which significantly improves the efficiency and manageability of virtual workloads.

Virtual Volumes virtualizes shared SAN and NAS storage devices, which are then presented to vSphere hosts, providing logical pools of raw disk capacity, called a *virtual datastore*. Then, Virtual Volume objects, which represent virtual disks and other virtual machine entities, natively reside on the underlining storage, making the object, or virtual disk, the primary unit of data management at the array level, instead of a LUN. As a result, it becomes possible to execute storage operations with virtual-machine, or even virtual-disk, granularity on the underlining storage system, and therefore provide native array-based data services, such as snapshots or replication, to individual virtual machines.

To facilitate a simplified and unified approach to management, all this is done with a common storage-policy-driven mechanism, which encompasses both Virtual SAN storage resources and Virtual Volumes external storage, into a single management plane. Virtual Volumes is covered in more detail in Chapter 8, "Policy-Driven Storage Design with Virtual Volumes."

Classic and Next-Generation Storage Models

This book refers to storage technologies as either *classic* or *next-generation*. Because these terms can have multiple meanings, this section provides an overview of each to clarify.

This book uses *classic storage model* to describe the traditional shared storage model used by vSphere. This typically includes LUNs, VMFS-based volumes and datastores, or NFS mount points, with a shared storage protocol providing I/O connectivity. Despite its constraints, this model has been successfully employed for years, and will continue to be used for some time by IT organizations and cloud service providers across the industry.

The *next-generation* storage model refers to VMware's software-defined solutions, Virtual SAN and Virtual Volumes, which bring about a new era in storage design, implementation, and management.

As addressed earlier in this chapter, the primary aim of VMware's software-defined storage model is to bring about simplicity, efficiency, and cost savings to storage resources. The model does this by abstracting the underlining storage in order to make the application the fundamental unit of management across a heterogeneous storage platform. With both Virtual SAN and Virtual Volumes, VMware moves away from the rigid constraints of the classic LUNs and volumes, and provides a new way to manage storage on a per virtual machine basis, through its more flexible policy-driven approach.

However, before addressing these *next-generation* storage technologies, you first need to understand the approach taken to storage over the last generation of vSphere-based virtualization platforms, and see how the VMware stack itself interacts with storage resources to provide a flexible, modern virtual data center.

This first chapter has addressed the VMware storage landscape, processes associated with storage design, and challenges faced by vSphere storage administration teams when maintaining complex, heterogeneous storage platforms on a daily basis for enterprise IT organizations and cloud service providers. The next chapter presents many of the essential design considerations based on the classic storage model previously outlined.

Chapter 2

Classic Storage Models and Constructs

This chapter covers the design considerations for deploying classic storage technologies in a VMware-based virtual data center, and addresses the primary storage concepts that impact the platform design of the storage layer.

Classic Storage Concepts

Storage infrastructure is made up of a multitude of complex components and technologies, all of which need to interact seamlessly to provide high performance, continuous availability, and low latency across the environment. For students of vSphere storage, understanding the design and implementation complexities of mixed, multiplatform, multivendor enterprise or service provider–based storage can at first be overwhelming. Gaining the required understanding of all the components, technologies, and vendor-specific proprietary hardware takes time.

This chapter addresses each of these storage components and technologies, and their interactions in the classic storage environment. Upcoming chapters then move on to next-generation VMware storage solutions and the software-defined storage model.

This classic storage model employs intelligent but highly proprietary storage systems to group disks together and then partition and present those physical disks as discrete logical units. Because of the proprietary nature of these storage systems, my intention here is not to address the specific configuration of, for instance, HP, IBM, or EMC storage, but to demonstrate how the vSphere platform can use these types of classic storage devices.

In the classic storage model, the logical units, or storage devices, are assigned a logical unit number (LUN) before being presented to vSphere host clusters as physical storage devices. These LUNs are backed by a back-end physical disk array on the storage system, which is typically served by RAID (redundant array of independent disks) technology; depending on the hardware type, this technology can be applied at either the physical or logical disk layer, as shown in Figure 2.1.

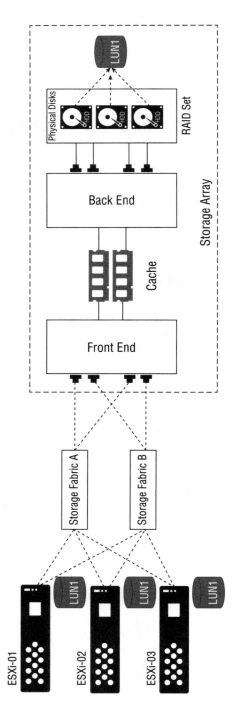

FIGURE 2.1
Classic storage model

The LUN, or storage device, is a virtual representation of a portion of physical disk space within the storage array. The LUN aggregates a portion of disk space across the physical disks that make up the back-end system. However, as illustrated in the previous figure, the data is not written to a single physical device, but is instead spread across the drives. It is this mechanism that allows storage systems to provide fault tolerance and performance improvements over writing to a single physical disk.

This classic storage model has several limitations. To start with, all virtual disks (VMDKs) on a single LUN are treated the same, regardless of the LUN's capabilities. For instance, you cannot replicate just a single virtual disk at the storage level; it is the whole LUN or nothing. Also, even though vSphere now supports LUNs of up to 64 terabytes, LUNs are still restricted in size, and you cannot attach more than 256 LUNs to a vSphere host or cluster.

In addition, with this classic storage approach, when a SCSI LUN is presented to the vSphere host or cluster, the underlying storage system has no knowledge of the hypervisor, filesystem, guest operating system, or application. It is left to the hypervisor and vCenter, or other management tools, to map objects and files (such as VMDKs) to the corresponding extents, pages, and logical block address (LBA) understood by the storage system. In the case of a NAS-based NFS solution, there is also a layer of abstraction placed over the underlying block storage to handle file management and the associated file-to-LBA mapping activity.

Other classic storage architecture challenges include the following:

- Proprietary technologies and not commodity hardware
- Low utilization of raw storage resources
- Frequent overprovisioning of storage resources
- Static, nonflexible classes of service
- Rigid provisioning methodologies
- Lack of granular control, at the virtual disk level
- Frequent data migrations required, due to changing workload requirements
- Time-consuming operational processes
- Lack of automation and common API-driven provisioning
- Slow storage-related requests requiring manual human interaction to perform maintenance and provisioning operations

Most storage systems have two basic categories of LUN: the traditional model and disk pools. The traditional model has been the standard mechanism for many years in legacy storage systems. Disk pools have recently provided compatible systems with additional flexibility and scalability, for the provisioning of virtual storage resources.

In the traditional model, when a LUN is created, the number and choice of disks directly corresponds to the RAID type and disk device configured. This traditional model has limitations, especially in virtual environments, which is why it was superseded by the more modern disk pool concept. The traditional model would often have a fixed maximum number of physical disks, which could be combined to form the logical disk. This maximum disk limitation was imposed by storage array systems as a hard limit, but was also linked to the practical considerations around availability and performance.

With this traditional disk-grouping method, it was often possible to expand a logical disk beyond its imposed physical limits by creating some sort of MetaLUN. However, this increased operational complexity and could often be difficult and time-consuming.

An additional consideration with this approach was that the amount of storage provisioned was often far greater than what was required, because of the tightly imposed array constraints. Provisioning too much storage was also done by storage administrators to prevent application outages often required to expand storage, or to cover potential workload requirements or growth patterns that were unknown. Either way, this typically resulted in expensive disk storage lying unutilized for a majority of the time.

On the plus side, this traditional approach to provisioning LUNs provided fixed, predictable performance, based on the RAID and disk type employed. For this reason, this method of disk provisioning is still sometimes a good choice when storage requirements do not have large amounts of expected growth, or have fixed service-level agreements (SLAs) based on strict application I/O requirements.

In more recent years, storage vendors have moved almost uniformly to disk pools. Pools can use far larger groups of disks, from which LUNs can be provisioned. While the disk pool concept still comprises physical disks employing a RAID mechanism to stripe or mirror data, with a LUN carved out from the pool, this device type can be built across a far greater number of disks. As a result of this approach, storage administrators can provision significantly larger LUNs without sacrificing levels of availability.

However, the sacrifice made by employing this more flexible approach is the small level of variability in performance that results. This is due to both the number of applications that are likely to share the storage of this single disk pool, which will inevitably increase over time, and the potential heterogeneous nature of disk pools, which have no requirement for uniformity, as it relates to the speed and capacity of individual physical disks (see Figure 2.2).

FIGURE 2.2
Storage LUN provisioning mechanisms

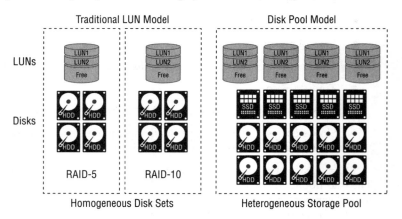

Also relevant from a classic storage design perspective are the trade-offs associated with choosing between provisioning a single disk pool or multiple disk pools. If choosing multiple pools, what criteria should a design use to define those pools?

We address tiering and autotiering in more detail later in this chapter, but this is one of the key design factors when considering whether to provision a single pool, with all the disk resources, or to deploy multiple storage pools on the array and to split storage resources accordingly.

Choosing a single pool provides simpler operational and capacity management of the environment. In addition, it allows LUNs or filesystems to be striped across a larger number of physical disks, which improves overall performance of the array system. However, it is also likely that a larger number of hosts and clusters will share the same underlying back-end disk system. Therefore, there is an increased possibility for resource contention and also an increased risk of specific applications not using an optimal RAID configuration, and maximizing I/O, which is likely to result in a degraded performance for those workloads.

Using multiple disk pools offers the flexibility to customize storage resources to meet specific application I/O requirements, and also allows operational teams to isolate specific workloads to specific physical drives, reducing the risk of disk contention. However, as the pools are inevitably smaller in this type of architecture, some systems may experience lower levels of performance than with a single larger pool. In addition, with multiple smaller pools, capacity planning becomes more complex, as growth across disk pools may not be consistent, and there is likely to be an increase in overall disk resources not being used.

Neither of these options is without its advantages and drawbacks, and there is no one perfect solution. However, designing a solution that uses multiple smaller pools over one universal disk pool will likely come down to one or more of the following key design factors:

- Disk pools based on function, such as development, QA, production, and so on. This option may be preferred if you are concerned with performance for specific environments, and want to isolate them from impacting the production system.

- In multitenanted environments, whether public or based on internal business units, each tenant can be allocated its own pool. However, depending on the environment and SLAs, each tenant might end up with multiple pools in order to address specific I/O characteristics of various applications.

- Application-based pools, such as database or email systems. This can provide optimum performance as applications of similar type often have similar I/O characteristics. For this reason, it may be worth considering designing pools based on application type. However, this also carries the risk of some databases, for instance, generating very high volumes of I/O and potentially impacting other databases residing on the same disk pool.

- Drive technology and RAID type. This allows you to place data on the storage type that best matches the application I/O characteristics, such as reads versus writes versus sequential. However, this approach can also increase costs and does not address any specific application I/O intensity requirement.

- Storage tier–based pools (such as Gold, Silver, and Bronze) could allow you to mix drive technologies and/or RAID types within each pool, therefore reducing the number of pools required to support most application types, configurations, and SLAs.

RAID Sets

The term *RAID* has already been used multiple times in different contexts, so let's address this technology next.

RAID (redundant array of independent disks) combines two or more disk drives into a logical grouping, typically known as a RAID set. Under the control of a RAID controller (or in the case of a storage system, the storage processors or controllers), the RAID set appears to the connected hosts as a single logical disk drive, even though it is made up of multiple physical disks. RAID sets provide four primary advantages to a storage system:

- Higher data availability
- Increased capacity
- Improved I/O performance
- Streamlined management of storage devices

Typically, the storage array management software handles the following aspects of RAID technology:

- Management and control of disk aggregation
- Translation of I/O requests between the logical and the physical entities
- Error correction if disk failures occur

The physical disks that make up a RAID set can be either traditional mechanical disks or solid-state flash drives (SSDs). RAID sets have various levels, each optimized for specific use cases. Unlike many other common technologies, RAID levels are not standardized by an industry group or standardization committee. As a result, some storage vendors provide their own unique implementation of RAID technology. However, the following common RAID levels are covered in this chapter:

- RAID 0–striping
- RAID 1–mirroring
- RAID 5–striping with parity
- RAID 6–striping with double parity
- RAID 10–combining mirroring and striping

Determining which type of RAID to use when building a storage solution largely depends on three factors: capacity, availability, and performance. This section addresses the basic concepts that provide a foundation for understanding disk arrays, and how RAID can enable increased capacity by combining physical disks, provide higher availability in case of a drive failure, and increase performance through parallel drive access.

A key element in RAID is redundancy, in order to improve fault tolerance. This can be achieved through two mechanisms, *mirroring* and *striping*, depending on the RAID set level configured. Before addressing the RAID set capabilities typically used in storage array systems, we must first explain these two terms and what they mean for availability, capacity, performance, and manageability.

NOTE Some storage systems also provide a JBOD configuration, which is an acronym for *just a bunch of disks*. In this configuration, the disks do not use any specific RAID level, and instead act as stand-alone drives. This type of disk arrangement is most typically employed for storage devices that contain swap files or spooling data, where redundancy is not paramount.

Striping in RAID Sets

As highlighted previously, RAID sets are made up of multiple physical disks. Within each disk are groups of continuously addressed blocks, called *strips*. The set of aligned strips that spans across all disks within the RAID set is called the *stripe* (see Figure 2.3).

Figure 2.3
Strips and stripes

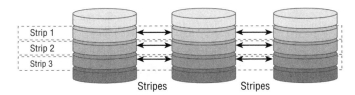

Striping improves performance by distributing data across the disks in the RAID set (see Figure 2.4). This use of multiple independent disks allows multiple reads and writes to take place concurrently, providing one of the main advantages of disk striping: improved performance. For instance, striping data across three hard disks would provide three times the bandwidth of a single drive. Therefore, if each drive runs at 175 input/output operations per second (IOPS), disk striping would make available up to 525 IOPS for data reads and writes from that RAID set.

Striping also provides performance and availability benefits by doing the following:

- Managing large amounts of data as it is being written; the first piece is sent to the first drive, the second piece to the second drive, and so on. These data pieces are then put back together again when the data is read.

- Increasing the number of physical disks in the RAID set increases performance, as more data can be read or written simultaneously.

- Using a higher stripe width indicates a higher number of drives and therefore better performance.

- Striping is managed through storage controllers, and is therefore transparent to the vSphere platform.

As part of the same mechanism, *parity* is provided as a redundancy check, to ensure that the data is protected without having to have a full set of duplicate drives, as illustrated in Figure 2.5. Parity is critical to striping, and provides the following functionality to a striped RAID set:

- If a single disk in the array fails, the other disks have enough redundant data so that the data from the failed disk can be recovered.

- Like striping, parity is generally a function of the RAID controller or storage controller, and is therefore fully transparent to the vSphere platform.

- Parity information can be

    - Stored on a separate, dedicated drive
    - Distributed across all the drives in the RAID set

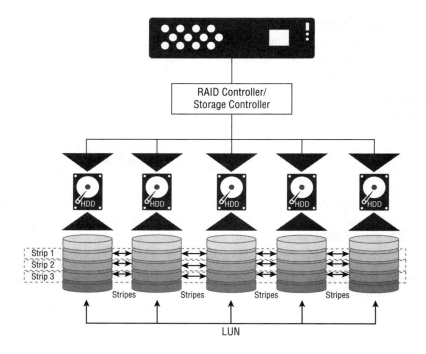

FIGURE 2.4
Performance in striping

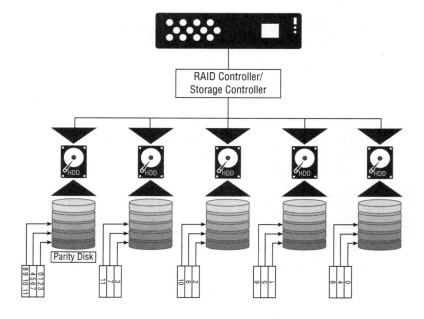

FIGURE 2.5
Redundancy through parity

Mirroring in RAID Sets

Mirroring uses a mechanism that enables multiple physical disks to hold identical copies of the data, typically on two drives. Every write of data to a disk is also a write to the mirrored disk, meaning that both physical disks contain exactly the same information at all times. This mechanism is once again fully transparent to the vSphere platform and is managed by the RAID controller or storage controller. If a disk fails, the RAID controller uses the mirrored drive for data recovery, but continues I/O operations simultaneously, with data on the replaced drive being rebuilt from the mirrored drive in the background.

The primary benefits of mirroring are that it provides fast recovery from disk failure and improved read performance (see Figure 2.6). However, the main drawbacks include the following:

- Degraded write performance, as each block of data is written to multiple disks simultaneously
- A high financial cost for data protection, in that disk mirroring requires a 100 percent cost increase per gigabyte of data

FIGURE 2.6
Redundancy in disk mirroring

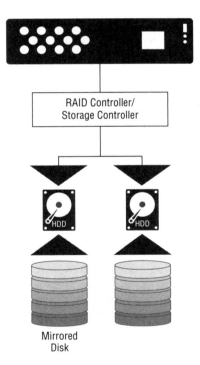

Enterprise storage systems typically support multiple RAID levels, and these levels can be mixed within a single storage array. However, once a RAID type is assigned to a set of physical disks, all LUNs carved from that RAID set will be assigned that RAID type.

Nested RAID

Some RAID levels are referred to as *nested RAID*, as they are based on a combination of RAID levels. Examples of nested RAID levels include RAID 03 (RAID 0+3, also known as RAID 53, or RAID 5+3) and RAID 50 (RAID 5+0). However, the only two commonly implemented nested RAID levels are RAID 1+0, also commonly known as RAID 10, and RAID 01 (RAID 0+1). These two are similar, except the data organization methods are slightly different; rather than creating a mirror and then striping the mirror, as in RAID 1+0, RAID 0+1 creates a stripe set and then mirrors it.

Calculating I/O per Second RAID Penalty

One of the primary ways to measure disk performance is input/output per second, also referred to as I/O per second or, more commonly, IOPS. This formula is simple: one read request or one write request is equal to one I/O.

Each physical disk in the storage is capable of providing a fixed number of I/O. Disk manufacturers calculate this based on the rotational speed, average latency, and seek time. Table 2.1 shows examples of typical physical drive IOPS specifications for the most common drive types.

TABLE 2.1: Typical average I/O per second (per physical disk)

Drive Speed	Typical Average IOPS/Drive
Solid-State Disk (SSD)	6,000
15,000 RPM	175
10,000 RPM	125
7,200 RPM	75
5,400 RPM	50

A storage device's IOPS capability is calculated as an aggregate of the sum of disks that make up the device. For instance, when considering a JBOD configuration, three disks rotating at 10,000 RPMs provide the JBOD with a total of 375 IOPS. However, with the exception of RAID 0 (which is simply a set of disks aggregated together to create a larger storage device), all RAID set configurations are based on the fact that write operations result in multiple writes to the RAID set, in order to provide the targeted level of availability and performance.

In a RAID 5 disk set, for example, for each random write request, the storage controller is required to perform multiple disk operations, which has a significant impact on the raw IOPS calculation. Typically, that RAID 5 disk set requires four IOPS per write operation. In addition, RAID 6, which provides a higher level of protection through double fault tolerance, also provides a significantly worse *I/O penalty* of six operations per write. Therefore, as the architect of such a solution, you must also plan for any I/O penalty associated with the RAID type being used in the design.

Table 2.2 summarizes the read and write RAID penalties for the most common RAID levels. Notice that you don't have to calculate parity for a read operation, and no penalty is associated with this type of I/O. The I/O penalty relates specifically to writes, and there is no negative performance or IOPS impact when calculating read operations. It is only when you have writes to disk that you will see the RAID penalty come into play in RAID calculations and formulas. This is true even though in a parity-based RAID-type write operation, reads are performed as part of that write. For instance, writes in a RAID 5 disk set, where data is being written with a size that is less than that of a single block, require the following actions to be performed:

1. Read the old data block.
2. Read the old parity block.
3. Compare data in the old block with the newly arrived data. For every changed bit, change the corresponding bit in parity.
4. Write the new data block.
5. Write the new parity block.

As noted previously, a RAID 0 stripe has no write penalty associated with it because there is no parity to be calculated. In Table 2.2, a no RAID penalty is expressed as a 1.

TABLE 2.2: RAID I/O penalty impact

RAID LEVEL	READ	WRITE PENALTY	EXAMPLE OF WRITE IOPS FOR A 15K DISK
RAID 0–Striping	1	1	175
RAID 1–Mirroring	1	2	85
RAID 3–Parallel transfer with parity	1	3	65
RAID 5–Striping with parity	1	4	40
RAID 6–Striping with double parity	1	6	30
RAID 10–Combining mirroring and striping	1	2	85

Parity-based RAID sets introduce additional processing overhead on the storage controllers, which results from the additional calculations required to determine the parity data. The higher the level of parity protection you provide to a RAID set, the more processing overhead you incur on the controllers, although, as you would expect, the actual overhead incurred is highly dependent on the workload's read/write balance.

In calculating the number of IOPS incurred by the RAID penalty, the following formula provides a good starting point, assuming you have derived the customer's workload balance between read and write operations from a current state analysis. However, you must also take into account peak and average workloads, to ensure that the storage device can deliver the required IOPS.

(total workload IOPS) × (% of workload that is read operations) + (total workload IOPS × % of workload that is read operations × RAID I/O penalty)

In this example calculation, the customer has provided the following workload I/O values:

- Total IOPS required: 250 IOPS
- Read workload: 50 percent
- Write workload: 50 percent
- RAID level required: 6 (I/O penalty of 6)

You would require a RAID 6 disk set that could support 875 IOPS, in order to meet the customer's requirement for 250 IOPS on a RAID 6 disk set, where the workload has 50 percent write operations.

As this example makes clear, the number of disks is far more important than the disk capacity. Based on the information provided by the customer, you would require twelve 7,200 RPM disks, seven 10K RPM disks, or five 15K RPM disks to support the required IOPS.

As you can see, determining the correct RAID type for a specific workload is key, and will come down to various design factors and compromises between cost, availability, and performance.

RAID Levels Explained

The RAID type chosen for a specific LUN determines the level of redundancy and data integrity that the LUN provides to the applications running on it. However, not all storage array vendors support all RAID types, and some have even developed their own. As part of ensuring that your storage design meets the customer's needs, you should establish the types of RAID available for the hardware vendor's storage devices. Tables 2.3 through 2.8 provide insight into the types of RAID that are most commonly employed in storage arrays, with illustrations in Figures 2.7 through 2.12.

Table 2.3: RAID 0—striped disk array without fault tolerance

Design Factor	Description
Data protection	None. RAID 0 stripes the information across the drives in the array without generating redundant data. By providing no parity or mirroring, there is no fault tolerance, making it extremely difficult to recover data.
Advantages	RAID 0 offers great performance, both in read and write operations. No overhead is created by parity controls, which also allows all storage capacity to be used. This technology is also easy to implement.
Drawbacks	RAID 0 is not fault-tolerant. If one drive fails, all data in the RAID 0 disk set will be lost. This RAID type should not be employed for business-critical systems.
Performance characteristics	RAID 0 is superior to a JBOD configuration, as it uses striping. All the data is spread out in chunks across all the disks in the RAID set. The I/O rate, or throughput, can be good when I/O sizes are small; however, larger I/Os will produce high bandwidth (data moved per second) with this RAID type. Performance can be further improved when data is striped across multiple controllers, with only one drive per controller.

Table 2.3: RAID 0—striped disk array without fault tolerance *(CONTINUED)*

Design Factor	Description
Minimum disks	At least two disks.
Cost per GB	Low.
RAID write penalty	1
Maintenance & operational considerations	RAID 0 provides zero fault tolerance, so recovering data from the disks is extremely difficult. Full data recovery is typically done from backups.
Typical application usage	Suitable applications include those that need high bandwidth or high throughput, but where the data is not critical, such as on an image-retouching or video-editing application. RAID 0 is ideal for high-bandwidth, noncritical storage of data that has to be read or written at a high speed in order to meet application requirements.
Key design characteristics	Not a *true* RAID type, as it provides no fault-tolerance. The failure of just one drive results in all data in the RAID set being lost. Therefore, this RAID type should never be used in mission-critical environments.

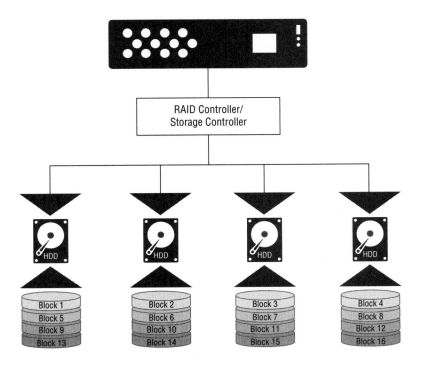

Figure 2.7
RAID 0 striped disk array without fault tolerance

TABLE 2.4: RAID 1—disk mirroring and duplexing

Design Factor	Description
Data protection	RAID 1 employs mirroring to provide fault tolerance. A RAID 1 group typically consists of two or more disks, although more than two is rarely seen, as in such configurations RAID 10 would generally be employed. Every write to a data disk is also written to the mirror disk. If a disk fails, the disk array controller can use either drive for data recovery, while also continuing normal I/O operations. All data on the replaced drive is rebuilt from the mirrored drive.
Advantages	RAID 1 offers excellent read speed, and a write speed comparable to that of a single drive. In case of a failed disk, the data does not have to be rebuilt; it just has to be copied to the replacement drive, making RAID 1 a simple technology to implement.
Drawbacks	The total number of disks in the array will equal twice the usable drives. This means that the overhead cost is equal to 100 percent, making a usable storage capacity of 50 percent of the total available across all disks in the RAID set.
Performance characteristics	RAID 1 improves read performance, as reads can be distributed across multiple disks, but degrades write performance. Write performance is the same as for single-disk storage.
Minimum disks	At least two disks.
Cost per GB	Expensive because of the extra capacity required to mirror data 1:1.
RAID write penalty	2
Maintenance & operational considerations	Low operational complexity.
Typical application usage	RAID 1 is ideal for mission-critical storage—for instance, for accounting, payroll, finance, and applications requiring high availability and fast read I/O. It is also suitable for small RAID sets, in which only two drives will be used.
Key design characteristics	RAID 1 offers the best all-around performance, and also offers good protection that can sustain double drive failures that are not in the same mirror set. Economy is the lowest of the RAID types, since usable storage is only 50 percent of the total available raw disk capacity.

FIGURE 2.8
RAID 1 disk mirroring and duplexing

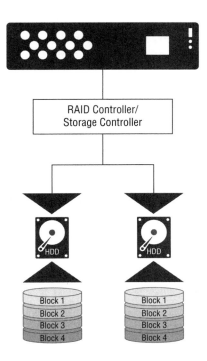

TABLE 2.5: RAID 1+0—mirroring and striping

DESIGN FACTOR	DESCRIPTION
Data protection	RAID 1+0 (also known as RAID 10, RAID 1/0, or RAID A) combines the speed of RAID 0 with the redundancy of RAID 1, but it is implemented in a different way. This is a nested RAID configuration, which provides availability by mirroring all data on secondary drives, while also employing striping across each set of drives to speed up data-transfer rates.
Advantages	RAID 1+0 provides high reliability, high data availability, high I/O rates, in small block sizes, and the ability to withstand multiple drive failures as long as they occur on different mirrors. If something goes wrong with one of the disks in a RAID 1+0 configuration, the rebuild time is fast, because all that is required is for all of the data to be copied from the surviving mirror to a new drive. This can take as little as 30 minutes for 1 TB drives.

TABLE 2.5: RAID 1+0—mirroring and striping *(CONTINUED)*

Design Factor	Description
Drawbacks	The total number of disks is twice that of the usable data disks, with an overhead cost equaling 100 percent.
Performance characteristics	High I/O rates are achieved using multiple stripe segments, with writes being slower than reads, as they are mirrored.
Minimum disks	4
Cost per GB	Expensive because of the high mirroring overhead, with half of the storage capacity going to mirroring. Therefore, when compared to large RAID 5 or RAID 6 arrays, this is an expensive way to provide redundancy.
RAID write penalty	2
Maintenance & operational considerations	Low operational complexity.
Typical application usage	Databases requiring high I/O rates with random data, and applications requiring maximum data availability.
Key design characteristics	Very high reliability, combined with high performance.

FIGURE 2.9
RAID 1+0 mirroring and striping

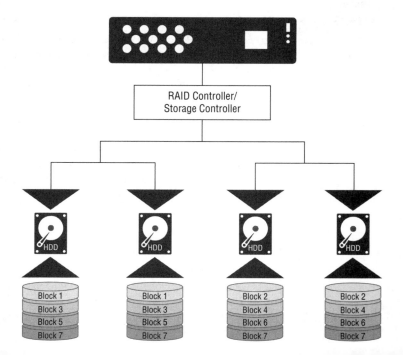

TABLE 2.6: RAID 3—parallel transfer with dedicated parity disk

DESIGN FACTOR	DESCRIPTION
Data protection	RAID 3 stripes data for high performance and uses parity for fault tolerance. In RAID 3, data is striped across all the disks in the RAID set, except one, with parity information stored on a dedicated drive, instead of striping it across all drives with the data, as in RAID 5, so that data can be reconstructed if a drive fails. RAID 3 always reads and writes complete stripes of data across all disks, as there are no partial writes, which update one out of multiple strips in a stripe.
Advantages	The total number of disks is less than in a mirrored solution. For instance, 1.25 times the data drives for a group of five disks. RAID 3 also provides good bandwidth for large data transfers.
Drawbacks	RAID 3 provides poor efficiency in handling small data blocks. It is not well suited to transaction processing applications, such as databases. Also, data is lost if multiple drives fail within the same RAID set.
Performance characteristics	RAID 3 provides high data read/write transfer rates, and a disk failure does not have a significant impact on the throughput. However, disk rebuilds can be very slow.
Minimum disks	RAID 3 requires a minimum of three physical disks, although some storage vendors' implementations of this RAID type might require five or nine disks.
Cost per GB	Medium.
RAID write penalty	3
Maintenance & operational considerations	Medium operational complexity.
Typical application usage	Since an I/O operation addresses all drives at the same time, RAID 3 cannot overlap I/O. For this reason, RAID 3 is best for single-user systems, with long record applications, and applications where large sequential data access is employed, such as medical and geographic imaging software.
Key design characteristics	This RAID level performs well under applications that just want one long sequential data transfer. Applications like video servers work well with this RAID type.
Key design characteristics	RAID 3 provides a very high read data-transfer rate, as well as a very high write data-transfer rate, and the low ratio of parity disks to data disks means a design gets high efficiency of capacity. In RAID 3, a disk failure has an insignificant impact on the throughput of the storage device.

FIGURE 2.10
RAID 3 parallel transfer with dedicated parity disk

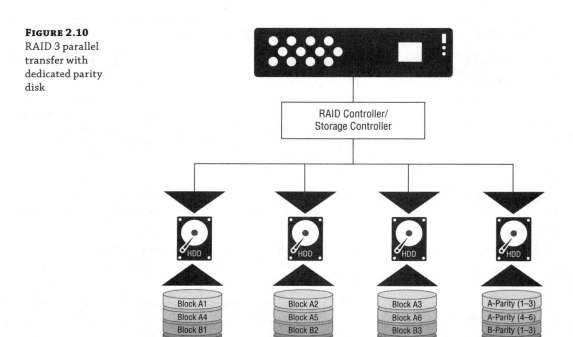

TABLE 2.7: RAID 5—independent data disks with distributed parity blocks

Design Factor	Description
Data protection	RAID 5 does not read and write data to all disks in parallel like RAID 3. Instead, it performs independent read and write operations. There is no dedicated parity drive, as data with parity information is distributed across all drives in the set. In RAID 5, a single disk failure puts the RAID set in degraded mode. RAID 5 is slower to rebuild (when compared to RAID 1), as it employs block-level or multiple-block-level striping.
Advantages	Read data transactions are very fast, while write data transactions are somewhat slower, because of the parity that has to be calculated. If a drive fails in a RAID 5 set, the storage controller rebuilds the data on the new drive from parity information gathered across all disks in the RAID set.

TABLE 2.7: RAID 5—independent data disks with distributed parity blocks *(CONTINUED)*

Design Factor	Description
Drawbacks	RAID 5 has a slower transfer rate than RAID 3. Small writes are slow, as they require a read-modify-write (RMW) operation. There is also degraded performance when the RAID set is in recovery or reconstruction mode, which can be prolonged. For instance, if one of the disks in a RAID 5 set, using 4 TB drives, fails and is replaced, the rebuild time may take a whole day or longer, depending on the load on the array, and the speed of the controller. A RAID 5 disk set can tolerate only a single drive failure, so data loss is likely to occur during this period if an additional drive, within the same RAID set, fails during a rebuild operation.
Performance characteristics	RAID 5 provides a high read data-transaction rate and a medium write data-transaction rate. RAID 5 also provides a low ratio of parity disks to data disks, and typically provides good overall performance for mixed virtual workloads.
Minimum disks	RAID 5 requires at least three drives, but can also work with far more disks in a modern storage array pool. Traditionally, five-disk and nine-disk sets were most popular, but storage arrays typically have a wide range of options.
Cost per GB	Significant cost savings result from the use of parity over mirroring. The cost per gigabyte economy of RAID 5 is excellent, with a usable capacity of 80 percent of the disks' raw storage.
RAID write penalty	4
Maintenance & operational considerations	RAID 5 has a medium level of operational complexity, although the understanding of this technology is generally good among IT professionals, because of its common use.
Typical application usage	RAID 5 is a good all-round option that combines efficient storage with good levels of availability and decent performance. It is ideal for file and application servers that have a limited number of data drives, and is good for parallel processing and multitasking application environments. RAID 5 is also good for file and application servers, some types of database servers, and WWW and email services.
Key design characteristics	RAID 5 offers the best mix of performance, protection, and economy, although it is generally considered a poor choice for use on write-intensive systems, because of the performance impact associated with writing parity information. It has a higher write performance penalty than RAID 1, since two reads and two writes are required to perform a single write operation. However, for large block sequential writes, internal optimization typically eliminates this penalty, as parity can be calculated in memory. RAID 5 is able to provide only single parity. Therefore, this RAID type is more susceptible to data loss after double drive failure, or an error during a disk rebuild operation.

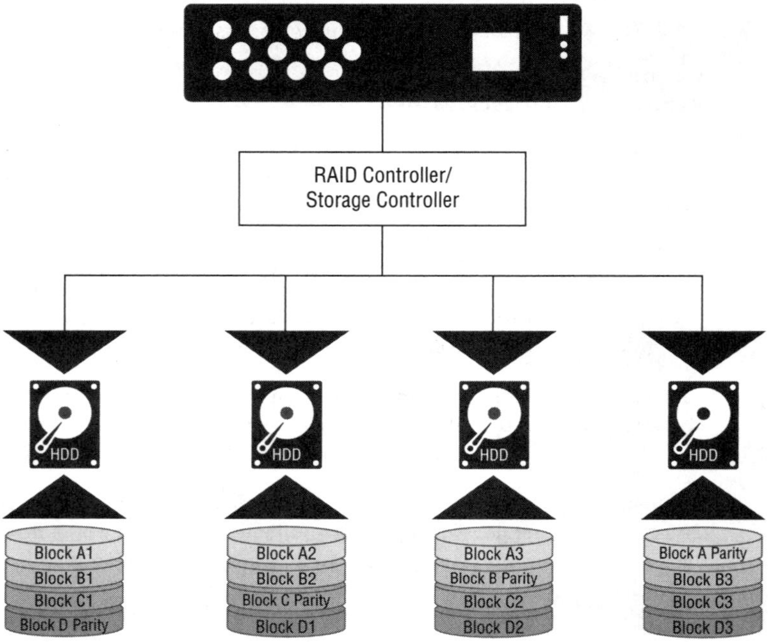

FIGURE 2.11
RAID 5 independent data disks with distributed parity blocks

TABLE 2.8: RAID 6—independent data disks with two independent parity schemes

DESIGN FACTOR	DESCRIPTION
Data protection	RAID 6 is similar to RAID 5, but the parity data is written to two drives. This means it requires at least four disks, but can withstand two drives dying simultaneously. Even though the chances of two drives failing at the same time are very small, if a drive in a RAID 5 system fails and is replaced by a new drive, it can take hours to rebuild the new disk. If another drive fails during that time, you would lose all data in the disk set. However, in a RAID 6 disk set, the data would be able to survive that second drive failure.
Advantages	As with RAID 5, read data operations are very fast. If two drives fail, you still have access to all data, even while the failed drives are being replaced. Therefore, RAID 6 is more secure, from an availability perspective, than RAID 5.
Drawbacks	Write data transactions are slowed down because of the parity that has to be calculated, and drive failures have an effect on throughput, although this is typically still acceptable. Also, rebuilding a RAID set in which one drive has failed can take a long time, and this is sometimes seen as a more complex technology than RAID 5, and is therefore less understood by IT personnel.

TABLE 2.8: RAID 6—independent data disks with two independent *(CONTINUED)*

DESIGN FACTOR	DESCRIPTION
Performance characteristics	Good random read and very good sequential read performance.
Minimum disks	4
Cost per GB	Economy is very good, with usable capacity at 75 percent of that of the raw disk aggregate storage.
RAID write penalty	RAID 6 is almost identical to RAID 5, except instead of calculating parity once, it has to do it twice. Therefore, we have three reads and then three writes, giving us a RAID penalty of 6.
Maintenance & operational considerations	Medium operational complexity, as this technology is often less understood by operational teams.
Typical application usage	RAID 6 is a good all-around system that combines efficient storage with excellent availability and decent performance. It is preferable over RAID 5 in file and application servers, which employ a large number of high-capacity drives for data storage.
Key design characteristics	RAID 6 offers the best protection and read performance, when compared to RAID 5. However, it does have a significant write performance penalty, since three reads and three writes are required to perform a single write operation.

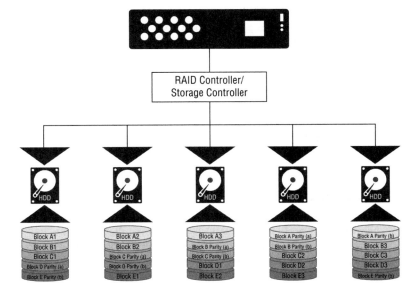

FIGURE 2.12
RAID 6 independent data disks with two independent parity schemes

In addition to these RAID levels, other less common RAID types are available on some storage systems. Also, some storage vendors provide their own variants on RAID sets, with additional vendor-specific features. For instance, RAID-DP from NetApp provides a similar feature set to RAID 6. However, in the NetApp implementation, rather than the parity being distributed across all disks, RAID-DP uses two specific disks for parity.

Despite the data protection being offered by RAID sets, an individual physical disk provides no protection at all. Therefore, hot spares are also required to provide protection for individual faulted disks.

Hot Spares

As good as they are, the performance of a single disk is limited by physics, electronic speeds, and in the case of mechanical spindles, physical movement. These devices still break down, with problems such as media failure and connectivity failures. A *hot spare* is an idle drive in a storage array that becomes a temporary or permanent replacement for a failed active RAID set participant disk within the storage array. This works as follows:

1. The hot spare takes the failed drive's identity in the storage array during a failure event.
2. Data is recovered onto the hot spare. How this occurs depends on the RAID type implemented.
3. If parity is being used, the data is rebuilt onto the hot spare from the parity and data from the surviving drives.
4. If mirroring is being used, the data is rebuilt on the hot spare disk by using the data from the surviving mirrored drive.

When the failed drive is replaced with a new disk by an engineer, one of the following scenarios occurs, which on most modern storage systems is a fully automatic mechanism:

1. The hot spare replaces the new drive permanently, meaning that this disk is no longer a hot spare, and a new hot spare might be required and reconfiguration carried out on the storage system.
2. When a replacement drive is inserted by the engineer, the new drive is added to the system, and data from the hot spare is copied to the new drive. The hot spare then returns to its previous idle hot-spare status, ready to act as a replacement for the next failed drive. It is an important consideration in storage design to ensure that the hot-spare drives are at least large enough to accommodate the data from the failed drives, and that different hot spares are provided for different drive types.

RAID Summary

As we have addressed, implementing RAID sets on storage arrays carries several key benefits. For instance, performance is increased as a result of the controllers having more physical disk devices to read from or write to, and availability is improved as the RAID controller can re-create lost data from parity information, or a mirror, should a physical disk drive fail.

As you have seen, drives can be grouped together in various ways to form RAID sets. The methods used to group drives are called RAID types, or RAID levels. RAID levels 0, 1, 5, 6, and 10 are the most common used by IT organizations and service providers, each providing optimum performance and/or availability, based on the specific requirements of the workload.

However, mirrored and nested RAID sets are more expensive to implement than native parity-based RAID sets, as they require a greater number of disks, and therefore the cost per gigabyte of storage is also higher, as so many of the drives are used for redundancy. Nested RAID, particularly RAID 1+0, has become popular despite its cost, as it helps to overcome some of the reliability problems associated with parity-based RAID.

Initially, when a new storage system is implemented, all of the drives in the RAID sets are installed at the same time, making all of the drives the same age, and typically subject to similar operating conditions and amounts of wear. Therefore, when a drive fails, there is a high probability that another drive in the array will also soon fail, because of the mean time between failures and increased workload, or maybe it's just *Murphy's law*. Nevertheless, the risk of multiple drive failures should be taken into account as part of any storage design. Because some RAID levels, such as RAID 5 and RAID 1, can sustain only a single drive failure, the storage system and the data it contains can be left vulnerable, until the failed drive is replaced and rebuilt on the newly swapped disk.

Even if a second disk failure does not occur while the failed disk is being rebuilt, there is still a chance that the remaining disks in the RAID set may contain bad sectors or unreadable data, which might make it impossible to fully rebuild the information on the failed disk. Nested RAID levels address this problem, to some extent, by providing a greater degree of redundancy and reducing the chances of a RAID set failure due to simultaneous disk faults, which are not part of the same mirrored set.

In addition to the common RAID levels addressed so far, several nonstandard RAID levels exist that can be infrequently implemented or provide specific storage-vendor hardware with proprietary options. Discussing all of these in detail goes beyond the scope of this book. However, some noteworthy, less common implementations of RAID technology include the following:

- RAID 03 (RAID 0+3, also known as RAID 53 or RAID 5+3), has a higher transaction rate than RAID 3, and offers all the protection of RAID 10, but there are disadvantages as well.

- RAID 50 (RAID 5+0) combines distributed parity (RAID 5) with striping (RAID 0).

- RAID-DP is a NetApp proprietary technology, which implements double parity protection in a RAID group, similar to that of a RAID 6 disk set.

The final consideration, which is addressed in more detail in Chapter 3, "Fabric Connectivity and Storage I/O Architecture," is the I/O transport mechanism. It is important for a storage architect to understand that raw I/O numbers do not take into account the transport choice, such as Fibre Channel or iSCSI. Even though the choice of storage I/O transport is an important consideration for many IT organizations and service providers, it does not directly impact the I/O per second, and is not considered directly as part of any IOPS formulas or calculations carried out as part of the design. This choice of transport is important, for reasons discussed in the following chapter, but it does not drive design requirements for fulfilling application I/O performance.

Virtual Provisioning

Also known as *thin provisioning*, modern storage system disk pools can typically provide the ability to present a LUN to the vSphere platform with more capacity than is physically allocated to the LUN by the storage system back end. Thin provisioning can also be applied at the hypervisor layer by vSphere administrators, but in this context, we are referring to thin provisioning specifically as it is applied at the storage-array level.

One of the biggest challenges faced by storage administrators is balancing the storage space required by the various applications being hosted in their data centers. Typically, storage administrators allocate space based on the predicted growth requirements of a given application. This is done in order to reduce the operational overhead and risk associated with application downtime, caused by insufficient storage being provisioned in the first instance. It is also the case that some applications may require a planned outage, in order to expand their storage capacity. As a result, this proactive behavior often sees storage being significantly overprovisioned, which results in higher costs, through increased power utilization, cooling, and data-center floor space. These operational and cost challenges can be largely addressed by employing virtual provisioning technology at the storage layer.

Virtual provisioning provides the storage administrator the ability to present a *thin LUN* to a host system or cluster, which appears to vSphere with more capacity than what is physically allocated to the LUN on the storage system (see Figure 2.13). This works by allowing the storage array to allocate physical storage to the LUN on demand, from a shared pool of physical disk capacity. This mechanism provides significantly increased efficiency of storage resources by using only the physical disk capacity actually required by the virtual machines, and as such, reducing the amount of allocated physical storage overall.

FIGURE 2.13
Virtual provisioning

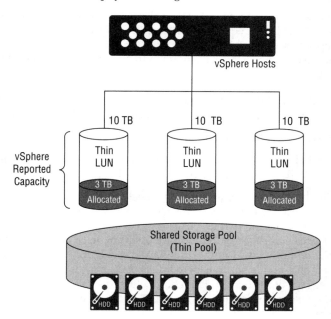

An additional benefit of virtual provisioning is the ability to overallocate storage resources, meaning that provisioning decisions are not necessarily bound by the array's currently available physical disk. For instance, in the example in Table 2.9, the storage administrator plans on provisioning three new LUNs to be presented to an existing vSphere cluster, using traditional thick-provisioning methods. The total current usable capacity of the storage system is 2 terabytes.

TABLE 2.9: Thick-provisioning example

LUN ID	LUN PROVISIONED CAPACITY	AMOUNT OF DATA STORED	UNUSED CAPACITY
LUN 1	500 GB	100 GB	400 GB
LUN 2	550 GB	50 GB	500 GB
LUN 3	800 GB	200 GB	600 GB

As shown in the example, the storage array contains 350 GB of data, although 1.5 TB of allocated capacity is unused, and 150 GB of storage space is unprovisioned.

Two weeks later, a new vSphere cluster is being configured and requires 400 GB of storage capacity for new virtual machines. However, the storage system has only 150 GB of unprovisioned capacity, so it is not possible to provide the 400 GB of storage the new cluster requires without new hardware being procured, even though 1.5 TB of capacity on the array is unused by the existing workloads. This example demonstrates the underutilization of storage when employing a traditional thick storage provisioning mechanism.

However, if you consider the same 2 TB storage array, employing virtual provisioning technology, the provisioning of capacity is not bound by the available storage. Even though the storage administrator creates the same three LUNs, no physical capacity is allocated by the storage system that is not used. Therefore, using the same values in a virtual provisioning environment, with 350 GB of data being used and 1.65 TB of available capacity, the storage array has more than enough resources available for the new vSphere cluster to be brought online. This is in contrast to the traditional storage thick-provisioning mechanism, where just 150 GB of disk remained available, and new storage resources would have been required for the new cluster (see Figure 2.14).

FIGURE 2.14
Traditional provisioning versus virtual provisioning

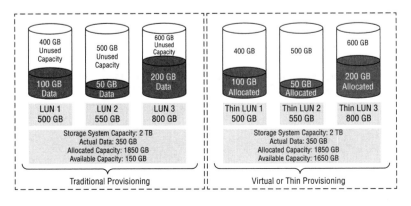

However, as capacity is used up by the virtualized workloads, the physical disk resources available in the pool will reduce. It is extremely important that the acceptable level of overprovisioning is defined by the storage operations team and monitored regularly. It is critical to the health of the environment that operational teams limit the level of overprovisioning, and ensure that the amount of available capacity does not fall below acceptable levels, before physical storage resources are exhausted or new disks can be added to the pool. If this situation does occur, running virtual machines and vSphere hosts will experience critical errors when trying to write data to the array, the result of which is likely to leave multiple applications unavailable.

VIRTUAL PROVISIONING DESIGN CONSIDERATIONS

When *thick* LUNs are provisioned on the storage array, the entire capacity is allocated up front, so typically a storage administrator will want to allocate the smallest amount of capacity possible in order to prevent wasted space. Therefore, by overprovisioning through thin provisioning at the vSphere layer instead, you can use the space more efficiently. However, when that capacity is used up, the LUN must be expanded, or additional LUNs provisioned.

With *thin* LUNs, a large amount of capacity can be allocated initially, with only a small fraction of it being used up front. From the host's perspective, it can consume the total amount of space allocated for its virtual machines. However, as the space on the LUN is consumed by virtual machines, the storage array dynamically adds capacity from the storage pool into the provisioned space.

Even though *thin* LUNs reduce the amount of storage consumed, they also present a new management challenge, especially if both the vSphere and storage layer are overprovisioned. As the two layers have no way to coordinate with one another, it is critical that both environments are monitored adequately. If the environment's storage resources on both the storage array and vSphere platform are not monitored for allocated and available capacity, the storage infrastructure can easily end up in a situation where one of the layers has exhausted its capacity, while the other layer has capacity available.

Typically, *thick* provisioned LUNs are best used for applications that cannot tolerate any performance variations on its back-end storage, or that require consistent high levels of performance in order to maintain their service-level agreements. In this type of tier 1 business-critical transactional application environment, space is not generally of concern. In this type of design use case, capacity is typically sitting idle because the storage administrator has allocated far more capacity than required, across a large number of disks, in order to meet application I/O requirements and not that of storage capacity. However, this approach is now changing, with more and more such use cases employing all-flash disk configurations.

In this type of classic storage environment, *thin* LUNs are best suited to environments where space efficiency is the primary concern, or where flexibility of storage resources is paramount. VMware's ESXi hypervisor can, in its own right, emulate the capabilities of a *thin* LUN through *thin provisioning* at the vSphere layer. When a virtual machine is deployed with this functionality, the hypervisor uses only the required capacity for each virtual disk, then dynamically expands the amount of consumed virtual disk space as the virtual machine grows in size. This technology was originally intended to be used on *thick* LUNs, but can also be used on *thin* LUNs, which are presented next.

Table 2.10 highlights typical use cases for both traditional *thick provisioning* and *virtual provisioning* at the storage array level.

TABLE 2.10: Virtual provisioning design considerations

USE TRADITIONAL *THICK* PROVISIONING WHEN	USE VIRTUAL *THIN* PROVISIONING WHEN
• Best and most predictable performance is required.	• Space efficiency is needed.
• Precise data placement is required.	• Minimal host impact to expanding resources is required.
• Less concerned about space efficiency.	• Reduced power consumption and capital cost savings are high priority.
	• Applications have space requirements and consumption that is difficult to predict.

VIRTUAL PROVISIONING LAYERING

It was referenced earlier in this chapter that *thin provisioning* could also be performed at the compute, vSphere layer, where the virtual machine disk files (VMDKs) are not allocated storage up front, but only zeroed out on demand when the virtual machine writes to the disk. Let's address the design considerations that come about as a result of layering thick and thin storage allocations at both the hypervisor and storage-array level, on top of one another. These are called *thick-on-thick*, *thick-on-thin*, *thin-on-thick*, and *thin-on-thin*.

In a vSphere environment, VMDKs can be deployed in three formats: thin, lazy zeroed thick (LZT), or eager zeroed thick (EZT). Both thin and thick disk files use lazy zeroing, in which the initial zeroing of the disk blocks is delayed until the first write of the virtual machine. However, eager zeroed thick disk blocks are pre-allocated with zeros at the time of disk provisioning, making it unnecessary to zero the disk on a first-write basis, during normal running virtual machine operations. This has traditionally provided approximately a 10 to 20 percent performance improvement over the other disk formats.

Figure 2.15 and Table 2.11 illustrate the outcomes associated with layering vSphere thin provisioning technology on top of both a storage array *thickly* and *thinly* deployed LUN. In effect, we're layering both of these mechanisms on top of one another at both the vSphere host layer and storage-array layer, in order to maximize disk efficiency.

Figure 2.15
Virtual provisioning layering

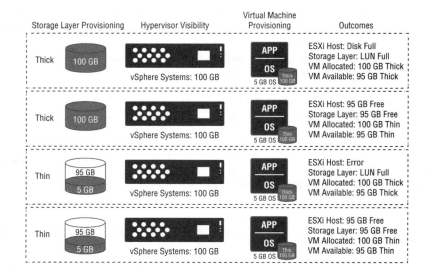

Table 2.11: Design factors of virtual provisioning

Capability	Thick VM / Thick Storage	Thin VM / Thick Storage	Thick VM / Thin Storage	Thin VM / Thin Storage
Array Capacity	Fully allocated at creation	Fully allocated at creation	Partially allocated at creation	Partially allocated at creation
Utilization	Full disk amount for each VM	Needed capacity for each VM	Full amount for each VM	Needed capacity for each VM
Overprovisioning	None	Hypervisor	Storage array	Hypervisor and storage array
Latency Cost	None	Initializing blocks	Array allocating storage	Initializing blocks and array allocating storage

The key consideration, as illustrated in Figure 2.15, is how the hypervisor and storage array handle zeros. In environments that use *thin provisioning* on the storage array, the vSphere administrator needs to be aware of how the hypervisor is allocating the blocks for an EZT virtual machine. With all the blocks being immediately zeroed upon creation, the full amount of the capacity from the *thin-provisioned* storage will be used. As you create multiple virtual machines in the same way, on that same storage, the consumed storage will approach the allocated storage, and if the thin storage is heavily oversubscribed across the pool, there is a substantial risk that disk resources might be exhausted quicker than anticipated.

That being said, some arrays will also perform zero reclamation on thin pools, in that they will examine blocks of data within the thin pool, and if they contain no data, will return that capacity to the pools' free resources. This can lead to performance variability on the array, because the storage capacity will have to be allocated from the pool when the hypervisor attempts to use it. This mechanism can also lead to capacity issues, as the hypervisor believes that it has allocated the full amount of disk to the virtual machines, but some of that space has been returned to the pool in the background, and therefore may have been consumed by other hosts.

In addition to the impact on the storage array, layering the two *thin* technologies also has some potential impact on the performance of virtual machines. In a *thin-on-thick* environment, every time the host writes data to the LUN on behalf of the virtual machine, the hypervisor has to write zeros to the blocks to initialize the space. This adds a small amount of latency to the write operation of the hypervisor, as each time a new block is needed, it first must be initialized.

In a *thick-on-thin* environment, a similar scenario occurs; the storage array has to allocate additional blocks when the allocated storage has been exhausted. In a *thin-on-thin* environment, both actions have to occur as the virtual disks expand. As the hypervisor writes data, storage must be allocated on the array from the pool of free capacity, and blocks must be initialized before first-write operations by the vSphere host.

Storage Tiering

For several years now, organizations have experienced unprecedented levels of data growth, which has increased their requirements for data-center storage resources. In turn, IT organizations have been seriously challenged by this tremendous growth in data, and the often long-term retention and management of this data for compliance reasons. The cost of storing this data to their businesses, while continuing to meet SLAs, can often be a problem. Continuing to buy more and more high-end storage systems to meet these business and regulatory requirements is not cost-effective, so IT organizations have needed a way to enable the storage of the right data, with the right availability, performance, and access, at the right cost per gigabyte.

Storage tiering has emerged as one potential solution to address this challenge. Storage tiering establishes a hierarchy of storage types and capabilities, and then helps to identify active or inactive data and relocate it to an appropriate storage type. This solution enables IT organizations to meet SLAs at an optimal cost per gigabyte of data.

Each tier of storage is designed to provide different levels of performance and availability. For instance, solid-state drives (SSDs) may be configured as tier 1 storage for frequently accessed data, where high performance is required by the SLA. At the same time, low-cost Near-Line SAS (NL-SAS) mechanical drives might be configured as a tier 4 storage solution, for less frequently accessed data, at a significantly lower cost per gigabyte. This allows storage teams to move the active (hot) data to SSD to improve application performance, and also move inactive (cold) data to the NL-SAS disks, to free up capacity on the high-performance drives, and as such, reduce the overall cost of the storage solution.

The IT organization's tiering policy might define the movement of data based on various factors. These might include the following:

- File type
- Frequency of access

- Performance
- Availability
- Data-center location

For instance, if the storage-tiering policy defines that files that are not accessed for 30 days be moved to a lower tier of storage, then all files matching this criterion will be moved to align with the policy. As illustrated in Figure 2.16, this has the advantage of lowering the cost per gigabyte, but also typically results in slower access to the data.

FIGURE 2.16
Tiered storage systems

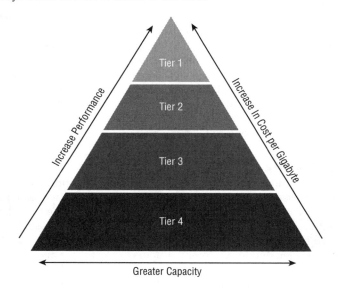

However, it is also important to recognize that simply representing the service levels, as they relate to specific hardware performance, might not constitute a complete tiered storage solution. Service levels are also influenced by other factors, such as people and processes.

Figure 2.17 illustrates an example of a manual tiering policy within an enterprise IT organization, and the component technologies that are pertinent to a particular tier of storage and its capabilities.

This example addresses typical enterprise customer requirements. However, manual tiering of storage is only one option available to IT organizations. Also available from most storage vendors are autotiering solutions, which are fully integrated into the array software and therefore can reduce the operational overhead associated with the manual tiering of data.

With manual tiering, perhaps the more traditional method, the storage team has to monitor workloads and move data between the tiers periodically. This can prove to be repetitive and operationally time-consuming, ranging from hours to days to complete. Automated storage tiering, on the other hand, addresses many of these same challenges around data retention and access, but in a more automated way, as you will see next.

	Specification	Attributes	Tier 1	Tier 2	Tier 3	Tier 4
	Service Levels					
	Hardware Platform		XtremIO	VMAX	VNX	ScaleIo
Primary Storage	Guaranteed Performance	Performance Throughput per Port (I/O sec.)	5,000+	3,500 – 5,000	1,500 – 3,500	500 – 1,500
Primary Storage	Guaranteed Performance	Response Time (ms)	< 8 ms	7 – 14 ms	12 – 30 ms	12 – 30 ms
Primary Storage	Availability	Maximum Unplanned Downtime per Year (mins)	<26.5	<26.5	<52.5	<263
Secondary Storage	Guaranteed Performance	Response Time	< 1 second	< 1 second	< 24 second	
Secondary Storage	Guaranteed Performance	Throughput	<= 300 Mbps	<= 700 Mbps	<= 280 Mbps	
Secondary Storage	Availability	Maximum Unplanned Downtime per Year (mins)	< 5.25 mins	< 52.56 mins	< 175.2 hours	
Secondary Storage	Retention & Disposition	Application Retention Period Required	< 10 years	< 7 years	< 3 years	
Secondary Storage	Retention & Disposition	Data-Shredding Compliance	Yes	No	No	
Secondary Storage	Accessibility	Read Access Frequency	< Hourly	> Hourly	Daily	
Secondary Storage	Data Integrity	Guarantee of Authenticity	Yes	No	No	
Secondary Storage	Offsite					
Operational Recovery (OR)	Recovery Granularity	Recovery Point Objective	< 1 minute	< 28 hours	< 38 hours	
Operational Recovery (OR)	Recovery Granularity	Recovery Granularity	Complete App. Restore	Complete App. Restore	File or Filesys. Restore	File or Filesys. Restore
Operational Recovery (OR)	Recovery Point Objective (RPO)	Amount of Data Loss	1 hour	24 hours	24 hours	30 hours
Operational Recovery (OR)	Recovery Time Objective (RTO)	Time to Restore Data	< 30 minutes	< 30 minutes	7 GB/minute	0.5 GB/minute
Operational Recovery (OR)	Recoverability	Ability to Recover Backed-Up Data	100%	100%	98%	95%
Operational Recovery (OR)	Retention Period	Time Data Is Retained	2 hours	24 hours	3 weeks	15 months
Disaster Recovery (DR)	Recovery Point Objective (RPO)	Amount of Data Loss	0 minutes	< 4 hours	24 – 48 hours	24 – 48 hours
Disaster Recovery (DR)	Recovery Time Objective (RTO)	Time of Restore Data	< 2 hours	< 12 hours	< 48 hours	< 72 hours

Figure 2.17
Storage tiering design example

Automated Storage Tiering

Automated storage tiering comes with two types of mechanism: intra-array (internal to the storage system) and inter-array (across different storage systems).

Intra-array automates the storage-tiering process within a single storage array by enabling efficient use of different mechanical and SSD technologies, to provide best performance at an optimal cost. The automated storage-tiering mechanism within the storage array monitors workloads and automatically moves active data to the higher-performing tiers, while simultaneously moving inactive data to slower and higher-capacity drives.

The goal, as you would expect, is to keep frequently accessed data on the highest-performing media, while moving less-active data down to the lower-tier capacity drives. This data movement is typically performed at the sub-LUN level, within the disk pool on the storage array.

To complement this automated tiering mechanism, many storage vendors also provide cache-tiering technology to improve array performance, by retaining frequently accessed data for longer periods of time in a dedicated SSD cache configured on the array. This enables a form of cache tiering between the DRAM (primary cache mechanism) and SSD (a secondary caching mechanism). As this technology allows the array to store significantly larger volumes of frequently accessed data on the cache tier, most reads are served directly from either the primary or secondary cache, providing much improved read and write performance during bursts of heavy utilization. Note that a caching tier offers no real benefit for pools or LUNs that are using SSD as a dedicated tier of storage, such as a *homogeneous* SSD pool.

NOTE Disk pools can be homogeneous or heterogeneous. *Homogeneous* pools have a single drive type (such as SSD, SAS, or NL-SAS), whereas *heterogeneous* pools contain different drive types.

Inter-array storage tiering identifies and automates the movement of active or inactive data in order to relocate it to different performance or capacity tiers, between different storage array systems. This type of solution typically optimizes storage to meet performance or capacity requirements, based on the cost to the business.

The inter-array storage-tiering solution uses a policy engine, which monitors and facilitates the movement of inactive or infrequently accessed data from the primary to secondary storage, and vice versa. The primary reason an IT organization would adopt this type of mechanism to tier data across different array systems is to meet archival or compliance requirements. For instance, the policy engine might be configured to relocate all data in the primary storage system to the secondary system that has not been accessed for six months. This technology works by the policy engine leaving a small stub file behind on the primary storage system that points to the actual data on the secondary system. When users try to access that data from its original location on the primary storage, the user is transparently provided with the actual data that the stub file points to, on the secondary storage system (see Figure 2.18).

Table 2.12 lists advantages and drawbacks of automated tiering technologies.

CLASSIC STORAGE CONCEPTS | 53

TABLE 2.12: Advantages and drawbacks of automated storage tiering

ADVANTAGES	DRAWBACKS
• Array-based automated tiered storage systems are typically more efficient than external software tiering solutions, appliances, or taking a manual approach. • This mechanism usually creates smaller block segments, providing more granularity. This means the technology is less likely to move data unnecessarily. • The ability of array-based automated tiered storage systems to move sub-LUN data segments between tiers automatically can be especially effective at optimizing the most expensive SSD storage resources. • As this automated storage tiering is an integrated part of the storage controller, the tiering function can be integrated with other storage functions that involve block data movement, such as snapshots and replication, which can help to reduce system overhead. There is also no requirement for an external automated tiering storage system. • Array-based automated tiering makes capacity expansion and storage management much simpler as this platform-based virtualization enables each tier to expand independently, without taking the array system down.	• Array-based automated tiered storage is typically limited to storage within a single array, and cannot easily be used to consolidate external storage platforms without migrating data. • As this solution is block-based, it cannot support applications requiring file services natively, which therefore cannot benefit the significant majority of data growth, which is file based. • Compared with an external block-based tiered storage solution, an internal storage array-based automated tiered storage system can also be less flexible, as it cannot tier data between different storage platforms, or offer a way to repurpose existing storage systems. • Auto-tiering isn't typically well suited to nonpersistent virtual desktops in a VDI environment.

FIGURE 2.18
Storage-tiering mechanisms

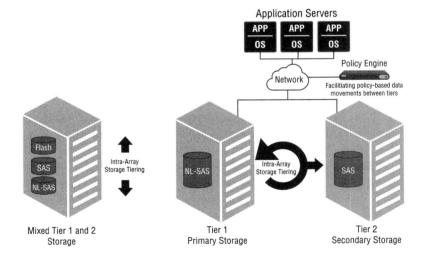

STORAGE TIERING DESIGN CONSIDERATIONS

Designing a tiered storage infrastructure requires an architect to consider these design factors before determining the most appropriate solution:

- What data do you plan to tier?
- Will you tier all data on the storage array or target specific applications?
- Will the solution employ autotiering or a manual approach?
- If the tiering is to be performed automatically on the array, what policy will it be based on?

For instance, some applications that are highly transactional in nature or do not store data for long periods of time may not be good candidates for tiering technology, as the data is always active. Performance is also a key design factor. When data is being moved on the array to a higher tier, that data has to be read from the slower disks, which may cause latency for the application. Further design factors may also include these:

- If the tiering is to be performed automatically, and is being performed at all times, what will be the impact on the performance of the storage array?
- Alternatively, if the tiering policy is to be maintained manually, what will the schedule be?

If the solution is to use inter-array tiering, many of the same considerations still apply. For instance, how will the policy engine determine what data to tier, and what are the performance considerations of users accessing data on slower storage array systems? In addition, some extra considerations for a multi-array tiering solution should be taken into account during the design and planning process:

- If the arrays are from different vendors, is there a requirement to maintain different management tools? If this is the case, can it even be implemented or managed?
- Is the policy-engine function embedded in the array's hardware, or do you require additional software or hardware components to be installed?
- How will the data traverse the fabric between systems? Will this be done using the front-end ports and network, and will this impact host performance, or is the data sent across an isolated back-end network? If so, what protocols are being used, and are there any security considerations?

Even though it is perfectly possible to isolate tiering traffic, this can increase operational and management complexity, and require additional hardware infrastructure to support it.

Storage Scalability Design

When addressing storage scalability requirements, it is essential to build a solution that is designed to meet the growth requirements from day one. Scalability in storage can mean different things to different organizations. For instance, one organization may be planning for data growth of 4 TB to 8 TB over three years, whereas large enterprises and service providers consider growth in terms of hundreds of petabytes (PB) of data year after year.

The terms *scale up* and *scale out* are used in multiple ways when referring to technology infrastructure growth. When addressing storage, scaling up refers to expanding a storage device by adding more capacity in terms of the following:

- Larger or faster disks to increase storage capacity or IOPS
- Additional DRAM cache to increase IOPS
- Additional storage controllers to increase throughput and IOPS
- Additional or faster front-end host ports to expand and increase throughput

For some enterprises and service provider environments, a single storage device cannot scale to the required capacity, throughput, or IOPS. For these use cases, it becomes necessary to scale beyond a single storage system in each data center and take a scale-out approach. However, this brings with it a range of operational management challenges.

For instance, the sample design illustrated in Figure 2.19 represents a possible building-block scenario where a service provider is employing a classic model approach to storage. From a physical platform perspective, each of these building blocks is made up of 96 rackmount vSphere compute hosts configured as four 24-node VMware vSphere clusters, divided equally across three server cabinets and a two-node vSphere local management component cluster. Each building block also houses two 48-port 10 GbE *leaf* switches, two 48-port 8 GB multilayer fabric switches, and two 1 GbE IPMI management switches, to provide out-of-band connectivity. Each of these building blocks is designed to provide a complete fault domain for compute, network, and storage.

The exact number of building blocks in this sample design would be scaled out accordingly, not only depending on the service provider's capacity requirements, but also based on the hardware, software, and power constraints at the data center.

In this example, the compute, storage, and network resources available from each component layer of this building-block architecture are provided in Table 2.13.

TABLE 2.13: Capacity scalability of building-block architecture example

RESOURCE	PER HOST	PER CLUSTER	PER BUILDING BLOCK
Memory	512 GB DDR3	10.5 TB (24 nodes with 3 reserved for HA)	42 TB
CPU	2 × Intel E5 8-Core 3.1 GHz = 49.6 GHz	1,041.6 GHz (24 nodes with 3 reserved for HA)	4,166.4 GHz
Fast storage	N/A	Flexible configuration	180 TB
Standard storage	N/A	Flexible configuration	300 TB
Network bandwidth	20 Gb/s	420 Gb/s	1,680 Gb/s (80 Gb/s MLAG to Spine)

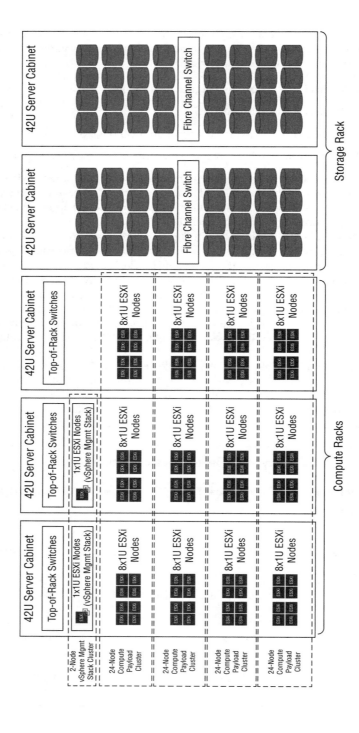

FIGURE 2.19
Scaling storage in a building-block approach

There is no one single correct solution to scaling storage for small or medium customers, nor for large service providers or enterprise IT organizations. However, during the design phase, several factors play an important role in choosing an approach to take, as illustrated in Table 2.14.

TABLE 2.14: Storage scalability design factors

REQUIREMENT	SCALING UP	SCALING OUT
Hardware scaling	Add more components to an existing storage system.	Add more storage systems.
Hardware limits	Expand to single storage system limits.	Expand to aggregate storage system limits.
Availability/resiliency	Protects against component failure	Protects against storage system failure
Storage management scalability/complexity	Single storage system to manage	Multiple storage systems to manage

When designing large-scale building blocks of storage, other factors may come into play and influence the approach taken, including, but not limited to these:

- Expectations regarding services and business growth
- Hardware availability and lead times
- Physical hardware scalability limitations (such as with storage management tools)
- Capital expenditure and hardware depreciation considerations
- Data center power, space, and cooling limitations

Storage Management Tools

When it comes to operational management of storage systems, many of the same considerations apply, which also are pertinent to hypervisor and network management, such as usability and single pane of glass. However, storage, especially large estates of storage systems—*SAN islands* as they are sometimes referred to—become far more complex to manage. You may have to deal with not only multiple vendors and their multiple management tools, but also different product lines from the same vendor, which also have a diverse set of management tools.

Managing storage systems such as this, from multiple vendors, quickly becomes an operational challenge. Employing multiple management tools typically requires expensive vendor training courses for those tasked with storage administration. In addition, different products from the same storage vendor can be just as complex to administer as a multivendor storage platform. The reason for this is that, almost perversely, it is common for different product lines from the same vendor to use a completely different management toolset.

In recent years, products such as EMC's ViPR have attempted to mitigate this challenge by providing multivendor support, in order to simplify storage operational management. However, this type of federated overlay software comes at a high cost, so you might have to consider other options, such as building your own in-house storage-as-a service (STaaS) offering, via an on-premises cloud management platform (CMP) tool. For instance, by taking advantage of the storage arrays' exposed APIs, it may be a viable solution to build custom workflows, which simplifies the required daily operational storage tasks being performed, a topic more fully addressed in Chapter 9, "Delivering a Storage-as-a-Service Design."

Scale-out environments also provide an additional set of operational complexities. For instance, some vendors' tools allow you to manage all storage systems globally, allowing them to appear to the storage administration team as a single node. Other management tools might show the storage systems as fully independent nodes that must be managed as separate entities, which can be complex, depending on how the environment is designed.

Table 2.15 addresses some of the operational management challenges of a large-scale, multivendor SAN environment.

TABLE 2.15: Multivendor SAN environment operational challenges

OPERATIONAL SITUATION	DESIGN FACTORS
Multivendor	• Do any of the vendors offer a solution to manage multiple vendors' storage systems from a *single pane of glass*?
	• Is there a third-party software or hardware option?
	• Are there storage APIs that can be used to build a custom storage-as-a-service solution?
Single vendor with multiple products	• Is there a single management plane offering for multiple products from that vendor?
	• Is there a third-party software or hardware option?
	• Are there storage APIs that can be utilized to build a custom storage-as-a-service solution?
Scale-out environment	• Are all the storage systems managed as a single entity or multiple entities?
	• Is information and storage system configuration shared across the platform?

Multitenanted Storage Design

Previously we addressed the types of scalability requirements required by cloud service providers. However, a multitenanted environment often carries an additional requirement to ensure separation of one tenant's data from all other tenants. This storage isolation can occur at various places within the environment, both at logical and physical layers. The specific design factors— such as which CMP is being used, hardware, disk pool architecture, or where the isolation is going to be employed— will make these design decisions clearer.

There are many options for designing a cloud solution that provides storage isolation for tenant applications. Table 2.16 illustrates (ordered in levels of increasing isolation and security) typical options for cloud service providers when employing a classic approach to their storage platform.

TABLE 2.16: Multitenanted storage design

Shared disk pools	This type of solution provides the logical isolation of data at all levels. Even though data across tenants is stored on the same physical disks, different LUNs or filesystems will provide logical segmentation. The hardware and software components of the storage system, such as front-end and back-end ports, DRAM cache, and storage controller software and management, are all shared.
Separate disk pools	This type of solution from a cloud service provider delivers physical separation of the data on the underlying disk subsystem, with each tenant given their own pool of disks. The other storage system components, such as front-end and back-end ports, DRAM cache, and storage controller software and management, remain shared.
Separate physical ports	In addition to separate disk pools, this design expands isolation by providing dedicated front-end ports to specific tenants, which can typically be achieved by also isolating the tenant onto dedicated compute hosts. Back-end port isolation would typically be far more complex to design. System DRAM cache, storage controllers, and management are still all shared among tenants.
Full isolation	The final example provides full component isolation. The design expands on the previous model by allocating a storage system dedicated to the tenant, meeting their specific platform requirements.

When architecting this type of solution in a multitenanted environment, also consider that it can drastically limit the scalability of a storage system and increase operational complexity significantly. For instance, consider a storage system with eight front-end ports available. If allocating ports to a specific tenant was a requirement of the design, and assuming a minimum of two ports per tenant, a maximum of four tenants can be configured across that storage system. As a result, the only design options would be to deploy multiple instances of that eight front-end port system, to scale out the design, or to deploy larger systems that provide higher levels of scalability and scale up the platform.

Quality of Service

Quality of service (QoS) provides an additional mechanism that can be used on a storage array to provide further service level–based performance guarantees, in either a multitenanted or multitiered environment. While not a feature provided by all storage vendors, QoS might be able to guarantee a certain response time, throughput, or bandwidth, to either a specific tenant or workload. This mechanism can be used in a design to ensure that specific applications can achieve the required performance levels, to meet required SLAs. In a multitenanted environment, in which SLAs are typically strictly enforced, QoS is a tool that can be used to enforce this design requirement. In addition, QoS can also be utilized to minimize the exposure to denial-of-service (DoS) attacks and noisy neighbor scenarios.

Data Deduplication and Data Compression

The process of *deduplication*, also referred to as *intelligent compression* or *single-instance storage*, can be performed at the controller level of some manufacturers' storage systems. Deduplication is the process used at the array level, to reduce storage requirements by reducing or even eliminating altogether redundant data, allowing space to be saved on the storage system.

During the deduplication process, unique chunks of data, or byte patterns, are identified. When multiple files have identical data, the filesystem stores only one copy of the data and shares that data between the multiple files. The redundant chunk is replaced with a small reference *stub* file, which points at the stored chunk, although different instances of the file can have different names, security attributes, and time stamps, with none of this metadata being affected by the deduplication process.

This technology can be used for various purposes, such as improving storage utilization or increasing levels of efficiency for network data transfers, across a WAN link, for remote backup, replication, or disaster recovery. The aim in this type of design scenario is that it reduces the overall number of bytes that must be sent over the site interconnect.

Data deduplication also offers other benefits, such as lowering storage space requirements, reducing costs, and facilitating the more efficient use of disk space, which can enable longer disk retention periods and provide improved recovery time objectives (RTOs), for a longer period of time, without a need for tape backups.

Most modern storage array systems perform all deduplication processing as a background, asynchronous operation, which acts on file data after it has been written into the filesystem. Typically, only backup or archiving disk systems process data as it is written into the filesystem.

In addition, most arrays also avoid processing active data, as active data is typically more likely to be accessed, modified, or deleted during a short window. It is the inactive, or cold, data that represents the largest component of the datasets, often as much as 80 percent. It is this data that can benefit the most, and is therefore targeted for deduplication within the array. It is these systems, which are designed not to process the files that are actively being used, that are able to provide the most efficient use of deduplication technology, thereby maximizing space utilization while also minimizing the impact on end-user applications. For most systems that support deduplication technology, storage administrators can tune the targeted data selection criteria if so desired, and optionally add filters to exclude specific file types by file extension, and also avoid files in specific directories or with specific names, or sequences of characters in their names, in order to optimize the process and performance of the storage platform in that specific environment.

Data deduplication is often used in conjunction with other forms of data reduction, such as conventional data compression. Compressing data can save additional storage capacity, further reduce file-transfer speeds, and decrease costs. The main disadvantage of data compression is the performance impact, resulting from the use of additional CPU and memory resources, to compress and decompress the data.

Compression and deduplication can be easily confused by customers, even though they are two very different technologies that can operate either independently or in conjunction with one another. Deduplication reduces data by looking for redundant chunks of data across the storage system or a filesystem, and replacing each duplicate chunk with a pointer to the original.

Compression, however, uses algorithms to decrease the size of the bit strings in the data stream, reducing it to fewer bits than the original.

Storage Device Security

Data that is stored on a storage array or any sort of backup media, depending on the business type, should be encrypted to provide data security as well as potential compliance with regulatory or other obligations.

The requirements for storing business-sensitive or personal customer data on a storage array should be addressed as part of any storage design. For instance, in the event that a disk is removed from the storage array, either maliciously or due to a hardware failure, the data stored on that drive could be accessed by unauthorized persons. When hardware reaches its end of usable life, the drives contained within that system can also expose data to loss or theft if not disposed of correctly.

Many storage vendors offer their own proprietary technologies to address these data security challenges. However, typically there are two key design considerations: data in flight and data at rest.

Encrypting data on the storage device either can occur after unencrypted data is sent from the host to the storage—*data at rest*, or can occur when data is encrypted in transit, prior to being committed to the storage system—*data in flight*. Each storage array vendor offers different types of solutions. Some solutions encrypt data on a per host or per LUN basis, whereas others encrypt data at the individual drive level, across the entire array, with each drive being encrypted with a unique key. Removing a drive from the array or even replacing the entire array renders the stored data unusable, as the decryption keys will typically be stored on an external device.

Hardware High Availability

High availability of hardware components is pretty much standard in most modern enterprise-class storage systems. This essential hardware layer availability is typically achieved by providing redundant components, such as front-end host ports, cache, back-end ports, power supplies, cooling fans, disk paths, and hot spare drives. For block-based systems, redundant controllers on the array allow the cache to be mirrored, so that no data is lost during a failure scenario. In the event of a controller or front-end port failure, the use of multipathing software, in the vSphere host, allows I/O to be moved to another available path, preserving access to the storage device.

For active/passive storage arrays, a controller failure often results in a momentary interruption of I/O, while access to the LUN is being transferred to the standby controller. This is not typically detectable, but some applications that are sensitive to latency variations might have an adverse reaction to this type of failure event with this hardware type.

Regardless of active/active or active/passive storage controllers, as part of the design, the architect should determine whether to preserve performance levels during an outage. For instance, in order to maintain service levels, in a dual storage controller system, it might be required that no processor should exceed 50 percent utilization, under normal operating conditions, in order to ensure that performance is not degraded during a hardware failure.

For file-based storage systems, many of the same principles continue to apply. However, instead of multiple storage controllers, multiple NAS heads are used to provide redundancy, which can again be either active/active or active/passive. In an active/active NAS environment, multiple NAS heads can read and write to the same filesystem simultaneously. If one NAS head fails or requires maintenance, access is still available through the remaining hardware. In an active/passive system, one NAS head typically owns a particular filesystem, and during a failure event, that filesystem is failed over to the remaining NAS head. The period of time required to complete a failover of the NAS head can vary significantly across storage vendors, so this factor should be considered when evaluating NAS-based solutions. However, unlike a block-based storage array, with NFS version 3, there has not been a multipathing mechanism native to vSphere hosts. Therefore, when this protocol is being used as part of a design, standard networking features, such as link aggregation across multiple hardware components, must be used to protect against interface or switch failure. However, vSphere 6, with NFS version 4.1, does provide a failover mechanism via session trunking. (Chapter 3 addresses this topic further.)

Storage Array–Based Disaster Recovery and Backups

When it comes to disaster recovery and backup mechanisms for storage arrays, and the data they hold, there are two main categories: replication and backups. Replication of data provides disaster-recovery capabilities, and also provides the ability to have data online at the recovery site or on the recovery storage system quickly.

The replication product, whether native to a storage vendor or an externally provided solution, will impact the networking design and may also influence the choice of storage protocol being used for the host to storage I/O fabric connectivity. Key design factors include whether the replication solution provides support for a heterogeneous storage environment, or whether it is a vendor-specific solution that requires all storage to be procured from that single vendor. In addition, in the event of a disaster-recovery scenario, how can the recovery site be activated? For instance, will the disaster-recovery solution integrate with VMware Site Recovery Manager and NSX, or is there cross-site integration at the hypervisor level, such as in a metro cluster design? In addition, is there an existing mechanism by which backups are replicated to a recovery site? How will application recovery be handled, how long will it take, and can the recovery point objective (RPO) and recovery time objective (RTO) be met? These, and many more, are all questions that must be addressed during design discovery and requirements gathering.

Backups enable you to recover lost data from either disk or tape by using a backup infrastructure. Even though backing up data to a local system is much faster, if the primary site goes down, this approach offers no benefit unless it can be restored to the recovery site. A system that allows you to back up data locally and then replicate it to the recovery site, facilitating both local data recovery and disaster recovery, can provide a solution to multiple challenges. However, copying data to a remote site could take a significant amount of time, and may potentially use WAN links for a prolonged period.

When designing a disaster-recovery solution, one of the most critical aspects to consider, relating to replication, is the bandwidth requirement between sites and the latency associated with the connectivity. Using synchronous replication provides a full mirrored copy of the data, as synchronous replication requires that data be written to both storage devices simultaneously. Because the data will be identical at both sites at all times, the design has an RPO of zero. However, synchronous replication requires sufficient bandwidth to transmit the data in real time, with extremely low latency between the sites, essentially treating both storage devices as if they were both local, and not acknowledging write operations to the applications or operating systems until both local and remote I/O is committed. Therefore, if the level of connectivity available between the locations does not meet the requirements, which are typically defined by the applications, degraded performance is inevitable, or even a complete replication failure may occur. For this reason, synchronous replication solutions are typically supported or viable in only a metro-style or campus-based design.

For asynchronous replication, the bandwidth and latency requirements are not as stringent, as data is not being written simultaneously or in real time. However, the design must ensure that the level of bandwidth and latency is able to support the defined RPO. An option that can be evaluated as part of a design, if the available connectivity cannot meet the design requirements, is a WAN accelerator. These solutions can be used to increase throughput and reduce latency across WAN interconnects. WAN acceleration products can significantly increase the throughput of WAN connections, and are typically ideally suited for sequential read or write functions, such as replication of backups offsite, or large data transfers across locations.

The level of transmitted data is directly impacted by the level of data changes that need to be replicated across the WAN interconnect, to the second data center. If the design requires a full mirrored copy of the data at the recovery site, you should design an identically configured storage system there also. This will provide not only identical performance in the event of a disaster occurring, but also the necessary performance required during normal replication operations, which do have the overhead of the storage array at the recovery location. If the design requires only a proportion of the environment to be replicated to the recovery site, such as tier 1 applications, or the customer is willing to run in a degraded state when failed over during a recovery scenario, then there is a potential to reduce CapEx by providing a smaller storage system or a lower-cost product from the same or a different storage vendor. However, by electing to introduce a different storage system at the recovery site, this may also then introduce a new requirement for an external, nonvendor replication product, which can facilitate the replication process across a heterogeneous storage environment.

Other key design considerations for a disaster-recovery solution based on a storage device might also include the following:

- How is the disaster scenario activated? Is it a manual or automated process to bring online the replica data and the attached vSphere hosts?

- Is VMware Site Recovery Manager being used in the design, or a third-party product, to automate the recovery process for the hypervisors and storage array?

- Once the environment has failed over, what is the process for failing back?

- How often will the process be tested after the environment is deployed?

Finally, when evaluating a backup solution for a design, deduplication is essential. Even when you simply consider how much redundant duplicated data exists only within guest operating systems, the need for a deduplication-based solution can be easily justifiable.

Storage Array Snapshots and Clones in a Classic Storage Environment

Using snapshots and clones in a classic storage infrastructure provides a mechanism for quickly restoring data to a previous point in time. Although this does not provide an end-to-end disaster-recovery solution, it can provide an effective solution to address data loss or data corruption, particularly in a dynamically changing or transient environment.

However, in a vSphere infrastructure, care must be taken when using a snapshot and cloning strategy as part of the environment's design. With classic block-based storage devices, snapshots and clones are taken at the individual LUN level and are therefore also restored or reverted to at the LUN level. Typically, a LUN is used by numerous virtual machines, and as a result, inadvertently rolling back virtual machines during a revert operation can be easy. For instance, in the classic storage environment, if a database administrator accidently deletes critical data from a database table, when the storage administrator reverts to a previous version, using a snapshot of that LUN, all virtual machines and data that exist on that datastore will be reverted, not just the database administrator's missing tables, which could result in loss of business-critical files. This situation is exacerbated in vSphere environments in which virtual machines use disks distributed across multiple datastores. Therefore, operationally, snapshots and clones should be used for only specific use cases, such as development environments or transient workloads, which may require the flexibility to quickly restore data to a previous point in time, but without increasing the risk to business-critical information (see Figure 2.20).

FIGURE 2.20
Snapshots and clones

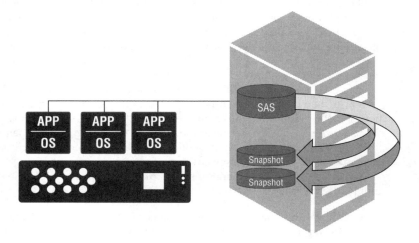

In a file-based environment, in addition to the full filesystem recovery described previously, there is often a mechanism to revert or recover individual files from snapshots. However, this may be complicated if data must be reverted from multiple incremental snapshots and then reassembled into the final state.

If storage-level snapshots or clones are a factor in a design, it is critical that operational procedures be implemented, in order to ensure that valuable business data is not inadvertently overwritten and destroyed during the revert process. For most use cases, vSphere-based snapshots are a far better mechanism to revert individual virtual machines to a previous state. This approach prevents the risk of accidental data loss, which can be associated with storage array–based snapshots.

Chapter 8, "Policy-Driven Storage Design with Virtual Volumes," addresses how the next-generation snapshot technologies, when used with vSphere Virtual Volumes, can be used to mitigate many of the risks associated with the classic LUN-based snapshot and clone mechanisms highlighted here.

vSphere Metro Storage Cluster

A vSphere Metro Storage Cluster (vMSC) is a design option that enables a vSphere cluster to be stretched across two geographical locations. By using this type of design, organizations can perform both load balancing of workloads and nondisruptive vMotion operations between the two data centers.

A vMSC can provide IT organizations and service providers with multiple benefits, which enhance single-site vSphere High Availability (HA) clusters with the added benefit of geographically separated sites for recovery and disaster avoidance. This is facilitated in a stretched cluster design by allowing an organization to live-migrate virtual machines between the two data-center locations for disaster avoidance or proactive load balancing. This type of disaster-avoidance design is typically most appropriately used in environments where downtime cannot be tolerated. However, it should not be used instead of an IT organization's disaster-recovery design, but rather employed alongside it to provide business continuity through a multitiered strategy.

A vMSC design, illustrated in Figure 2.21, requires a high-bandwidth and low-latency connection between the two data centers. Typically, the maximum supported network latency between the two sites is a round-trip time (RTT) of 10 milliseconds, which limits the geographic distance between sites in which a stretched cluster can be implemented to less than 100 kilometers, in most cases. A vMSC supports Fibre Channel, iSCSI, NFS, and Fibre Channel over Ethernet (FCoE) protocols, and requires a stretched storage solution, such as EMC's VPLEX or NetApp's MetroCluster System. A vMSC also requires cross-data-center links with low-latency and high-bandwidth network connectivity, as outlined previously, which can provide the required layer 2 extension technology across the two data-center locations.

Stretched clusters are addressed again in Chapter 5, "Virtual SAN Stretched Cluster Design," which focuses on offering this technology through Virtual SAN–enabled, distributed vSphere clusters.

All-Flash Disk Arrays

An all-flash storage array provides 100 percent solid-state storage devices, in order to deliver high-speed, low-latency storage to vSphere workloads, as opposed to mechanical spinning hard disk drives, or a hybrid array, which combines both solid-state storage and mechanical disks.

Solid-state drives, or SSDs, are storage devices that have no moving parts and are made up of nonvolatile memory that can be erased and reprogrammed in units, referred to as *cells*. Nonvolatile memory is a form of Erasable Programmable Read-Only Memory (EPROM), so called as the memory cells can be erased in a single action. Typically, an all-flash array can transfer data to and from SSDs significantly quicker than mechanical disks, making them the ideal choice for high-performance, low-latency workloads.

Even though enterprise flash drives are still more expensive than mechanical disks, the development of multi-level cell (MLC) technology has made flash more affordable for data center use. However, MLC flash is slower and less durable than single-level cell (SLC) flash, although SSD manufacturers have now developed software that improves their life expectancy, to the point where MLC is now acceptable for typical enterprise application workloads. However, SLC flash remains the optimum design choice for applications that require the highest level of durability and for intensive I/O workloads. These aspects of flash-based storage are detailed further in Chapter 4, "Policy-Driven Storage Design with Virtual SAN," which addresses how they relate to VMware Virtual SAN design.

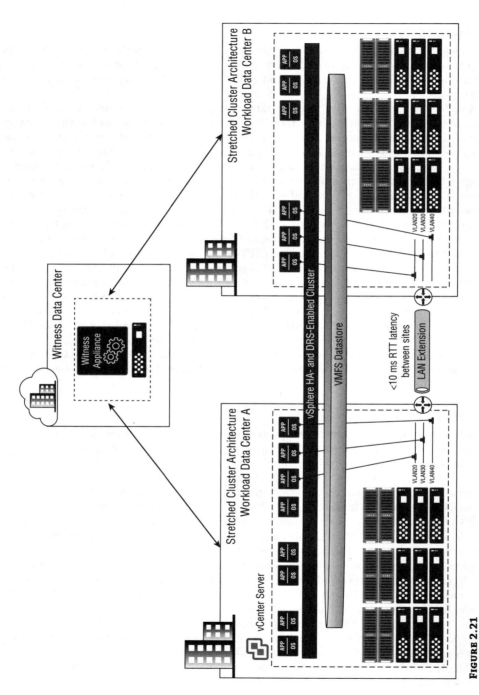

FIGURE 2.21
vSphere Metro Storage Cluster design

vSphere Storage Technologies

The vSphere platform provides vSphere and storage administrators with a wide range of storage component technologies. It is crucially important that both vSphere and storage operational teams know what features are offered, and how they interact with traditional as well as next-generation storage solutions. The remainder of this chapter discusses vSphere storage technologies and how these components interact with the storage layer. As you move forward in this book, you will notice a general trend, and concise moves by both VMware and storage vendors, toward far more direct interaction and communication between the vSphere layer and the storage hardware layer. However, from a classic storage model perspective, we draw this demarcation line, shown in Figure 2.22, pretty much down the middle, between virtual and physical storage constructs.

FIGURE 2.22
Identifying the demarcation line between the vSphere layer and the storage array layer

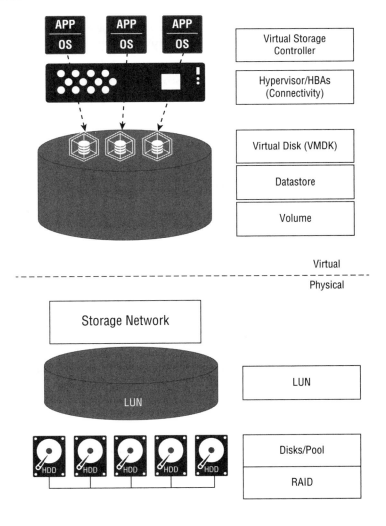

Virtual Disks

When you deploy a new virtual machine, the vSphere platform components automatically create a certain set of files that are associated with the new virtual machine. This virtual machine is then stored as this set of files in a directory created by the host during the provisioning process.

The component files, which make up that virtual machine in the newly created directory, are described in Table 2.17. However, note that some of the files listed are not created until other events, such as a Power On, or Snapshot Taken, occur.

TABLE 2.17: Virtual machine component files

FILENAME	DESCRIPTION
`<vm-name>.vmx`	The `.vmx` file provides the virtual machine configuration. This is the primary configuration file, which stores all chosen settings for the virtual machine.
`<vm-name>.vmdk`	The `.vmdk` is an ASCII format file, which stores the contents of the virtual machine's hard disk drive. There could be one or more virtual disk files, depending on the virtual machine's configuration.
`<vm-name>-flat.vmdk`	The `-flat.vmdk` is a single, pre-allocated disk file, which contains the virtual machine's data.
`<vm-name>.vswp`	The `.vswp` is the swap file. This file gets created when a virtual machine is powered on, and is deleted when powered off. This file size is equal to the configured memory, less any virtual machine reservation.
`<vm-name>.nvram`	The `.nvram` file is nonvolatile RAM that stores virtual machine BIOS information.
`<vm-name>.vmem`	The `.vmem` is the virtual machine memory mapped to a file.
`<vm-name>.vmss`	The `.vmss` is the virtual machine suspend state file, which is created when the virtual machine is suspended, and stores the current state of the suspended virtual machine.
`<vm-name>.vmsd`	The `.vmsd` is a centralized file for storing information and metadata about snapshots. This file is created as part of a snapshot operation.
`<vm-name>-snapshot.vmsn`	The `.vmsn` is the snapshot state file, which stores the running state of a virtual machine at the time you take the snapshot.
`vmware-0.log, vmware-1.log, etc.`	The `.log` files are the virtual machine's log files. This file can be useful for troubleshooting and is always stored alongside the configuration (`.vmx`) file of its associated virtual machine.

The virtual disk, which in vSphere is represented with the file extension `.vdmk`, is a disk-image file format for storing the complete contents of the virtual machine's hard drive. The virtual disk represents and replicates a physical hard drive, and includes all data and structural elements, as if it were a physical disk, located inside a physical computer. A virtual disk can be stored anywhere that the vSphere host can access it and hides the physical storage layer from the virtual machine's operating system. This allows virtual machines to run various operating systems, including operating systems that are not certified for storage area networks (SANs).

From a storage perspective, the VMDKs are the most important file, as they provide the virtual machine's virtual disks. In vSphere 5, VMDKs have a maximum size of 2 TB, specifically 2 TB minus 512 bytes. However, with vSphere 5.5, VMware introduced a new absolute maximum limit for VMDKs of 62 TB on VMFS5, with the limit on NFS depending on the underlying filesystem.

In reality, however, the virtual disk is made up of two files: a descriptor file with the extension .vmdk provides metadata and a link to the second file, the file containing the data and using the extension -flat.vmdk.

When a virtual machine is deployed, an administrator can choose from three disk types: lazy zeroed thick, and eager zeroed thick and thin. The advantages and drawbacks of each option are presented next.

Lazy Zeroed Thick

In the LZT format type, the size of the VMDK on the datastore is the same size as the virtual disk that you create during the provisioning operation. However, within the file, the space is not *pre-zeroed*. For instance, if you create a 2 TB virtual disk and place 100 GB of data on it, the VMDK will appear as the full 2 TB on the datastore, but contain just 100 GB of data on disk. As I/O occurs on the guest operating system, the VMkernel zeroes the space required immediately before the guest I/O is committed, but the VMDK file size does not grow, as the full 2 TB is already allocated. Table 2.18 lists the advantages and drawbacks of lazy zeroed thick disks.

Table 2.18: Advantages and drawbacks of lazy zeroed thick disks

Advantages	Drawbacks	Typical Use Case
Provides fast deployment with the space being allocated to the VMDK on provisioning, so after creating a virtual disk with this format, the datastore will show that the space is no longer available.	The performance of a lazy zeroed disk is not as good as an eager zeroed disk. For each write to a new block, the block is zeroed out before the data is written. For each new write operation, there is a small amount of overhead.	Applications with low levels of write I/O or where very low levels of write latency are not an issue. Also where the allocation of datastore capacity up front is important.

Eager Zeroed Thick

The EZT format type provides a VMDK file size on the datastore that is the full size of the virtual disk that you created during the provisioning operation. The virtual disk file in this format type is *pre-zeroed*. For instance, if you create a 2 TB virtual disk and place 100 GB of data in it, the VMDK will appear as 2 TB on the datastore but contains only 100 GB of data and

1,900 GB of zeros. As I/O occurs in the guest operating system, the VMkernel does not need to zero the blocks prior to the I/O being committed. As a result, I/O latency is reduced, and less back-end storage I/O occurs during normal operations. However, significantly more back-end storage I/O operations occur up front, during the creation of the virtual machine. Although, with compatible hardware, this process can be speeded up and offloaded to the array via the Block Zeroing VAAI primitive. Table 2.19 lists the advantages and drawbacks of eager zeroed thick disks.

TABLE 2.19: Advantages and drawbacks of eager zeroed thick disks

ADVANTAGES	DRAWBACKS	TYPICAL USE CASE
Reduces latency associated with zeroing blocks dynamically at first write. Some deduplication storage devices work better when the VMDK has been zeroed, as the data can be de-duplicated more cleanly.	Takes longer to create as the virtual disk is fully zeroed during the provisioning process. Zeroes fill the capacity of the VMDK in advance; therefore, all space on a thin LUN is pre-allocated.	Latency-sensitive applications and write-intensive applications.

THIN DISKS

This format type ensures that the size of the VMDK file on the datastore uses only the amount of used space within the virtual machine's guest operating system itself. For instance, if you create a 2 TB virtual disk and place 100 GB of data on the virtual machine, the VMDK file will be 100 GB in size. The disk size is increased only when I/O occurs within the guest operating system, with the VMkernel zeroing the space required immediately before the guest I/O is committed. Table 2.20 lists the advantages and drawbacks of thin disks.

TABLE 2.20: Advantages and drawbacks of thin disks

ADVANTAGES	DRAWBACKS	TYPICAL USE CASE
Fast provisioning	Increases write latency associated with zeroing blocks dynamically at first write. Therefore, write performance is not as good as other disk types.	Applications with low levels of write I/O.

If the array supports thin provisioning, you'll generally get more efficiency using the array-level thin provisioning in most operational models. However, if you're using thin provisioning at the storage level, you will want to avoid using eager zeros, as this will completely defeat the purpose of thin provisioning on the storage array.

Virtual Machine Storage Controllers (vSCSI Adapters)

When a virtual machine is deployed, a storage controller adapter is created to provide connectivity for the guest operating system, to manage I/O requests, to and from the virtual disk or disks. When provisioning a new virtual machine, you select the guest operating system for the new virtual machine. At this point in the process, vCenter Server automatically selects which storage controller is to be added, based on what drivers are available in that specific operating system distribution. The default controller type for each operating system is optimized for best performance, with typical virtual workload use cases in mind.

It is possible, even recommended, for specific workloads and configurations to add additional controllers or change the controller type after the virtual machine has been deployed. A virtual machine can use a maximum of four SCSI controllers and four SATA controllers, with the default SCSI or SATA controller numbered as 0. When you provision a virtual machine, the default hard disk is assigned to the default controller 0, at bus node (0:0). If the vSphere administrator adds storage controllers to the virtual machine, they are numbered sequentially 1, 2, and 3.

When designing virtual machines for optimal performance, it is important to understand the characteristics, controller limitations, and compatibility of controller types, in order to understand the impact of adding or changing controller devices. Not doing so risks causing potential boot problems and inadvertently creating a bottleneck in terms of the storage controller queue, or lack thereof.

Figure 2.23 illustrates where the virtual machine controller, also sometimes referred to as the vSCSI adapter, is integrated into the storage stack, end to end.

FIGURE 2.23
vSphere storage controller stack

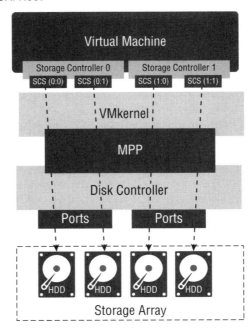

Storage controllers appear to a virtual machine as different types of SCSI controllers, which are optimized for different types of workload and use cases. Essentially, vSphere administrators have five storage controller options:

- BusLogic
- LSI Logic Parallel (formerly known as just LSI Logic)
- LSI Logic SAS
- VMware Paravirtual (also known as PVSCSI)
- AHCI SATA

The BusLogic adapter type was one of the first emulated vSCSI controllers available on the VMware platform. The earlier versions of Microsoft Windows had this driver available by default, embedded into its distribution, making it easy to install. However, the Windows driver for this adapter type was limited to a queue depth of 1. Therefore, knowledgeable vSphere administrators would often manually replace the BusLogic adapter with the LSI Logic driver instead. This adapter type should now be considered as legacy, unless the IT organization is still running Microsoft Windows 2000.

The LSI Logic Parallel (formerly known as just LSI Logic), was the other early emulated vSCSI controller available on the VMware platform. Most legacy guest operating systems included an embedded driver that supported a queue depth of 32, and for this reason, it quickly became a common choice, if not the default option for most vSphere administrators. For most modern use cases, this adapter type should also be considered legacy.

The LSI Logic SAS adapter type is the next generation in evolution of the parallel driver, which supports a new future-facing standard. This adapter type grew in popularity after the release of Microsoft Windows Server 2008, as it was required for its new clustering technology, but is still among the most common choices for general workload types today.

The VMware Paravirtual, or PVSCSI, is a virtualized SCSI controller, which is virtualization aware and has been designed from the ground up to support high-throughput workloads with minimal processing cost on the host CPU. The PVSCSI adapter is now the most efficient driver in low, medium, and high I/O deployments. Previous issues with low I/O virtual machines and this adapter type are no longer found in the most recent releases of vSphere.

In reality, the Paravirtual and LSI Logic Parallel SAS adapters are essentially the same when it comes to overall performance capability. However, the Paravirtual adapter is more efficient in the number of host compute CPU cycles required, which are necessary to process the same number of I/O operations per second. Therefore, in designs that employ I/O-intensive storage workloads, this controller type will ensure that you reduce the overall number of CPU cycles on the host servers.

Finally, the AHCI SATA adapter, a new storage controller that became available with vSphere 5.5, requires Virtual Hardware 10. This controller type allows the connection of large amounts of storage capacity to a virtual machine, but is not designed to be as efficient at handling large amounts of I/O as the Paravirtual or LSI Logic Parallel SAS controllers. For this reason, this adapter type is not recommended for use with workloads that are performance sensitive.

As previously outlined, it is common for vSphere administrators to use multiple virtual storage controllers on I/O-intensive virtual machines, to ensure that the virtual storage controller does not become a bottleneck, as illustrated in Figure 2.24. In this example, the Oracle workload has been split over four PVSCSI adapters, to ensure that sufficient I/O can be achieved across the storage controllers and disk devices.

FIGURE 2.24
Example of a multiple storage controller virtual machine design, for splitting workload across storage controllers

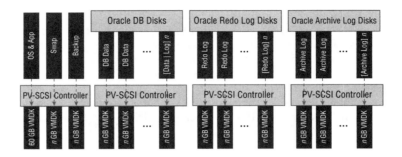

Datastore

A *datastore* is a volume that has been mapped to the vSphere hosts and has been formatted with either the Virtual Machine File System (VMFS), or the Network File System (NFS). In a non-hyper-converged infrastructure (HCI), a shared datastore is typically located on a physical storage device system.

There can often be much confusion about the terms *LUN*, *volume*, and *datastore*. To avoid this confusion, we will clarify these terms before proceeding.

LUN refers to a logical unit number. This means that a logical space has been carved from the storage array by the storage administrator. For ease of identification, the storage administrator will assign it a number, known as a LUN ID. In the following example, shown in Figure 2.25, the LUN ID is 10 and it has 2,000 GB of space. A single LUN can be created from the entire space on the storage array, or from part of the available space in a disk pool. For instance, as addressed earlier in this chapter, it is typical that a disk pool is carved into multiple LUNs.

FIGURE 2.25
Volume, datastore, and LUN

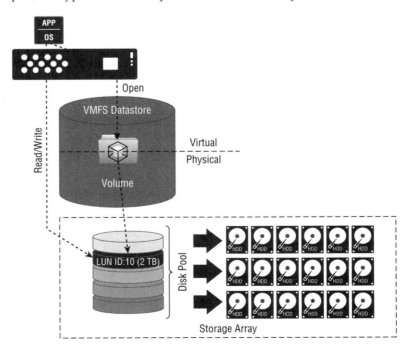

When the LUN is mapped to vSphere hosts, it can then be referred to as a *volume*. When a volume is formatted with a filesystem, either VMFS or NFS, it can then be referred to as a *datastore*.

Datastores are logical containers, equivalent to filesystems, that mask the specifics of each storage device, and therefore provide a consistent model for storing virtual machine files. VMware strongly advises its customers that there should always be a one-to-one mapping between the VMFS datastore and a LUN.

When it comes to sizing datastores, there is no one correct answer. The architect should make recommendations based on some of the key design factors listed here:

- Design each LUN to have the correct RAID level and read/write characteristics for the applications running on the virtual machines that use that LUN.
- VMware strongly advises that each LUN should contain only one VMFS datastore.
- If multiple virtual machines access the same datastore (which is the most common approach), employ disk shares and Storage I/O Control to prioritize virtual machines, based on service-level requirements.

With vSphere supporting datastore sizes of up to 64 TB, Table 2.21 further addresses the scale-up or scale-out approach, and how it relates to datastore-sizing considerations.

TABLE 2.21: Making LUN sizing decisions

Fewer, Larger LUNs (Scale Up)	More, Smaller LUNs (Scale Out)
• It's more flexible. You can create new virtual machines without asking the storage administrator to provision more space. • It's more flexible when you are required to resize virtual disks, perform snapshots, etc. • It lowers storage operational overhead, if fewer VMFS datastores exist.	• Improved performance, as there is less contention per volume. • Different applications typically require different RAID read/write characteristics. • Less wasted storage space, if not virtually provisioned on the storage system. • Additional flexibility, as multipathing policies and disk shares are configured per LUN. • Microsoft cluster services require physical RDMs, where each cluster disk resource is configured as its own LUN.

vSphere allows you to use up to 256 VMFS datastores per system, with the maximum volume size of 64 TB. By default, you can have up to 8 NFS datastores per system, but this number can easily be expanded to 256 via the advanced configuration within the vSphere Web Client. Datastores can also be used for storing ISO disk images, virtual machine templates, and even floppy disk images.

The type of datastore that a design uses largely depends on the type of physical storage systems you have, or plan to have, in the data center. The vSphere platform provides various ways to present disk devices to hosts. Typically, shared storage devices are presented cluster wide, via SAN-based solutions, such as Fibre Channel disk arrays and iSCSI disk arrays, or network-attached storage (NAS) offering up NFS filesystems. Local disks, which can be used as boot devices in some designs, can also provide local SCSI-attached storage for the temporary or permanent storage of virtual machines.

SAN- and NAS-based storage systems store virtual machine files on datastores mapped to external shared storage devices located outside the vSphere host. In the most common design scenario, illustrated in Figure 2.26, ESXi hosts communicate with the storage devices through the storage network. When used in a classic storage architecture, utilizing external storage devices facilitates vSphere availability technologies, such as HA and vMotion, which are not available with locally attached storage. However, as detailed in future chapters, this limitation changes when the architecture being employed follows the HCI model instead.

Storage volumes, presented from Fibre Channel and iSCSI SANs, are formatted with the VMFS filesystem type, whereas NAS-based arrays are formatted with the NFS filesystem, both allowing the host servers to create the datastore logical container on top of them. One key differentiator between NFS- and SAN-based storage is that SAN-based storage also allows the use of raw LUNs, referred to as raw device mappings (RDMs). Through the raw data mapping of a volume, it is possible to present the raw LUN directly to the virtual machine. This is addressed in more detail later in this chapter, and the concepts of SAN- and NAS-based storage protocols are addressed in far more depth in Chapter 3.

The number of virtual machines that can share a single datastore will depend on the workload's I/O activity, and also on the underlying capability of the storage system. Typically, several design factors govern the virtual machine consolidation level on a VMFS datastore. These might include, but are not limited to, the maximum LUN queue depth, backup capacity, workload distribution for high availability, workload capacity, workload I/O, and disaster-recovery capabilities. In the absence of any specific constraints or guidance from the customer, a general guideline is to place 10, 15, or 20 workloads per datastore, where 10 represents virtual machines that have high demands on storage, 15 represents virtual machines with average demand, and 20 represents workloads with low I/O demand.

Datastore Signatures

Every VMFS datastore that gets created is assigned a universal unique ID (UUID) in the header metadata, to identify which LUN is associated with that datastore. This ID must be unique; if two datastores are accidently assigned the same UUID, the vSphere hosts are unable to determine on which volume to perform read and write operations, resulting in data corruption. For this reason, each UUID is generated using a hash of four variables: date, time, ESXi MAC address, and LUN ID. The VMFS datastore UUID applies to block-based VMFS volumes presented by Fibre Channel, Fibre Channel over Ethernet, and iSCSI storage arrays only, and does not apply to NFS volumes.

When a snapshot or clone is taken of a LUN, or the LUN is replicated, the LUN copy that is created is 100 percent identical to the original, including its UUID. In order to use this LUN copy in the same environment as the source LUN, it must be associated with a new UUID. When the LUN is identified by vSphere, the administrator is presented with three options:

- Keep the existing signature
- Assign a new signature
- Format the disk

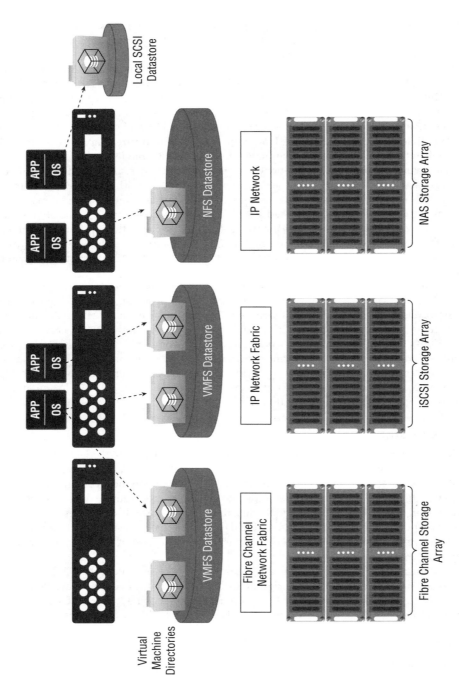

Figure 2.26
Types of datastore and storage network

Applying a new signature has significant consequences for the virtual machines that reside on that volume. Each virtual machine references the datastore UUID in its VMX, VMSD, and VMDK files, and therefore, a volume re-signature results in this UUID no longer identifying the correct datastore. This situation can easily be resolved by reregistering the virtual machines in vCenter Server, but depending on the scenario, this may not always be the correct course of action.

The datastore UUID becomes particularly significant when it comes to disaster-recovery planning, where the replicated volumes change signature, resulting in the problem outlined previously. In this type of scenario, all virtual machines must be removed from the vCenter Server inventory and reregistered with the new UUID, which can be operationally cumbersome if performed manually, and difficult to automate. Fortunately, VMware Site Recovery Manager can automate this process as part of a workflow to recover virtual machines, which simplifies the process significantly and helps avoid errors. With Site Recovery Manager in the design, the replicated volume is created with a different signature on the backup site, and all placeholder virtual machine configuration files, located at the recovery site, reference the correct recovery site UUID.

VIRTUAL MACHINE FILE SYSTEM

The VMware proprietary Virtual Machine File System (VMFS) is a clustered filesystem that allows multiple physical servers to read and write to the same storage device simultaneously. This clustered filesystem enables virtualization-based, distributed, infrastructure technologies, such as vMotion, to provide the live migration of running virtual machines from one physical server to another, by allowing multiple vSphere hosts to access the shared virtual machine storage concurrently. Likewise, VMFS also allows virtualization-based services such as vSphere Distributed Resource Scheduler (DRS) and vSphere HA to operate across a cluster of vSphere hosts. The vSphere HA mechanism works as the VMFS on-disk distributed locking mechanism ensures that the same virtual machine is not powered on by multiple servers at the same time. If a host fails, the on-disk lock for each virtual machine is released, so that the virtual machines can be restarted on other hosts within the cluster that have access to the same storage. All of this is possible only because of the unique properties of VMFS.

A VMFS-formatted datastore can be deployed on only SCSI-based storage devices, such as Fibre Channel, Fibre Channel over Ethernet, and iSCSI SAN equipment, as well as local storage or direct-attached storage (DAS), and is fundamentally the foundation that allows virtualization to scale beyond the boundaries of a single guest virtual machine per host.

In addition, VMFS allows IT organizations to significantly simplify the virtual machine–provisioning process, as illustrated in Figure 2.27, by storing the entire machine state in one central location, although this does not necessarily need to be the case.

Each of the three host servers has two virtual machines residing on it. The connecting lines from the virtual machine to the virtual machine disks (VMDKs) are logical representations of the connection between the two entities. The VMFS datastore represents a single LUN provisioned from a shared storage system and accessed across the SAN.

The virtual machine sees the VMDK as a local SCSI target within the guest operating system. The virtual machine itself does not recognize any VMDK files. As described previously, when addressing the virtual disk, the virtual machine contents on the VMFS datastore are really just a group of files. For the guest operating system running inside the virtual machine, VMFS preserves the internal filesystem semantics, which ensures the correct behavior and data integrity for running applications. In addition, it is a simple and standard operating procedure for guest operating system virtual disks to be *hot-added* or *hot-removed* from a virtual machine, if supported by the guest, without the need to power it down.

FIGURE 2.27
VMFS datastores

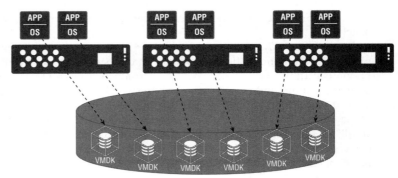

VMFS Datastore

Newly presented LUNs can also be added from the physical storage layer's subsystems, and can be discovered and made available to hosts and clusters without powering down physical servers. Storage capacity on a previously created VMFS datastore can also be *hot-extended* without powering down physical hosts. Virtual disks residing on VMFS datastores can also be *extended*, if supported by their guest operating system.

Equally, if any of the VMFS datastores fail or become unavailable, such as in an All Paths Down (APD) or Permanent Device Loss (PDL) event, only the virtual machines that touch that LUN are affected. All other virtual machines with virtual disks residing in other LUNs continue to function normally.

NOTE In Permanent Device Loss (PDL), ESXi considers the loss of the storage device as permanent. This event is typically caused by making a LUN inaccessible to the vSphere host. The PDL state is inferred from SCSI sense codes returned for a LUN by the storage array, with ESXi interpreting certain codes as a permanent failure.

However, in an All Paths Down (APD) failure scenario, ESXi considers the loss of connectivity as transient. This might occur if a host is unable to access the storage array via the SAN.

NFS Volumes

NFS is a file-based protocol used to establish a client-server relationship between the vSphere hosts and a NAS device. Unlike SCSI-based block storage, the NAS system itself is responsible for maintaining the layout and structure of the files and directories on the physical storage.

An NFS datastore provides most of the same features and functionalities as SCSI presented VMFS volumes. For instance, NFS also allows volumes to be accessed simultaneously by multiple hosts running multiple virtual machines, and also provides support for vSphere DRS, and vSphere HA. In addition, like VMFS datastores, once the storage has been mounted by the hosts, the vSphere administrator is able to use the storage to deploy virtual machines.

Additional benefits of NFS datastores include the lowest per port cost, when compared to Fibre Channel solutions, high performance in 10 GB environments, and storage savings provided by VMware thin provisioning, which is the default format for VMDKs created on NFS. Chapter 3 addresses these storage protocols extensively, and provides insight into the advantages and drawbacks of each transport mechanism as a design choice.

Raw Device Mapping

For some application use cases, the virtual machine needs to see a raw LUN directly. In a vSphere environment, this can be achieved via a raw device mapping (RDM).

Unlike VMFS and NFS datastores, which provide shared storage in a global pool, an RDM provides LUN access directly to one or more virtual machines. This is achieved by the RDM providing a symbolic link from a raw volume to the virtual machine. The mapping file, and not the raw volume, is referenced in the virtual machine configuration file (VMX file), which contains the symbolic link to the raw device (see Figure 2.28).

FIGURE 2.28
Raw device mapping connection topology

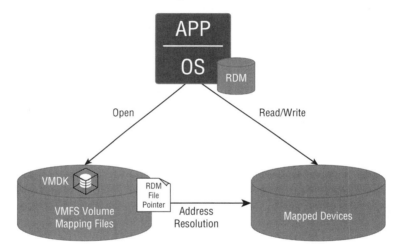

In the design illustrated in Figure 2.29, the hypervisor acts as a connection proxy between the virtual machine and the LUN, presented by the storage array. The primary use case for this solution type is support for Cluster Across Boxes (CAB) virtual machine clustering, and Physical and Virtual Machine (Physical and $n + 1$ VM) clustering technologies, such as Microsoft Windows Server Failover Clustering (WSFC) and Oracle RAC.

When using raw device mappings, you can still do the following:

- Use vMotion to migrate virtual machines that have raw disks attached.
- Add raw volumes to virtual machines using the vSphere Web Client.
- Use filesystem features such as distributed file locking, permissions, and naming.

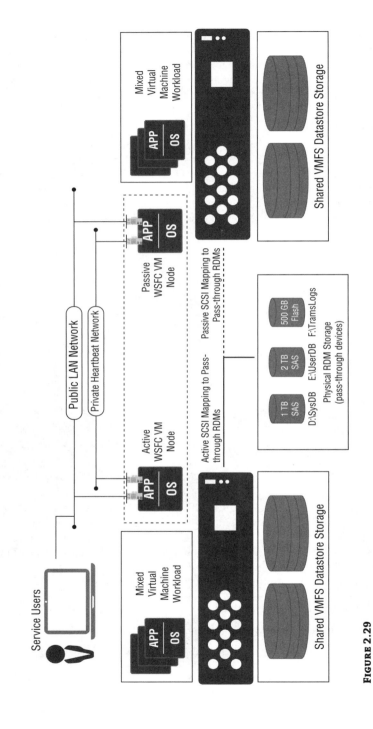

FIGURE 2.29
Cluster Across Boxes, Windows Server Failover Clustering example

In vSphere, raw device mappings operate under two compatibility modes:

- Virtual compatibility mode, which allows the mapping to act exactly like a virtual disk file, including the use of storage array snapshots
- Physical compatibility mode, which allows direct access of the SCSI device, for those applications that require lower-level control

Both of these compatibility modes support key VMware availability features, such as vMotion, HA, and DRS clusters. The key differentiator between the two mechanisms is the amount of SCSI virtualization that occurs at the virtual machine level, which results in some limitations around WSFC and virtual machine snapshots.

When to Use RDMs over VMFS or NFS?

When should you use RDMs and when should you use VMFS or NFS datastores? The answer is, it depends on the workload. Recent performance testing has shown that RDMs give little performance benefit, except for small block-size transactions.

However, working with templates and provisioning virtual machines is much more complicated with RDMs. When using RDMs, you must interact with the storage management team every time you want to provision a virtual machine that requires a raw volume. In addition, there are memory considerations when you map large numbers of LUNs to a single host.

For most applications, VMFS or NFS is the clear choice. VMFS provides the automated filesystem that makes provisioning and managing storage for virtual machines running on a vSphere cluster easy. In addition, VMFS has an automated hierarchical filesystem with user-friendly file-naming. It automates the directory and subdirectory naming processes to make vSphere administration much simpler. VMFS also enables higher levels of disk utilization by facilitating the process of provisioning the virtual disks from a shared pool of storage, and optimizing the I/O handling of multiple virtual machines through using a common Storage I/O Control mechanism.

When using RDMs in the design, each LUN must be sized based on the needs of each specific virtual machine's application or workload, which usually requires more-frequent dependence on the storage administration team. For these reasons, raw device mappings are recommended in modern virtual data-center environments only when a virtual machine must directly interact with the raw disk on the storage system. In typical use cases, you use storage-system-based snapshots of data, or you have a large amount of data that you don't want to move onto a virtual disk. Also, as illustrated previously in Figure 2.29, RDMs are required for single-copy Microsoft Windows Server Failover Clustering, and some in-band storage systems for management purposes.

Storage vMotion and Enhanced vMotion Operations

Storage vMotion (SvMotion) allows you to live-migrate a running virtual machine to a new storage environment or device, with little or no performance impact on the running applications. With SvMotion specifically, as many enhanced options are now available, the virtual machine remains running on the same host while its disks are individually moved to a different storage location. In the classic storage model described in this chapter, hosts are typically presented with multiple volumes, potentially of different performance capability and/or capacity.

Therefore, SvMotion can facilitate the migration of disk resources for a virtual machine from one datastore to another. The destination can be of different storage capability, capacity, or type, as SvMotion is protocol agnostic; you can migrate a virtual machine from one physical storage type, such as Fibre Channel, to a different storage type, such as iSCSI or NFS.

When the SvMotion migration occurs, the process does not significantly impact the virtual machine's performance, and there is no downtime. In most instances, it is transparent to the guest operating system and running applications. As part of the SvMotion operation, vSphere administrators can choose to place the virtual machine and all its files in a single datastore or select separate locations for the virtual machine configuration file and each of its virtual disks.

Introduced with vSphere 5.1, *X-vMotion*, more commonly known as *Enhanced vMotion*, combines both compute-based vMotion and Storage vMotion into a single operational task. This allows administrators to fully relocate a virtual machine, and its storage resources, from one vSphere host to another, while running and without impact to services, even if the storage is not shared between hosts.

There are a number of use cases for Storage vMotion, and when Enhanced vMotion technologies are used, such as long-distance vMotion, the use case options increase exponentially. However, traditionally vSphere administrators have used SvMotion for standard operating tasks in administering the data center. For instance, common tasks or scenarios for SvMotion might include the following:

- During an upgrade of the ESXi host platform, from a legacy release to the latest version, vSphere operational teams can migrate running virtual machines from a VMFS3 datastore to a VMFS5 datastore, and upgrade the old datastore without any impact on virtual machine workloads. For several reasons relating to block and volume size, creating a new VMFS5 volume is preferential to upgrading. The administrator can then use SvMotion to migrate virtual machines back to the original location without any virtual machine downtime.

- When performing storage maintenance tasks, vSphere administrators can use SvMotion to migrate virtual machines to another datastore to allow maintenance, reconfiguration, or retirement of the storage device, without virtual machine downtime.

- When redistributing storage load, vSphere administrators can use SvMotion to redistribute virtual machine workloads, or individual virtual disks, to different storage volumes with differing capabilities to balance capacity and improve performance.

- To meet service-level requirements, vSphere administrators can migrate virtual machine workloads to different tiered storage, with different levels of service, to address the changing business requirements for those applications.

Datastore Clusters

A datastore cluster can be easily compared to a vSphere compute cluster. A datastore cluster pools storage resources into one single logical entity, which then becomes the managed object. This object allows vSphere operational teams to manage multiple datastores as a single unit of storage resource, which, depending on the datastore cluster features being used, can provide optimized balancing of virtual machine resources across datastores within the cluster.

In effect, the datastore cluster provides a logical management layer that is abstracted from the storage itself. This being the case, the design does have to consider the relationship between hosts, clusters, virtual machines, and virtual disks (see Figure 2.30). It is often the case that datastore separation is carried out by enterprises and service providers based on business requirements. This new logical abstraction layer of the storage devices may impact this and even disrupt existing organizational policies and processes relating to workload placement across the existing storage platform.

Implementing vSphere datastore clusters can also impact other design factors, such as datastore sizing, number of datastores, and number of virtual machine disks per datastore. However, in reality, without Storage Distributed Resource Scheduler (SDRS) being implemented, a datastore cluster is simply a group of datastores, abstracted into a single logical management construct. With SDRS enabled, the datastore cluster becomes the unit by which to load-balance virtual machine storage resources across the cluster, based on application workload, in the same way that vSphere DRS balances compute workload across the physical hosts, based on CPU and memory load and resource availability.

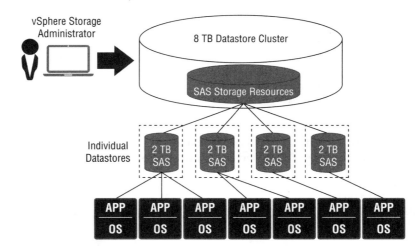

FIGURE 2.30
Datastore cluster design example

Storage Distributed Resource Scheduler

Building further on the Storage vMotion mechanism and the logical constructs of the datastore cluster, the Storage Distributed Resource Scheduler (SDRS) allows the datastore cluster to become a load-balanced storage domain,. vSphere operational teams can manage the SDRS datastore cluster as a single entity for provisioning virtual machines, instead of having to make individual decisions on virtual disk placement.

Without SDRS enabled, a vSphere administrator who wants to deploy a new virtual machine would typically have to manually try to find a datastore with available space, and one that did not suffer from high levels of latency. Additionally, the administrator would have to periodically check whether the virtual machines were suffering from high levels of latency, and monitor the datastores for low-disk-space alerts. This approach is, of course, inefficient and costs operational teams time with all the required manual intervention. In addition, all of this operational overhead leaves a lot of room for error, which could result in poor performance or application failures.

However, with SDRS, virtual machines and their virtual disks are automatically placed on the datastore with the lowest latency and the most capacity available. Virtual machines are then automatically balanced by being migrated, via SvMotion, to the most appropriate datastore within the cluster, if they are receiving higher levels of latency than they need to. The initial placement of a workload on the most appropriate datastore in the datastore cluster occurs in the following scenarios:

- The virtual machine is first deployed.
- A clone of the virtual machine is created.
- The virtual machine is relocated.

In addition to initial placement, SDRS load balancing can trigger a migration based on either high latency or low capacity. Low capacity is calculated based on capacity information that is continuously collected, with a default migration threshold of 80 percent utilization. The I/O latency is evaluated every 8 hours, when the Storage DRS algorithm does a cost and benefit analysis based on the previous 24-hour trend.

As you can see from the prolonged trend-based calculation, the SDRS algorithm does not make quick or snap decisions, but instead evaluates latency over a period of time before changes are made, or are recommended to be made, to a virtual disk's placement within the storage cluster. SDRS also works in conjunction with Storage I/O Control, by taking into account the percentage of shares assigned to a virtual machine. This is addressed in more detail in the next section, "Storage I/O Control."

Additional features of the SDRS mechanism that must be addressed include support for virtual disk level affinity and anti-affinity rules, illustrated in Figure 2.31.

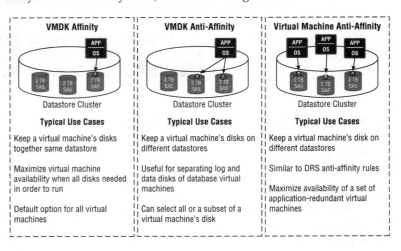

Figure 2.31 Storage DRS affinity rules

In addition, when a datastore is put into maintenance mode, all registered virtual machine disks will be evacuated from that datastore onto other datastores within the same datastore cluster. However, templates, ISO images, and unregistered virtual machines will not be migrated, and will therefore be unavailable during the maintenance period.

The SDRS mechanism improves overall operational management of vSphere storage resources and saves storage administrators and vSphere administrators a significant amount of time, which in turn can reduce operational costs.

Finally, the following are additional design factors that are pertinent to the implementation of SDRS:

- It will take at least 16 hours of I/O statistics being gathered before SDRS will make its first recommendations based on latency.

- It is possible to create a scheduled task to modify SDRS settings, to stop collecting statistics for a fixed period of time—for instance, during a backup cycle, to prevent unnecessary migrations resulting from skewed data.

- Design considerations that specifically relate to storage systems (such as tiering, replication, deduplication, and thin provisioning) should also be taken into account. For instance, all datastores within a single SDRS cluster should provide the same capabilities and characteristics to workloads, in order for the load balancing across datastores to provide meaningful benefits to application performance.

Storage I/O Control

There is a general consensus that the vast majority of virtualization performance issues are caused by latency in shared storage. The aim of Storage I/O Control is to provide a *quality of service* mechanism to ensure that all virtual machines get the storage performance they require, and prevent contention issues before they occur. Therefore, by taking this proactive approach, Storage I/O Control is able to mitigate a situation that can cause one virtual machine from impacting multiple other virtual machines' storage resources, the so-called *noisy neighbor* scenario.

In the typical noisy neighbor scenario, a small number of low-priority virtual machines impact and slow down a larger number of higher-priority virtual machines, with their continuous high demand for storage resources. Storage I/O Control, once enabled, helps alleviate these scenarios by monitoring the latency on each datastore. If the latency reaches a certain threshold, the Storage I/O Control mechanism will begin to throttle high I/O from some workloads, based on the *share* values assigned to each virtual machine.

The Storage I/O Control feature achieves this by monitoring and identifying when latency exceeds the set threshold. This feature works actively to relieve the congestion by ensuring that each virtual machine accessing the datastore is allocated an appropriate level of I/O resources, in proportion to their assigned share value. As such, this technology dynamically allocates storage I/O resources to ensure that virtualized workloads are able to maintain their performance service levels, even during periods of contention, which is the result of peak load. The net result of this mechanism is that operational teams can have increased productivity by reducing the operational overhead associated with manually providing this ongoing storage I/O management.

Storage I/O Control is configured at the virtual machine level by setting a specific level of shares and the maximum number of I/O per second. If the limit you want to configure for a virtual machine is measured in terms of megabytes per second (MB/s) instead of IOPS, it is also possible to convert MB/s into IOPS, based on the typical I/O size for that virtual machine.

For instance, to limit a specific application with 64 KB I/Os to 10 MB/s, configure a maximum limit of 160 IOPS for that virtual machine.

To translate MB/s into IOPS or vice versa:

IOPS = (MB/s throughput / KB per I/O) × 1,024

Or

MB/s = (IOPS × KB per I/O) / 1,024

Storage I/O Control does not take any action during normal operational conditions. The Storage I/O Control queue-throttling mechanism intervenes only when one of two specific thresholds, configurable by the vSphere administrator, are broken:

- The explicit congestion latency threshold set is broken for a datastore; the observed latency or response time in milliseconds exceeds the configured threshold.

- The percentage of peak performance is being broken for the datastore.

Storage I/O Control is enabled on each datastore individually. When enabled, ESXi begins to monitor the device latency observed from hosts that are connected to that storage device. A file called `iormstats.sf` is created directly on the datastore in question, and all hosts accessing that datastore are allowed read/write access to the file. If the device's latency exceeds the value defined in the threshold configuration (30 ms by default), the datastore is considered congested, and the Storage I/O Control queue-throttling mechanism intervenes by giving each virtual machine accessing that datastore only its allocated I/O resources in proportion to their percentage of *shares*.

By default, all virtual machines accessing the datastore will have the same number of shares. In this scenario, each virtual machine will be granted equal access to the datastore, regardless of its size or application workload. For this reason, if used as part of a classic storage design, it is critical that virtual machines be configured with shares proportionate to their SLA, if Storage I/O Control is to be of benefit.

It is pertinent to note that Storage I/O Control is not just a cluster-wide setting. It applies to any vSphere host that is connected to an enabled datastore. All connected hosts write to the `iormstats.sf` file on the datastore, regardless of whether they are a member of a cluster or not. However, it is considered good practice not to share datastores across different clusters, although this is sometimes done in order to meet specific the design requirements of a certain customer environment.

Without Storage I/O Control enabled, all vSphere hosts have equal device queue depths. Therefore, if a single virtual machine can generate enough I/O to max out the datastore, it is possible that critical workloads will be throttled by the hypervisor, to the benefit of less-important virtual machines. Therefore, it is always a good idea to enable Storage I/O Control with a high-enough threshold so as not to inhibit performance, regardless of the back-end disk configuration.

It is also recommended to enable the Storage I/O Control feature across all datastores, unless the storage vendor has issued guidance otherwise. For this reason, it is a good idea to check with the storage vendor before including it in a design or recommending changes across an entire environment, which could affect hundreds or even thousands of virtual machines.

For instance, many storage system vendors recommend disabling Storage I/O Control when implementing autotiering technologies, because of the Storage I/O Control injector not being able to determine the capabilities of tiered storage.

With Storage I/O Control enabled, the total disk shares available are managed globally, with the proportion of the datastore resources that each host receives depending on the total sum of share values of the virtual machines running on that host, which is then relative to the total sum of shares across all the virtual machines accessing that datastore. For instance, as illustrated in Figure 2.32, Storage I/O Control has been configured with the following values:

1. Virtual Machine A is given 75 percent of the space in the device queue, as it has 1,500 shares and therefore 60 percent of the storage queue.

2. Virtual Machine B is given 25 percent of the space in the device queue, as it has 500 shares and therefore 20 percent of the storage queue.

3. Virtual Machine C is given 100 percent of the device queue, on a dedicated ESXi host, but subsequently only 20 percent of the storage array queue through its 500 shares.

FIGURE 2.32
Storage I/O control mechanism

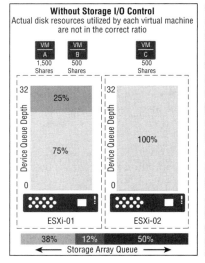
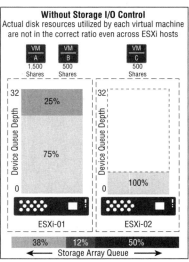

In vSphere 5, Storage I/O Control defaults to a 30 ms average latency threshold before limiting I/O on the noisy neighbor. However, in reality, with all the different performance characteristics of disks, disk tiers, and storage vendors, it is far more difficult to deterministically set a threshold. As such, this process was made dynamic in vSphere 5.5. Therefore, the latency threshold is now set to a value as determined by the I/O injector. This mechanism works by allowing the I/O injector to calculate the peak throughput; from this, it is able to determine the 90 percent throughput value and measure the latency at that point to determine the threshold. This is a configurable value that vSphere administrators can change, allowing them to set the throughput value to a different percentage; alternatively, they can continue to input a millisecond value.

For instance, if a vSphere administrator wishes to determine particular latency values for specifically configured tiered storage, based on predefined business service levels, they can do so as illustrated in the example shown in Table 2.22.

Even though in the past VMware has suggested the default threshold of 30 ms as a value that should meet the requirements for most virtual machines, the characteristics of the workload should, in most designs, play a factor in determining the optimum configuration, in allowing the vSphere platform to utilize the resources of the underlying storage system most effectively.

TABLE 2.22: Tiered Storage I/O Control latency values example

TIER/QOS	DISK TYPE	THRESHOLD
Tier 1	SSD	10–15 ms
Tier 2	SAS (15K)	15–20 ms
Tier 3	SAS (10K)	20–25 ms
Tier 4	SATA (7.2K)	30 ms

In autotiered environments, where the autotiering mechanism being used operates at the LUN level, make sure to use the vendor's recommended configuration. If one is not provided, use the threshold value defined previously for the slowest tier in the array. If the autotiering mechanism operates at the block level, also known as *sub-LUN tiering*, then again, use the vendor-recommended configuration. If no guidance is provided, combine ranges of the fastest and slowest media types in the array.

The Storage I/O feature should be considered an absolute design requirement for IT organizations that aim to achieve high consolidation ratios across their storage platform, such as cloud service providers, who host tens of thousands of virtual machines for multiple tenants in public cloud environments. Using Storage I/O Control also helps reduce the effect of a noisy neighbor scenario, in which a small number of workloads are trying to hog storage resources from the other, well-behaving, virtual machines.

The following are additional key design factors that are pertinent to employing Storage I/O Control as part of a storage design:

- Storage I/O Control is an Enterprise Plus feature only, and is therefore not available on other editions of the vSphere platform.

- Storage I/O Control–enabled datastores must be managed by a single vCenter Server.

- Just because the business is doing a storage refresh does not necessarily mean that vSphere will be updated to the latest release also. The design must ensure that vCenter Server, and all hosts connected to the datastore, are installed with vSphere 4.1 or later.

- Storage I/O Control supports both block-based storage, such as iSCSI and Fibre Channel, and NFS storage protocols. However, RDMs or multiple-extent LUNs are unsupported.

Classic Storage Model—vStorage APIs for Array Integration

VMware vSphere Storage APIs for Array Intergration (VAAI) is a feature that provides hardware acceleration functionality through the advanced features of storage arrays. The VAAI enables vSphere hosts to offload specific virtual machine and storage management operations to compliant storage systems when deploying storage configured to support this classic storage model.

Using these vStorage APIs provides an easy way to use the advanced storage capabilities of VAAI-enabled storage arrays from within the vSphere interface. These features allow vSphere to move the I/O load from the vCenter or host platform onto the storage controller. Instead of slowing execution time by consuming processing power, memory, and bandwidth on vSphere components, these operations are offloaded, and therefore speed up the completion of tasks and shift potential bottlenecks off hosts or vCenter Server to free up virtualization resources for more-critical operations.

These APIs define a set of *storage primitives*, aimed at reducing resource overhead on the vSphere hosts and vCenter Server, and as such, significantly improve the performance of storage-intensive operations, such as cloning and zeroing. The primary aim of these vStorage APIs is to allow storage vendors to provide hardware assistance to speed up these vSphere-intensive I/O operations that are more efficiently accomplished on the storage system.

The exact list of features offloaded to the storage hardware will vary, depending on the vSphere version and storage vendor integration. However, the following capabilities can typically be offloaded in most modern storage-array systems using the following primitives:

Full Copy   The time it takes to deploy or migrate a virtual machine is greatly reduced with the use of the *full copy* primitive, as the process is entirely executed on the storage array, and not on the ESXi server. In addition to being used when deploying new virtual machines from a template or through cloning, *full copy* is used when doing a Storage vMotion migration. When a virtual machine is migrated between datastores on the same array, the live copy is entirely performed on the storage system. Without VAAI, such copy operations must use the VMkernel software Data Mover driver. If the files being cloned or copied are hundreds of gigabytes in size, these operations can last for hours. On a VAAI-enabled array, the *full copy* primitive makes a request to the array that it performs the copy of the data blocks on behalf of the Data Mover. The *full copy* primitive's primary use cases are cloning and migration operations. Not only does *full copy* save time, but it also saves significant server CPU cycles, memory, DMA buffers, SCSI commands in the HBA queue, IP or SAN network bandwidth, and storage front-end controller I/O.

Block Zeroing   Having the storage array system complete the *block zeroing* out of a disk is far more efficient and much faster than the hypervisor. Typically, *block zeroing* is used when creating virtual disks that are EZT in format. Without the *block zeroing* primitive, the vSphere host must complete all the zero writes of the entire disk, before it reports that the task is complete. For large disk sizes, this can be time-consuming. When using the *block zeroing* primitive, however, the disk array returns the cursor to the requesting service as though the process of writing the zeros has been completed. It then finishes the job of zeroing out those blocks, without the need to hold the cursor until the job is done, as is the case with software zeroing at the hypervisor level. The following provisioning use cases are accelerated with the use of the *block zeroing* primitive:

- Cloning operations for EZT target disks
- The allocating of new file blocks for thin-provisioned virtual disks
- Initializing previous unwritten file blocks for zeroed thick virtual disks

Hardware-Assisted Locking   Also known as atomic test-and-set (ATS), hardware-assisted locking enables the offloading of the lock mechanism to the array. This permits a significant increase in scalability in a vSphere cluster sharing a datastore, without compromising the integrity of the VMFS shared storage-pool metadata. The ATS lock is a mechanism used to modify a disk sector, which once successful, allows a vSphere host to do a metadata update on the VMFS datastore. For instance, this includes the allocation of space to a VMDK during provisioning, as specific characteristics need to be updated in the metadata to reflect the new size of the file. It is noteworthy that in the initial VAAI release with vSphere 5, the ATS primitives had to be implemented differently on each storage array; this meant you had a different ATS opcode, depending on the storage vendor. However, ATS is now a T10 standard SCSI command and uses a common opcode of 0×89, COMPARE AND WRITE.

Thin-Provisioning Stun   This primitive facilitates a storage system to notify the vSphere host when a thin-provisioned volume reaches a certain capacity utilization threshold, in how it relates the available physical space on the storage array. When enabled, this primitive allows the host to take preventative measures to maintain virtual machine integrity, by pausing the virtual machines that will be impacted.

Delete Status (Dead Space Reclamation)   Also known as the UNMAP primitive, this allows a vSphere host to inform the storage array that space can be reclaimed that had previously been occupied by virtual machines that have been either deleted or migrated to another datastore. The SCSI UNMAP command, by performing space reclamation, allows the storage array to accurately report its space consumption of thin-provisioned datastores, and enables administrators to monitor and correctly forecast future storage requirements. This primitive has changed significantly since it was first introduced in vSphere 5.0. In the initial release, the operation was automatic, so when a virtual machine was deleted or migrated from a datastore, the UNMAP primitive was called immediately and space was reclaimed on the array. However, several problems were quickly identified with this approach, predominantly concerning the array performance and the storage system's ability to reclaim the space within an optimal time frame. For this reason, the UNMAP operation is now delivered as a manual process.

NOTE   Several important considerations associated with the use of the UNMAP primitive should be addressed before recommending it for a design. Please VMware Knowledge Base article 2014849 for more details.

These VAAI primitives need to be enabled on both the storage array and on the vSphere host. By default, vSphere enables a number of these primitives natively. However, others may require user intervention.

Classic Storage Model—VASA 1.0

The VASA feature was first released with VMware's vSphere 5. vSphere APIs for Storage Awareness (VASA) provide a set of APIs that enables vCenter Server to recognize the capabilities provided by supported storage arrays.

These capabilities might include characteristics such as supported RAID type, storage layer thin provisioning, or deduplication, which are published within vCenter, making it easier for vSphere and storage administrators to make informed decisions about the placement of virtual machines. For instance, providing published capabilities that can allow an administrator to identify a specific disk device, which provides RAID 5 protection.

The VASA feature is also able to provide this capability information about the storage array to vSphere's Profile-Driven Storage mechanism, which determines whether the storage device that a virtual machine is stored on complies with the workload's service-level storage requirements. In addition, VASA provides SDRS with information on storage arrays, allowing SDRS to provide optimal integration with a supported storage system.

The VASA 1.0 plug-in provides visibility from vCenter Server into the storage array, and can enable the cross-platform features highlighted previously. However, as we address in forthcoming chapters on Virtual SAN and Virtual Volumes, the published capabilities supported by VASA in version 2.0 are able to provide unprecedented coordination between vCenter Server, vSphere hosts, and storage devices.

Figure 2.33 illustrates the VASA 1.0 feature, in providing vCenter Server insight into a supported storage array's published capabilities.

FIGURE 2.33
VASA 1.0 vCenter server and storage array integration

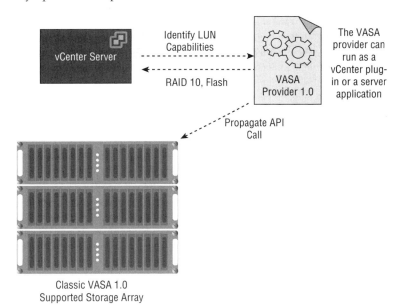

VADP and VAMP

Two further vStorage API features, which should be highlighted in order to ensure full coverage of the topic, are VMware vSphere Storage APIs for Data Protection (VADP) and vStorage API for Multipathing (VAMP).

VADP is an API that can be used to perform backup and restore operations on vSphere virtual machines. VADP was first introduced in vSphere 4 to replace the VMware Consolidated Backup (VCB) framework. However, VADP does not perform the data-protection operations itself. It is a framework that provides the mechanics for either VMware or third-party applications to perform backup and recovery operations. Therefore, VADP requires the installation of an additional VMware or third-party software backup product.

The primary purpose of VADP is to back up vSphere virtual machines from a centralized backup server, without requiring backup agents to be installed onto the guest operating systems. In addition, VADP does not require the backup processing to be performed inside each guest virtual machine which, as a result, allows the backup processing to be offloaded from the hypervisor, freeing up host resources.

One of the key features of VADP is Changed Block Tracking (CBT). Changed Block Tracking allows the VMkernel to track individual data blocks that have changed on the virtual machine's disks, allowing more-efficient incremental backups to be performed by identifying and targeting only the data blocks that are required to be backed up. This approach to backing up data helps to reduce backup times, and shortens the time required to replicate recovery data to a different site. However, this is likely to mean that virtual machine restoration will take longer as a result.

VAMP provides a set of APIs from VMware that aim to help control the I/O path selection between the vSphere host and storage array. These multipathing APIs provide storage-path failover and optimize storage I/O throughput by adding intelligence beyond that of the default pathing policies used by storage devices.

In order for a storage design to use VAMP, the storage vender's array must have built-in support for it. The vStorage APIs for Multipathing are provided through the vSphere framework in the VMkernel, called the Pluggable Storage Architecture (PSA), which is addressed in more detail in Chapter 3.

Boot from SAN

When you configure a host to boot from SAN, the hypervisor's boot image is stored on a single LUN in the SAN-attached storage system. When the host is powered on, it boots from the LUN through the SAN, rather than from any local media. A Boot from SAN environment can provide numerous benefits to the infrastructure, including providing a completely stateless compute environment. However, it can be complex to support and requires a specific implementation configuration. In addition, in certain use cases, you should not use Boot from SAN for ESXi hosts (for instance, where Virtual SAN is being used on the same hardware).

Before you decide whether Boot from SAN is appropriate for the environment, consider the advantages and drawbacks outlined here:

Boot from SAN Advantages

Less Power, Less Heat, Less State   Removing internal hard drives from servers means they consume less power and generate less heat. Therefore, they can be packed more densely, and the need for localized cooling is reduced. Without local storage, the servers effectively become *stateless* compute resources that can be pulled and replaced without having to worry about locally stored data.

Reduced Server CapEx   Boot from SAN enables organizations to purchase less-expensive diskless servers. Further savings can be made through reduced storage controller costs, although servers still need bootable HBAs.

More-Efficient Use of Storage   Whatever the footprint of the ESXi operating system, it will always be overprovisioned in terms of internal storage to accommodate it. Using Boot from SAN, the boot device can be configured to match the capacity it requires. This means a large number of host servers can boot from a far smaller number of physical disks.

High Availability   Spinning hard drives with moving internal components have limitations in terms of reliability, so removing reliance on internal hard drives should provide higher server availability. The servers will still rely on hard drives, but SAN storage arrays are much more robust and reliable, with far more redundancy built in to ensure that servers can boot.

Rapid Disaster Recovery   Data, including boot information, can easily be replicated from one SAN at a primary site to another SAN at a remote disaster-recovery site. This can mean that in the event of a failure, servers should be up and running at the remote site very rapidly.

Lower OpEx through More Centralized Server Management   Boot from SAN provides the opportunity for greatly simplified management of operating system patching and upgrades. For example, upgraded operating system images can be prepared and cloned on the SAN, and then individual servers can be stopped, directed to their new boot images, and rebooted, with very little down time. New hardware can also be brought up from SAN-based images without the need for any Ethernet network requirements. LUNs can be cloned and used to test upgrades, service packs, and other patches or to troubleshoot applications.

Better Performance   In some circumstances, the rapidly spinning, high-performance disks in a SAN might provide better operating performance than is available on a lower-performing local disk.

Boot from SAN Drawbacks

Compatibility Problems   Some operating systems, system BIOS, and especially HBAs, might not support Boot from SAN. Upgrading these components might change the economics in favor of local boot or vSphere Auto Deploy.

Single Point of Failure   If a server hard drive fails, the system will be unable to boot. However, if a SAN or its fabric encounters major problems, it is possible that no servers will be capable of booting. Although the likelihood of this happening is relatively small because of the built-in redundancy in most SAN systems, it is nevertheless noteworthy.

Boot Overload Potential   If a large number of servers try to boot at the same time (after a power failure, for example), this might overwhelm the fabric connection. In these circumstances, booting might be delayed or, if time-outs occur, some servers might fail to boot completely. This can be prevented by ensuring that the design distributes boot LUNs across as many storage controllers as possible, and that individual fabric connections are never overloaded, beyond the vendor's recommended limits.

Boot Dependencies   The SAN and array infrastructure must be operational to boot vSphere hosts. After a complete data-center outage, these components must be started and be operational prior to restarting the host servers.

Configuration Issues   Diskless servers can easily be pulled and replaced, but their HBAs have to be configured to point to their SAN-based boot devices before they boot. Unexpected problems can occur if a hot-swappable HBA is replaced in a running server. Unless the HBA is configured for Boot from SAN, the server will continue to run but fail to boot the next time it is restarted.

LUN Presentation Problems   Depending on the hardware, you might find that some servers can boot from SAN only from a specific LUN ID, such as LUN 0. If this is the case, you must have a mechanism in place to present the unique LUN that you use to boot a given server, with the LUN ID it expects to see. This is typically considered a legacy issue, which should not affect a new design and implementation.

Additional Complexity   There is no doubt that Boot from SAN is far more complex than providing a local boot device, and that adds an element of operational risk. As IT staff become accustomed to the procedures, this risk should diminish. However, the potential for problems in the early stage of Boot from SAN adoption should not be discounted. For example, Boot from SAN configurations require individual fabric zoning for each server, and potentially a much more complex HBA/CNA configuration.

Cost   SAN storage is typically more expensive than local storage, so any savings on server storage is lost on the extra storage array disks required.

Storage Team Overhead   A SAN LUN must be provisioned and managed for every host server, which can create significant additional work for the storage operations team.

Performance   Periods of heavy VMkernel I/O disk swapping can affect the virtual machine's disk performance, as they share the same disk I/O channels to the storage array.

Microsoft Clustering   In vSphere 4, virtual machines configured with Microsoft clustering (MSCS or failover clustering) are not supported on Boot from SAN configurations.

Scratch Partitions   ESXi does not automatically create scratch partitions in a Boot from SAN environment, as it sees the disks as remote. The creation of scratch partitions can be easily configured manually or scripted but must not be overlooked.

Classic Storage Model—vSphere Storage Policies

vSphere storage profiles were first introduced with vSphere 5, then renamed storage policies with the release of vSphere 5.5.

This technology forms part of VMware's profile-driven storage mechanism, which aims to logically separate storage by using storage capabilities and storage profiles in order to guarantee a predesignated quality-of-service of storage resources to a virtual machine.

The storage capability is a storage feature that may provide availability, redundancy, capacity, performance, or any other characteristic to storage resources. Two methods can publish storage capabilities within vCenter Server. The first are system-defined capabilities, provided if the storage array supports VASA 1.0, as outlined previously. The second option is to create and present a user-defined capability manually, based on the features supported by the storage system. Note, however, that a datastore can have only a single user-defined and system-defined capability associated at once.

The storage policy is used to map the defined storage capabilities to a virtual machine, and specifically, its virtual disks. The policy is based on the storage requirements for each virtual disk that guarantees a certain storage feature or capability.

This functionality allows vSphere administrators to identify storage capabilities, such as capacity, performance, availability, and redundancy, based on either system-defined, via

VASA 1.0, or user-defined storage features, and associate these with datastores. Once these capabilities are defined, storage policies can be created that reflect their associated technologies, such as tier 1, tier 2, and tier 3, before being associated with virtual machines. This mechanism allows vSphere to automatically place the workload on a datastore that meets its requirements, and allows vCenter Server to report in the Web Client user interface if it finds a virtual machine that is not placed on the correct storage type.

As highlighted earlier in this chapter, this feature has been enhanced significantly with the release of Virtual SAN and Virtual Volumes, to facilitate the Storage Policy–Based Management mechanism, which is fundamental to these software-defined storage offerings.

Figure 2.34 illustrates how this mechanism fits together, in order to offer this policy-driven storage solution, which is addressed in far greater depth in later chapters.

Figure 2.34
Classic storage policies

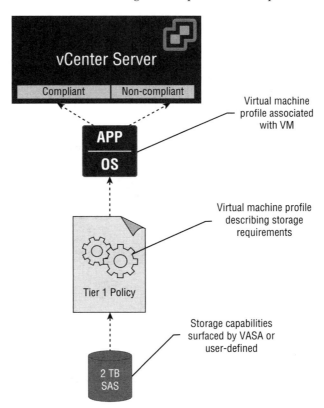

Tiered Storage Design Models in vSphere

Having now addressed storage tiering from both the storage array and a vSphere perspective, this section shows how these two separate levels of configuration should be brought together in an enterprise or service provider virtual data center design.

Several storage design models are available to users of classic storage architectures. By using a combination of the technologies already addressed (such as vSphere storage policies to categorize the storage capabilities of the backing storage subsystem, and SDRS to reflect the storage categorization such as tier 1, tier 2 and tier 3), operational simplification can be achieved. This combining of technologies will typically constitute a flexible solution, which can meet the needs of most enterprise and service provider storage architectures.

Next we present three example models that demonstrate how a combination of these technologies and customer design factors can provide various architectural solutions, each with advantages and drawbacks that require individual assessment.

Tiered Storage Model 1: Static Storage Tiering

The first model, shown in Figure 2.35, provides static storage tiers, allowing vSphere to correctly align each virtual machine to a guaranteed storage tier and level of service. This design requires that each storage tier will include specific storage capabilities, or drive type, that is consistent throughout the disk pool. Each pool in this typically multipool architecture will then be used to present a corresponding tiered storage capability, which will be consistent throughout the underlying subsystem.

In this storage model, there is no dependency on any array-based autotiering mechanism, and all blocks exist on the correct storage tier, throughout the life cycle of the virtual machine. The primary advantage of this storage model is that it allows a set of LUNs to be presented to the vSphere hosts from a pool, each with a consistent and predictable capability.

Figure 2.35
Static storage tier presentation model

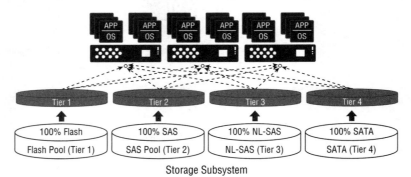

Tiered Storage Model 2: Mixed Storage Tiers

In this second model, we use a mix of storage tiers concurrently, including both static and dynamic tiering, within the array's disk subsystem. As illustrated in Figure 2.36, applications designated as either *tier 1* or *tier 2* employ LUNs from a statically tiered pool, ensuring that service levels are being maintained by the guaranteed block performance of the underlying storage. However, as depicted, *tier 3* and *tier 4* storage is made up of a single dynamically *autotiered* pool. which allows the storage system's autotiering mechanism to manage multiple storage tiers as a single entity.

FIGURE 2.36
Mixed storage tier presentation model

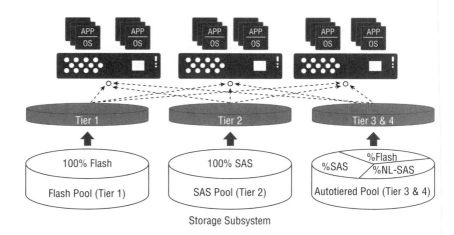

TIERED STORAGE MODEL 3: FULLY AUTOTIERED ENVIRONMENT

This final model, illustrated in Figure 2.37, uses a fully automated architecture for all presented storage. By using this technology in the design, the storage system's autotiering mechanism dynamically moves infrequently referenced blocks to lower-cost tiers of disk, while at the same time moving the more frequently accessed data to the faster drives. This eliminates much of the required user management and intervention associated with static storage tiers, while still aiming to maintain peak performance for the majority of workloads.

FIGURE 2.37
Fully auto-tiered presentation model

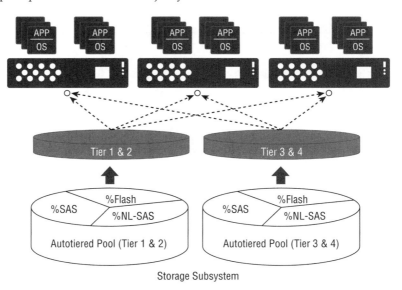

NOTE For LUNs provisioned from the autotiered storage, the storage vendor may recommend that Storage I/O Control should not be used to provide I/O control across virtual machines, particularly with older releases of vSphere. This is due to the Storage I/O Control injector not being able to determine the capabilities of autotiered storage.

Table 2.23 outlines key design factors that should be evaluated when carrying out an assessment of each solution, and when mapping customer requirements to the most appropriate choice of classic storage architecture.

TABLE 2.23: Storage tiering design factors

DESIGN FACTOR	DESCRIPTION/COMMENT
Availability	All design options have no impact on availability, as the storage array provides equal protection for both options.
Manageability	Model 1 provides simpler troubleshooting of storage performance issues, but increases daily management operational overhead.
Performance	Model 1 provides guaranteed performance, based on drive choice.
	For the other two models, performance could be impacted by Storage I/O Control not being available for use in an autotiered environment, depending on storage vendor and vSphere release.
Recoverability	All options provide equal data-recoverability features.
Security	All options provide equal data-recoverability features.

Sub-LUN System Access

Sub-LUN system access refers to how the storage subsystem is designed, and whether the disk pool is exclusively used by vSphere workloads, managed within a single vCenter Server, or is shared with other, non-vSphere workloads.

These are the only two options relating to storage subsystem access, and we refer to them as either VMware *dedicated* or *shared*. In this context, *shared* refers to a disk subsystem that is being written to and read from by other non-vSphere hosts, such as physical Microsoft Windows or Linux-based workloads, where the operating system is installed directly on the bare-metal hardware.

OPTION 1: VMWARE DEDICATED DISK SUBSYSTEM

With the disk subsystem being restricted to virtual workloads and managed by a single vCenter Server, all I/O can be observed as a single entity. This is considered the optimal design choice, a good practice and the recommended configuration by VMware. This design means that vCenter Server can accurately gauge the latency and I/O statistics of the platform, in order to allow Storage I/O Control to target and adjust the disk queue length of specific virtual machines, as and when required (see Figure 2.38).

FIGURE 2.38
VMware *dedicated* disk subsystem

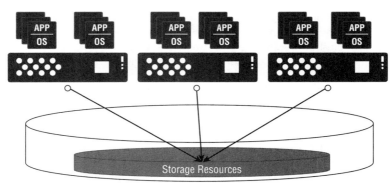

OPTION 2: VMWARE SHARED DISK SUBSYSTEM

In this second example, we can see that the underlying disk subsystem is being used to provide shared storage to other external workloads, which are unmanaged by vCenter Server. This approach is likely to have a negative impact on the virtual workloads while utilizing Storage I/O Control, as a result of the unmanaged I/O being detected on the system. This typically results in virtual machine workloads being throttled, allowing the unmanaged external workloads to consume more I/O, and will typically result in error messages, notifying administrators with a `Non-VI workload detected on the datastore` message appearing in the vCenter Server Web Client user interface (see Figure 2.39).

FIGURE 2.39
VMware *shared* disk subsystem

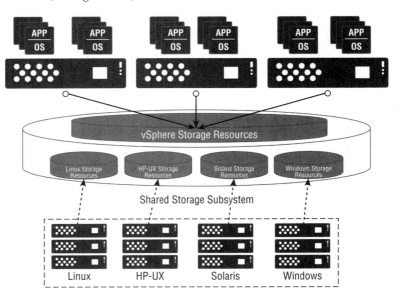

Another design factor to consider is that the sharing of the underlying subsystem presented to vSphere hosts, with other external workloads, will significantly increase the complexity of troubleshooting storage performance issues.

The primary reason for this is that VMware recommends enabling Storage I/O Control on all datastores. If the storage media (either flash or mechanical disks) on which the datastore is located is shared with volumes used by non-vSphere workloads, this will result in vSphere detecting that the datastore response time has exceeded the threshold, and issue the warning message shown previously. However, as Storage I/O Control is able to detect the presence of external workloads, and as long as they are present while the threshold is being exceeded, Storage I/O Control will intelligently compete with the interfering workload, by reducing its usual throttling activity.

It is also noteworthy that in older releases of Storage I/O Control, this behavior was not as intelligent, and as a result, the vSphere platform would begin throttling the ESXi workloads, assessing the datastore to ensure that the applications with the highest shares got priority over I/O access, and as such, a lower I/O response time. However, it is likely that this throttling would not have been required or have been applicable, as the workload causing the contention was not being managed by Storage I/O Control. Therefore, in these earlier releases of Storage I/O Control, this throttling would result in the external workload getting more I/O and more bandwidth, even though the vSphere workloads continue to get less and less. This is important to remember, as just because a customer is refreshing their storage platform or storage design, they aren't necessarily going to be deploying the latest vSphere software releases.

Despite the improved intelligence in newer releases of Storage I/O Control, the clear design recommendation is to always provide dedicated VMware subsystems for vSphere virtual machine workloads in order to avoid external interference and simplify performance troubleshooting activities.

Chapter 3

Fabric Connectivity and Storage I/O Architecture

Fabric connectivity refers to the interconnection between vSphere hosts or between hosts and storage devices, depending on the architecture. This chapter focuses only on the connectivity between the host and shared storage devices, as required by both the classic storage infrastructure (covered in Chapter 2, "Classic Storage Models and Constructs") and Virtual Volumes (covered in Chapter 8, "Policy-Driven Storage Design with Virtual Volumes"). Intra-host connectivity, as required for VMware Virtual SAN, is addressed in detail in Chapter 4, "Policy-Driven Storage Design with Virtual SAN."

Connectivity and communication between the ESXi hosts and the storage devices is enabled using physical components and interface protocols. The physical components of connectivity are the hardware elements that connect the hosts to the storage. There are typically just three physical components of connectivity between the host and storage device: the host interface devices themselves, switch ports, and cables.

A *host interface device*, or *host adapter*, connects the host to the fabric and its attached storage devices. Examples of host interface devices include the host bus adapter (HBA) and the network interface card (NIC). A host bus adapter is an application-specific integrated circuit (ASIC) board that performs I/O interface functions between the host and storage, off-loading the additional CPU I/O processing overhead associated with the storage protocol. A typical vSphere host includes multiple HBAs to provide increased performance and redundant connectivity to the fabric, and onto the storage devices.

A *port* is a specialized outlet that enables connectivity between the host and external devices. The HBA contains one or more ports to connect the host to the switching fabric, and then on to the storage device. *Cables* are used to connect hosts to external devices via either copper or fiber-optic media.

A *protocol* enables communication between the host and storage. Protocols are implemented using interface devices, or controllers, at both source and destination. The most common interface protocols employed for host-to-storage communications are based on Fibre Channel or Internet Protocol (IP). The most common options deployed in the data center are native Fibre Channel (FC), Fibre Channel over Ethernet (FCoE), Internet Small Computer System Interface (iSCSI) and the Network File System (NFS).

Storage connectivity and fabric architecture, when referring to block-based storage protocols, is more commonly referred to as the *storage area network* (SAN). A SAN provides a high-speed dedicated network of shared storage devices, which facilitates storage consolidation and enables storage to be shared across multiple host servers. This improves the utilization of storage resources when compared to a direct-attached storage architecture, and reduces the total amount of physical disk capacity an organization is required to purchase and manage. With this model, storage management becomes centralized and less complex than maintaining large amounts of directly connected storage, which further reduces the operational cost of managing data. A SAN also enables organizations to connect geographically dispersed servers and storage together. Furthermore, it helps to meet the storage demands efficiently with better economies of scale, and also helps provide effective maintenance and data protection.

The two most common SAN deployments are Fibre Channel SAN and IP SAN. Fibre Channel SAN uses the Fibre Channel Protocol as the transport mechanism for data, commands, and status information between vSphere hosts and storage devices. Likewise, IP SAN uses the iSCSI protocol for storage communication between host and storage device.

Despite the growth in hyper-converged infrastructure (HCI) solutions in our data centers, the requirement for a SAN and shared storage in most data centers is here to stay for the foreseeable future. Therefore, these technologies must be understood in depth by storage architects and storage operational teams alike. The following sections of this chapter cover each of the supported storage protocols, so you can gain a better understanding of how they are integrated with vSphere, and the key design factors that you must take into account when architecting solutions based on their use.

Fibre Channel SAN

In the majority of enterprise and storage provider environments, the *nonlocal* storage for vSphere is typically previsioned via Fibre Channel–accessible, storage area network LUNs. Of course, that does not rule out the use of iSCSI, NAS, FCoE, and raw disk storage implementations, which can be employed in native or mixed-protocol storage environments.

Fibre Channel Protocol

Fibre Channel is a well-established block protocol employed in enterprise and service provider data centers. It provides a high level of performance, typically running at speeds of 4, 8, 16, or even 20 Gb/s, and offers the solid foundation required for reliable storage access. The Fibre Channel Protocol requires a separate dedicated network architecture, which is typically deployed in a highly available redundant fashion, making it costly, as well as adding operational complexity to data-center management.

However, despite the cost and added complexity, Fibre Channel SAN meets the high storage demands of enterprise IT organizations and service providers efficiently, and with far better economies of scale than IP-based storage. Fibre Channel SAN also provides effective maintenance, operations, and protection of data.

Fibre Channel high-speed network technology runs on high-speed optical fiber cables or serial copper cables, and was developed to meet the demand for faster data transfer between servers and high-speed intelligent storage systems. The responsibility for the Fibre Channel interface standard lies with Technical Committee T11, which is a working group within the International Committee for Information Technology Standards (INCITS).

The latest common implementation of the Fibre Channel standard, at the time of writing, offers a raw bit rate of 16 Gb/s with a throughput of 6,400 MB/s. With its credit-based flow-control mechanism, Fibre Channel delivers data as fast as the destination buffer is able to receive it, without dropping frames. In addition, Fibre Channel has very little transmission overhead, and the architecture is highly scalable, with approximately 15 million devices theoretically being accommodated on a single Fibre Channel network.

Fibre Channel Protocol Layers

Many readers will already be familiar with the International Organization for Standardization (ISO) Open Systems Interconnect (OSI) model, which standardizes the communication functions of a telecommunication or computing system into seven layers: Physical, Data-Link, Network, Transport, Session, Presentation, and Application.

Breaking a communication protocol in this way into layers often makes it easier to understand. While Fibre Channel does not follow the same seven layers, it does follow a similar protocol model of five layers, shown in Figure 3.1. Each layer is briefly described in Table 3.1, along with the main functions it defines.

Table 3.1: Fibre Channel Protocol layers

Fibre Channel	Layer Function
FC-0	FC-0 defines the physical media including connectors, cables, transmitter, and receiver technology, and supports a variety of data rates. Signaling, media specifications, receiver, and transmitter specifications are also included in this layer.
FC-1	Defines transmission protocol and link maintenance, utilizing the 8B/10B code/decode method, which improves the transmission characteristics and enhances error recovery.
FC-2	Defines the rules by which nodes communicate, sequence management, exchange management, including data framing, frame sequencing, flow control, and class of service. In addition, login/logout, topologies, segmentation, and reassembly are all handled at FC-2.
FC-3	Defines a set of common services, such as services for multiple ports on one node, and support for advanced functions such as RAID. FC-3 is still under construction.
FC-4	Defines the interface model between the upper-level protocols (ULP): • Small Computer System Interface (SCSI-3) • Internet Protocol (IP) • High-Performance Parallel Interface (HIPPI) • Asynchronous Transfer Mode—Adaptation Layer 5 (ATM-AAL5) • Intelligent Peripheral Interface 3 (IPI-3) (disk and tape) • Single-Byte Command Code Sets (SBCCS) • Other, future ULPs

FC-0 and FC-1 can be compared with the Physical layer of the OSI model, in defining the physical media. FC-2 can be considered what other protocols define as a Media Access Control (MAC) layer, which is the lower half of the Data-Link layer. FC-3 is a largely undefined set of services for devices having more than one port (for instance, striping, whereby data is transmitted from all ports at the same time in order to increase bandwidth). Finally, FC-4 defines how the other, more established higher-layer protocols are mapped onto and transmitted over Fibre Channel.

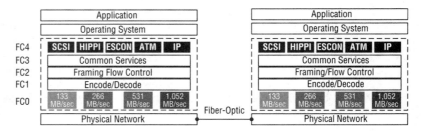

FIGURE 3.1
Fibre Channel Protocol layers

FIBRE CHANNEL ARCHITECTURE

Just as corporate data networks have become standardized with Ethernet and IP, for the enterprise organizations and cloud service providers, storage area networks are generally implemented using Fibre Channel. The Fibre Channel Protocol maps many existing protocol frames for transmission. These protocols include the following:

- Small Computer Systems Interface (SCSI)
- High-Performance Parallel Interface (HIPPI)
- Enterprise Systems Connection (ESCON)
- Fibre Connection (FICON), which is replacing ESCON as the Fibre Channel implementation for the IBM Z series mainframes
- Asynchronous Transfer Mode (ATM)
- Internet Protocol (IP)

COMPONENTS OF FIBRE CHANNEL SAN

In a Fibre Channel SAN environment, ESXi servers access the disk array through the dedicated Fibre Channel network, which is typically a dual fabric design, as illustrated by the components shown in Figure 3.2.

While the SAN topology may differ, depending on design requirements, essentially all hosts connect to two SAN fabrics using internal server HBAs. The fabrics are isolated from each other and are managed as two separate entities. Legacy configurations with four or more HBAs are far less common in modern designs, due to the availability of 4, 8, and 16 Gb/s hardware options.

As shown in Figure 3.2, the components of a Fibre Channel SAN can be divided into three categories, each of which is described in more detail next:

FIGURE 3.2
Fibre Channel component topology

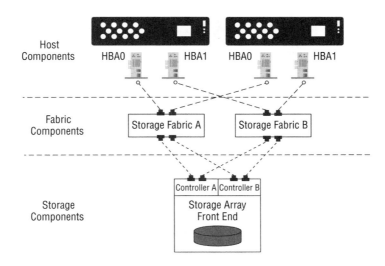

- Host components
- Fabric components
- Storage components

Host Components

The *host components* of a SAN consist of the host servers, or more specifically, the HBA component, which allows the servers to be physically connected to the SAN fabric. HBAs are located inside individual host servers and connect each host to the fabric ports. This mechanism employs vendor-specific HBA drivers running on the host ESXi servers, which enable the hypervisor to communicate with the HBA device.

The HBA provides the I/O processing and physical connectivity between the host server and storage. The storage can be attached via a direct-attached connection or a variety of storage-networking technologies. HBAs also provide server CPU off-loading, which is critical for freeing vSphere host servers' CPU and memory to perform processing related to virtual machines and applications. The HBA is the only part of a storage area network that resides inside the host server, and enables a range of high-availability and storage management capabilities, including SAN administration, load balancing, failover, and storage management. In effect, the host adapter card provides the interface between the server's internal bus and the external storage network.

In addition to the adapter card itself, the vendor supplies a device driver that allows the card to be recognized by the ESXi operating system, which is loaded as a kernel module. This device driver may also perform protocol translation or other similar functions, if these are not already executed on the host's motherboard. It is critical to the storage infrastructure that the Fibre Channel HBA must provide reliable communication at the Physical and Data-Link level at all times.

For this reason, in order to achieve HBA high availability, most designs use a minimum of two host ports in each server. Depending on the server hardware chosen, and its form factor, two design options are typically considered:

- A single-, dual-, or quad-port HBA
- Two or more single-port HBAs

Blade systems, however, typically use proprietary hardware, which forces the use of a single adapter card, which might be unable to provide redundancy at the hardware layer. However, where possible, by employing two or more independent HBA devices, an additional level of redundancy and availability can be achieved within the design.

Fabric Components

Fabric components consist of the network and interconnecting devices portion of the SAN, with all hosts connecting to the storage devices on the SAN through the SAN fabric. The SAN fabric can consist of the following fabric components:

- Fibre Channel hubs
- Fibre Channel edge switches
- Director class switches
- Data routers
- SAN cables
- Communication protocols

Fibre Channel hubs are now legacy devices, which you will rarely see in a modern fabric design but were previously used as communication devices in FC-AL implementations. Hubs physically connect nodes in a logical loop or a physical star topology. All of the nodes share the loop, as the data must always travel through all of the connection points. Because of the availability of low-cost and high-performance fabric switches, hubs are no longer generally part of a Fibre Channel SAN design.

A *Fibre Channel switch* is more intelligent than a hub, and can directly route data from one physical port to another. For this reason, SAN switches are employed to connect to servers, storage devices, and other switches, and provide the connection points for the modern SAN fabric. The type of SAN switch, its design features, port capacity, throughput, performance, fault tolerance, and the manner in which they are interconnected all contribute to define the fabric topology and overall architecture.

Fibre Channel switches function in a manner similar to traditional network switches, in that they provide increased bandwidth, and scalable performance through an increased number of interconnected devices. Also, just like Ethernet switches, Fibre Channel devices vary in the number of ports and media types they support. Multiple switches can be connected to form a switch fabric capable of supporting a large number of host servers and storage systems. *Edge switches*, which are effectively top-of-rack switches, are available with a fixed port count or in a modular design. In a modular switch, the port count is increased by installing additional line cards to open slots, in much the same way as for IP network hardware.

Director class Fibre Channel switches are high-end switches with a higher port count and better fault-tolerance capabilities. The architecture of a director is always modular, and its port count is increased by inserting additional line cards or blades into the director's chassis. Director-class switches always contain redundant components to provide high availability. Standard Fibre Channel and director switches both have management ports, typically Ethernet and serial, for connectivity to SAN management servers.

Data routers are used to provide an intelligent bridge between SCSI and Fibre Channel devices on the SAN fabric. Host servers, which are connected to the SAN, can access SCSI disks or other devices such as tape libraries through the data routers on the fabric.

SAN cables are typically special fiber-optic cables, although copper cables are occasionally employed for connecting devices over short distances. Optical cabling can be used for connections over much longer distances, because of their immunity from external interference, referred to as *noise*. Each link connects two Fibre Channel ports: a transmitting port (Tx) at one end and a receiving port (Rx) at the other end. The type of SAN cable, fiber-optic signal, vendor, and fabric switch licensing all go toward determining the maximum distances supported between SAN components, and contribute to the total bandwidth rating of the fabric.

Finally, the *communication protocol* is the special set of rules that endpoints use when they communicate with one another. As we have already highlighted, the Fibre Channel SAN is the storage-interface protocol used by most enterprise IT organizations and service providers. Fibre Channel was developed as a protocol for transferring data at high speeds between two ports over a serial I/O bus cable, making it the ideal choice for storage communication.

However, keep in mind that iSCSI connectivity over an IP network is also considered a SAN protocol, as is Fibre Channel over Ethernet. These other protocol options, as well as NFS, are covered later in this chapter.

Storage Components

The *storage components* of a SAN are the storage arrays themselves. Storage controllers provide the vSphere hosts with front-end connectivity to the storage devices, either directly or more commonly through a switched fabric. The storage controllers also provide the internal access to the disks themselves, through their back-end connectivity, which can use either a switch or a bus architecture, to provide the RAID, storage container, and Virtual Volumes functionality of the storage system.

An intelligent storage system typically consists of four key components: front end, cache, back end, and physical disks, as illustrated in Figure 3.3.

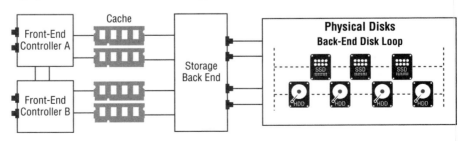

Figure 3.3
Physical storage array architecture

Figure 3.3 illustrates one example of a physical storage array architecture, which can vary significantly from vendor to vendor. Disk arrays are managed by vendor-proprietary operating systems, which include built-in intelligence. Therefore, disk arrays can vary significantly in their design, capacity, performance, and advanced capabilities.

When an I/O request is received from the host server on the front-end port, it is processed through the cache and back-end disk system in order to facilitate the storage and retrieval of the required data to and from the physical disks. A read request might also be serviced directly from the cache, if the requested data is being held there. In most modern storage systems, the front-end cache and back-end components are typically integrated on a single board, within the storage controller's hardware.

In addition, in an enterprise storage system, back-end disks are normally connected in loops. The back-end loop technology employed by the storage controllers can provide several key benefits:

- High-speed access to the disks
- The ability to add more drives to the loop
- Redundant access to a single drive from multiple loops when drives are dual-ported and attached to both back-end loops

Fabric Services

All Fibre Channel switches, regardless of the hardware vendor, provide a common set of services, which are defined in the Fibre Channel standards. These services are available at certain predefined addresses, shown in Figure 3.4. These services are described in Table 3.2.

Table 3.2: Fabric services

Fabric Services	Description
Fabric Login Server	The Fabric Login Server is located at the predefined address of FFFFFE and is employed during the initial part of the node's fabric login process.
Name Server	The Name Server is located at the predefined address of FFFFFC and is responsible for name registration, and management of node ports. Each switch exchanges its Name Server information with other switches in the fabric to maintain a synchronized, distributed name service.
Fabric Controller	Each switch has a Fabric Controller located at the predefined address of FFFFFD. The Fabric Controller provides services to both node ports and other switches in the fabric. The Fabric Controller is responsible for managing and distributing registered state change notifications (RSCNs) to the switch ports registered with the Fabric Controller. If there is a change in the fabric, RSCNs are sent out by the switch to the attached ports. The Fabric Controller also generates switch registered state change notifications (SW-RSCNs) to every other domain (switch) in the fabric. These SW-RSCNs keep the name server up-to-date on all switches in the fabric.

TABLE 3.2: Fabric services *(CONTINUED)*

FABRIC SERVICES	DESCRIPTION
Management Server	FFFFFA is the predefined address in the Fibre Channel standards for the Management Server. The Management Server is distributed to every switch within the fabric. The Management Server enables the Fibre Channel SAN management software to retrieve information and allow the administration of the devices within the fabric.

NODE PORTS

In a Fibre Channel network, the end devices, such as hosts, storage arrays, and tape libraries, are all referred to as *nodes*. Node ports provide the physical interface for communicating with other nodes, with each node being a source or destination of information. Each node requires one or more ports to provide a physical interface for communicating with other nodes on the same fabric. These ports are integrated components of HBAs and storage arrays' front-end controllers. In a Fibre Channel environment, all ports operate in full-duplex data-transmission mode, with both a transmit (Tx) link and a receive (Rx) link.

FIBRE CHANNEL ADDRESS MECHANISM

A Fibre Channel address is dynamically assigned when a node port logs into the fabric. The Fibre Channel address has a distinct format, shown in Figure 3.4. The first field of the address contains the domain ID of the switch that is logged into. The domain ID is a unique number provided to each switch in the fabric. Although this is an 8-bit field, there are only 239 available addresses for domain IDs, as some addresses are deemed special and are reserved for the fabric services. For instance, as shown in Table 3.2, FFFFFC is reserved for the Name Server, and FFFFFE is reserved for the Fabric Login Service. The area ID is used to identify a group of switch ports used for connecting host servers. An example of a group of ports that would have a common area ID is a line-card, providing additional ports on the fabric switch. The last field, the port ID, identifies the unique port within the group.

As a result of this switched fabric addressing scheme, the maximum possible number of node ports in a switched fabric is calculated as follows: 239 domains × 256 areas × 256 ports = 15,663,104.

FIGURE 3.4
Fibre Channel address mechanism

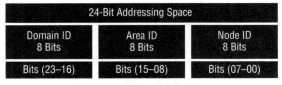

Fibre Channel Port Naming

Ports in a switched fabric can be one of the following types (see Figure 3.5), depending on their connectivity to other devices:

N_Port An endpoint in the fabric. This port is also known as the *node port*. Typically, it is a host port (HBA) or a storage array port that is connected to a switch in a switched fabric.

E_Port A port that forms the connection between two Fibre Channel switches. This port is also known as the *expansion port*. The E_Port on a Fibre Channel switch connects to the E_Port of another Fibre Channel switch in the same fabric. This port type creates Inter-Switch Links (ISLs).

F_Port A port on a switch that connects to an N_Port within the Fibre Channel topology. It is also known as a *fabric port*.

G_Port A generic port on a switch that can operate as an E_Port or an F_Port, and determines its functionality automatically during initialization.

NL_port A port used with an FC-AL topology, sometimes known as a *node loop port*.

FL_port A port on the switch that connects to an FC-AL loop. Also known as a *fabric loop port*.

Login Types in a Switched Fabric

Fabric services define three login types, depending on their connectivity to other devices:

Fabric Login (FLOGI) Performed between an N_Port and an F_Port. To log in to the fabric, a node sends a FLOGI frame with the WWNN and WWPN parameters to the login service at the predefined Fibre Channel address FFFFFE (Fabric Login Server). In turn, the switch accepts the login and returns an Accept (ACC) frame with the assigned Fibre Channel address for the node. Immediately after the FLOGI, the N_Port registers itself with the local Name Server on the switch, indicating its WWNN, WWPN, port type, class of service, and assigned Fibre Channel address. After the N_Port has logged in, it can query the Name Server database for information about all other logged-in ports.

Port Login (PLOGI) This login type is performed between two N_Ports to establish a session. The initiator N_Port sends a PLOGI request frame to the target N_Port, which accepts it. The target N_Port returns an Accept (ACC) to the initiator N_Port. Subsequently, the N_Ports exchange service parameters relevant to the session.

Process Login (PRLI) Also performed between two N_Ports. This login type relates to the FC-4 upper-level protocols (ULPs), such as SCSI. If the ULP is SCSI, N_Ports exchange SCSI-related service parameters.

World Wide Name Device Addressing

All Fibre Channel devices, such as a host adapter (initiator) or storage device (target), have a unique 64-bit identity, derived from IEEE OUI and vendor-supplied information, which is called the *World Wide Name* (WWN). This addressing system can be compared with Ethernet cards and MAC addresses. Each node port has its own WWN, but it is possible for a device with more than one Fibre Channel port to have its own WWN as well.

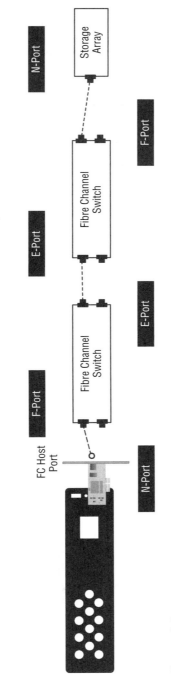

Figure 3.5
Fibre Channel port naming

There are two WWN addressing schemes. The older one starts with 10:00 followed by company ID and vendor-specific information. The newer scheme starts with hex 5 or 6 in the first half-byte, followed by the vendor information in the next 3 bytes, as shown in Figure 3.6.

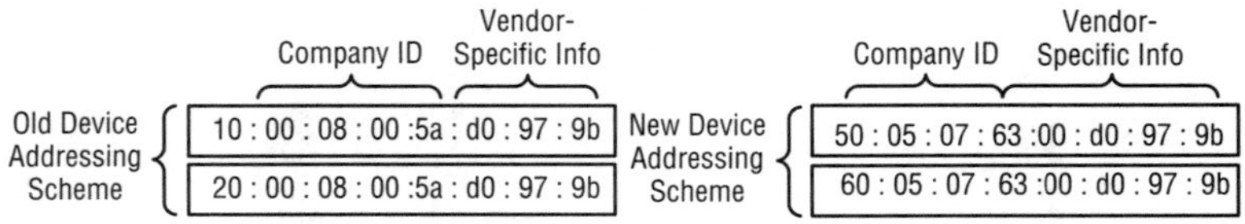

FIGURE 3.6
WWW device addressing

	Company ID	Vendor-Specific Info		Company ID	Vendor-Specific Info
Old Device Addressing Scheme	10 : 00 : 08 : 00 :5a : d0 : 97 : 9b		New Device Addressing Scheme	50 : 05 : 07 : 63 :00 : d0 : 97 : 9b	
	20 : 00 : 08 : 00 :5a : d0 : 97 : 9b			60 : 05 : 07 : 63 :00 : d0 : 97 : 9b	

Unlike a Fibre Channel address, which is assigned dynamically, a WWN is a static name for each node on the Fibre Channel network. The WWN was traditionally burned into the hardware, although more recently, with some vendors' equipment, it can also be assigned through software. A number of configuration definitions within the SAN use WWNs for identifying storage devices and HBAs. The name server in a Fibre Channel environment keeps the association of the WWN to the dynamically created Fibre Channel addresses for nodes.

The Fibre Channel environment uses two types of WWN:

World Wide Node Name   The WWNN is a unique identifier that identifies a server.

World Wide Port Name   The WWPN uniquely identifies a specific physical port on the HBA.

The WWNN is a globally unique 64-bit identifier that is assigned to each Fibre Channel node or device. For instance, for servers and hosts, the WWNN is unique for each HBA. A server with two physical HBA cards would have two WWNNs. For a SAN fabric switch, the WWNN is a common identifier for the chassis. For storage devices, the WWNN is common for each storage controller in the array, although with some vendor hardware, it could also be unique for the entire array.

The WWPN is a unique identifier for each Fibre Channel port associated with any Fibre Channel device. For instance, a host server has a WWPN for each port on the HBA, typically 1, 2, or 4. For a SAN fabric switch, a WWPN is available for each port in the chassis, and on the storage device, each front-end port has an individual WWPN.

The WWNN and WWPN addresses consist of 16 hex values. Figure 3.7 illustrates how the WWN is structured for an array and an HBA.

SAN Management Software

The final fabric component to be addressed is SAN management software. The *SAN management software* is employed by storage operational teams to manage the SAN fabric interfaces between hosts, interconnected devices, and storage arrays. This software typically provides an overall view of the SAN environment, and enables management of various resources from one central console (see Figure 3.8).

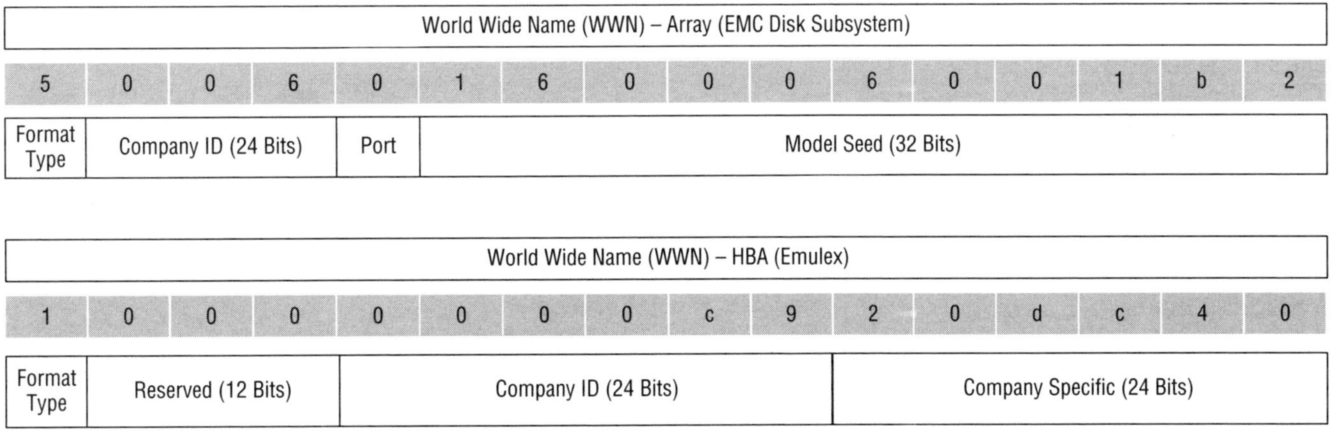

World Wide Name (WWN) – Array (EMC Disk Subsystem)

5	0	0	6	0	1	6	0	0	0	6	0	0	1	b	2
Format Type	Company ID (24 Bits)						Port	Model Seed (32 Bits)							

World Wide Name (WWN) – HBA (Emulex)

1	0	0	0	0	0	c	9	2	0	d	c	4	0
Format Type	Reserved (12 Bits)			Company ID (24 Bits)					Company Specific (24 Bits)				

FIGURE 3.7
World Wide Name (WWN) device addressing

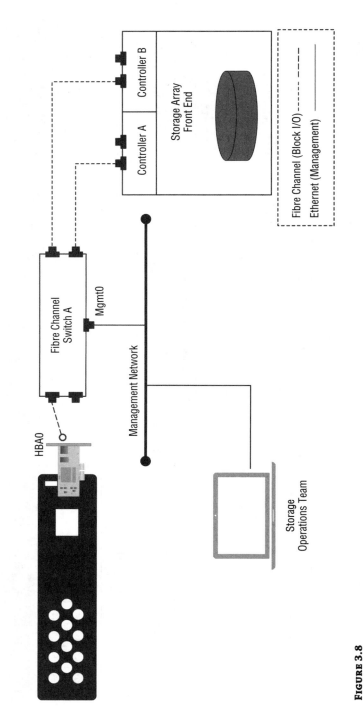

Figure 3.8
SAN management topology

In most cases, SAN management software provides key management functions (such as zoning, monitoring, and alerts) for discovered fabric devices via a web-based or command-line interface.

Fibre Channel Topologies

The ANSI Fibre Channel standards define three topologies, which describe how ports can be connected:

Point-to-Point (FC-P2P)   Two devices are connected directly to each other.

Arbitrated Loop (FC-AL)   A legacy shared topology in which all devices are interconnected in a loop, or ring.

Switched Fabric (FC-SW)   A shared topology in which all devices, or loops of devices, are connected to the common interconnect. The switched fabric topology is typically the basis for most modern SAN fabrics.

Point-to-Point Connectivity

The *point-to-point* (FC-P2P) topology is the simplest Fibre Channel configuration. As shown in Figure 3.9, two devices are connected directly to each other. This configuration provides a dedicated connection for data transmission between nodes. However, the point-to-point configuration offers limited connectivity and scalability, because only two devices can communicate with each other at any given time. Therefore, this topology is typically employed only to provide standard direct-attached storage (DAS) point-to-point connectivity between a host and storage device (see Figure 3.9).

FIGURE 3.9
Point-to-point (FC-P2P) topology

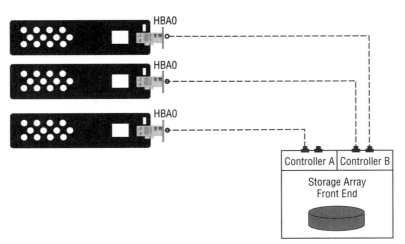

Arbitrated Loop Connectivity

The *arbitrated loop* (FC-AL) topology is a low-cost connectivity solution. It does not require a high CapEx investment for switching devices, as lower-cost hubs are used to scale server and storage connectivity. Hubs are largely employed in JBOD environments, and just as JBODs cost less than enterprise storage, hubs cost less than switches.

In the arbitrated loop topology, devices are attached to a shared loop. The arbitrated loop has the same characteristics as a token ring topology and a physical star topology in IP networks. In an arbitrated loop configuration, each device contends with every other device to perform I/O operations. Devices within the loop must *arbitrate* to gain control of the loop. At any given time, only one connected device can perform I/O operations on the loop.

An arbitrated loop topology, illustrated in Figure 3.10, can potentially be implemented without any interconnecting devices, by directly connecting one device to another, with the two devices in a ring through cable. However, arbitrated loop implementations typically use hubs, physically connected in a star topology. The arbitrated loop topology has the following limitations in terms of scalability and performance:

- The loop is shared, and only one device can perform I/O operations at any given time. Each device in a loop must wait for its turn to process an I/O request. As a result, the overall performance in an arbitrated loop environment is poor.

- An arbitrated loop uses only 8 bits of the 24-bit Fibre Channel addressing scheme (the remaining 16 bits are masked out), which allows for the assignment of up to 127 valid addresses to the ports. However, one address is reserved for optionally connecting the loop to a Fibre Channel switch port. Therefore, up to 126 nodes can be connected to the loop.

- An arbitrated loop has a maximum speed of 8 Gb/s Fibre Channel.

- Employs both NL_ports and FL_ports within the FC-AL loop topology.

- Adding or removing a device from the loop results in the loop reinitializing, which can cause a momentary pause in loop traffic.

FIGURE 3.10
Arbitrated loop (FC-AL) connectivity

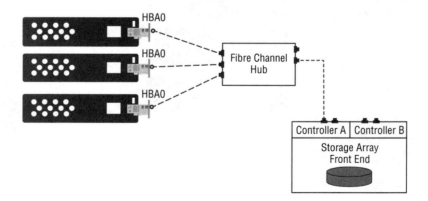

SWITCHED FABRIC CONNECTIVITY

The *switched fabric* topology, also referred to as a *fabric connect*, provides full-bandwidth performance, scalability, and flexibility between all the nodes on the SAN fabric, and is the basis for almost all modern Fibre Channel SAN designs. In a switched fabric, the device, or port, gains access to the fabric through a point-to-point connection with a port on an edge or director switch. At any one time, there can be $n/2$ full-bandwidth connections between nodes in the fabric: one connection for the initiator and one connection for the target.

This topology has evolved beyond the preceding two connection topologies to solve several distinct data-center problems, including these:

Proximity Extension The distance challenge when using shortwave-to-longwave conversion to extend server-to-storage distances, beyond the shortwave 500-meter limitation.

Capacity Expansion The ability to expand the storage capacity supported by a host port, by allowing a host port to connect to more than one storage array node.

The fabric is a logical space, in which nodes communicate with one another over the Fibre Channel network. This virtual space can be created with a single switch or a network of switches, typically in a redundant dual-fabric architecture. Each switch in the fabric maintains a unique domain ID, which is employed by the fabric's addressing scheme. In a switched fabric topology, nodes do not share a loop. Instead, data is transferred through a dedicated path between the nodes. In addition, each port in the fabric employs the full unique 24-bit Fibre Channel address scheme for communication.

In a switched fabric, illustrated in Figure 3.11, the connecting link between any two Fibre Channel switches is called an Inter-Switch Link (ISL). An ISL allows switches to be connected together to create a single larger fabric. The ISL is used to transfer host-to-storage data and fabric management traffic from one switch to another. This enables a switched fabric to be expanded and connected to a large number of nodes. This is possible because a switch-based fabric uses intelligent switch devices that can switch data traffic between nodes directly through switch ports, allowing frames to be routed between the source and destination by the fabric.

Unlike in an arbitrated loop topology, in a switch-based fabric, the network provides a dedicated path to data and provides the scalability, performance, and flexibility required by enterprise data-center customers and service providers. The addition or removal of a Fibre Channel device in a switched fabric is normally nondisruptive and does not affect the ongoing traffic flows between other devices.

Switch-Based Fabric Architecture

As discussed, a switch-based fabric provides a flexible and scalable approach to data-center fabric design. To this end, numerous standard topologies exist and are presented next.

However, it is important to recognize that *flexible* means just that. There is not a single best practice or approach that is better than another when it comes to SAN fabric design. Once again, it simply comes down to requirements, vendor recommendations, and understanding any constraints that may exist within the customer's storage environment.

CORE-EDGE FABRIC TOPOLOGY

The first topology example found in the enterprise or service provider data center is the *core-edge* fabric topology, which has two types of switch tiers. The *edge tier* is typically made up of top-of-rack (ToR) type switches, which offers a relatively inexpensive approach to adding more hosts to the fabric. Each edge-tier switch in this architecture is connected to a switch at the core tier through ISLs.

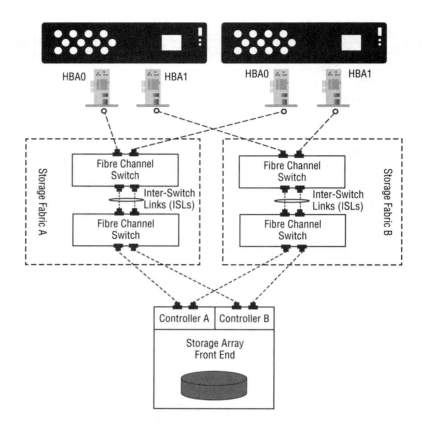

Figure 3.11
Switched fabric (FC-SW) connectivity

The *core tier* is typically made up of director-class switches that ensure high fabric availability and performance. Typically, in a *core-edge* fabric topology, all traffic must either traverse this tier or terminate at this tier. In this topology, all storage devices are connected to the core tier, allowing host-to-storage traffic to traverse only one ISL. It is also common that in specific use cases, hosts that require exceptionally low latency and higher performance can be directly connected to the core-tier director-class switches, and therefore avoid any potential ISL latency.

In addition, as illustrated in Figure 3.12, in this topology, the edge tier switches are not connected to each other via ISLs. The core-edge architecture increases connectivity within the SAN, while at the same time conserving the overall port count for hosts. If the fabric is required to expand, additional edge-tier switches can be connected to the director core layer. This topology has different variations, such as single-core architecture or dual-core architecture, as illustrated in Figure 3.12 and Figure 3.13, respectively. Therefore, if the fabric needs to be extended into a dual-core architecture, this can be achieved by adding directors at the core tier, depending on the ratio of edge-tier switches to core-tier switches required by the design. To extend the fabric from a single core to a dual-core design, new ISLs are required to connect each edge-tier switch to the new core-tier director switches.

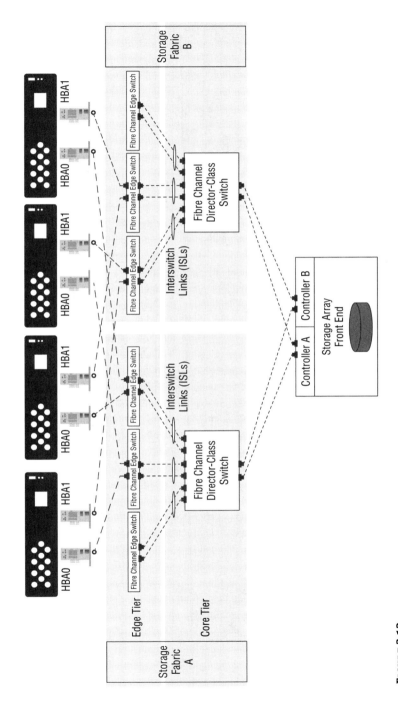

Figure 3.12
Single-core, core-edge fabric topology

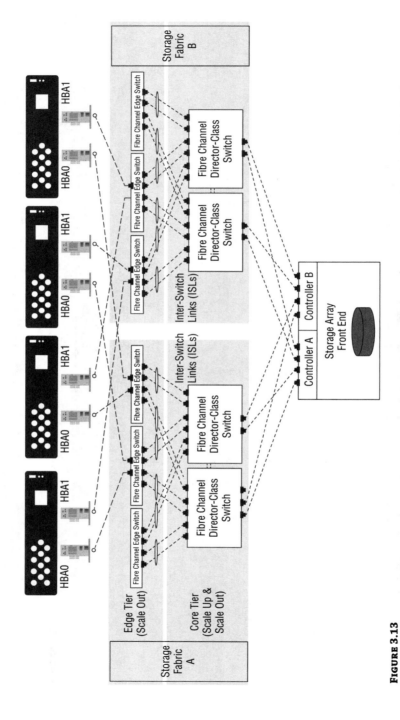

FIGURE 3.13
Dual-core, core–edge fabric topology

The key design factors associated with a *core-edge* fabric topology include the following:

- Consists of edge- and core-switch tiers.
- All network traffic traverses the core tier or terminates at the core tier.
- Storage devices are usually connected to the core tier.
- High availability.
- Medium data-center scalability.
- Medium to maximum data-center connectivity.

EDGE-CORE-EDGE FABRIC TOPOLOGY

The *edge-core-edge* fabric topology, shown in Figure 3.14, provides additional scalability for a scale-out storage strategy, by implementing an additional layer of edge-tier switches for storage device connectivity. This addresses the port limitation and cost-per-port factor on director-class switches when connecting directly to storage arrays. This fabric topology is typically found in large-scale enterprises or service provider data-center implementations, where a single unified fabric approach is being employed.

MESH TOPOLOGY

Within a fabric architecture, the *mesh topology* can be one of the following two types:

- Full mesh topology
- Partial mesh topology

In a *full mesh topology*, every switch is connected to every other switch via ISL. This topology is typically most appropriate when the number of switches included in the design is relatively small—for instance, a deployment that includes up to four edge or director switches, with each of them servicing highly localized host-to-storage traffic.

As illustrated in Figure 3.15, in a full-mesh topology, a maximum of one ISL, or network hop, is required for host traffic to reach the storage device. However, if the number of switches increases, , the number of switch ports required for ISLs also grows, which in turn reduces the number of ports available for host connectivity.

The following are key design factors associated with a full mesh topology:

- Each switch is connected to every other switch within the fabric.
- A maximum of one ISL, or hop, is required for host-to-storage device traffic.
- Hosts and storage can be connected to any switch within the fabric.

However, as illustrated in Figure 3.16, in a *partial-mesh topology*, several hops, or ISLs, might be required for the host traffic to reach the storage device. Despite this, a partial mesh offers a more scalable solution than the full-mesh topology. Therefore, without considered placement of host and storage devices, traffic management in a partial-mesh topology might be complicated and introduce latency. In addition, ISLs could become overloaded as a result of high levels of traffic aggregation between switches.

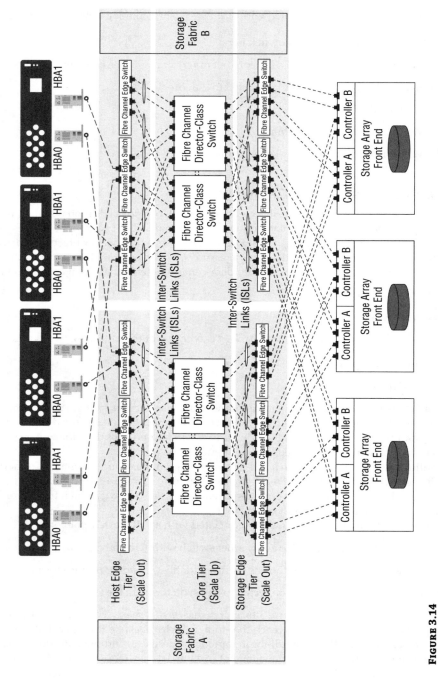

FIGURE 3.14
Edge-core-edge, dual-core, fabric topology

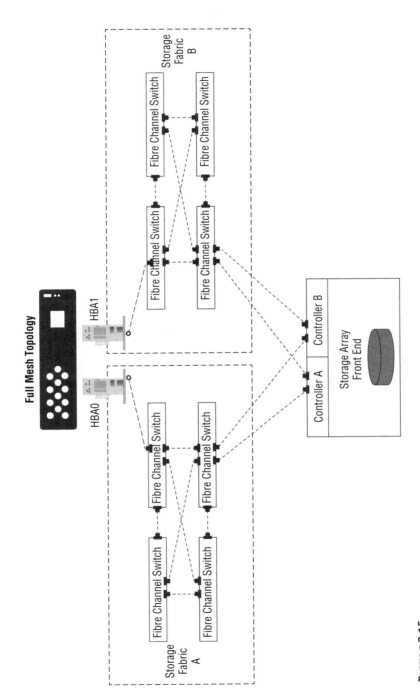

FIGURE 3.15
Full mesh topology

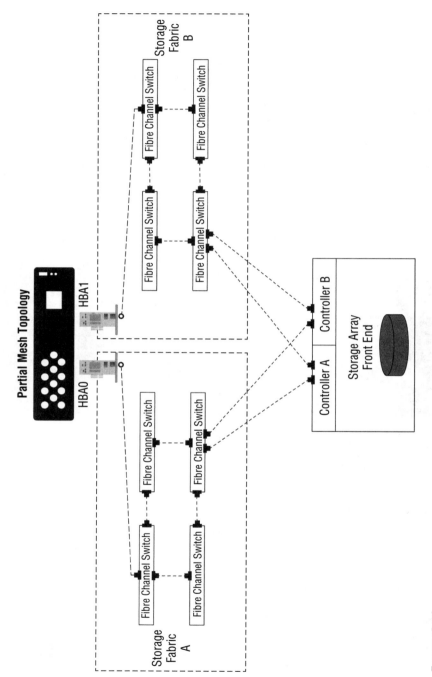

FIGURE 3.16
Partial mesh topology

The following are key design factors associated with a partial mesh topology:

- Not all the switches are connected to every other switch.
- Potential for increased complexity of traffic flows.
- More scalable than full mesh topology.
- Potential for ISLs to become overloaded.

Security and Traffic-Isolation Features

Securing data from accidental or malicious disclosure, whether it is data in transit or data at rest, is critical to the mission of most IT organizations. SAN security should be carefully considered as part of any storage fabric design, and then be implemented in accordance with all applicable security policies. In a Fibre Channel SAN, various technologies can contribute to securing the data.

Fabric Switch Zoning

Zoning is a function performed on Fibre Channel switches to enable node ports within the fabric to be logically segmented into groups and to communicate only with other node ports within the same group (see Figure 3.17).

There are several reasons that zoning should be implemented as part of any fabric design. First, whenever a change is implemented by the Name Server database, the fabric controller sends a registered state change notification (RSCN) to all the nodes impacted by the change. If zoning has not been configured, the fabric controller sends the RSCN to all the nodes in the fabric, including the nodes that are not impacted by the change. This results in an increase in the amount of fabric-management traffic. In a large fabric architecture, the amount of Fibre Channel traffic generated because of this process can be significant and may impact the host-to-storage data traffic. Therefore, the implementation of zoning helps to limit the number of RSCNs in a fabric; when zoning has been implemented, the fabric sends the RSCNs to only those nodes within the zone where the change has occurred.

In addition, zoning provides a level of access control, along with other storage-associated access-control mechanisms, such as LUN masking. Zoning provides access control by allowing only members in the same zone to establish communication with each other.

Zone members, zones, and zone sets form a hierarchy defined in the zoning process, which varies depending on switch vendor. A *zone set* is made up of a group of zones, which can be activated or deactivated as a single entity within the fabric. Even though multiple zone sets can be defined within a fabric, only one zone set can be active at any one time. Members of zones are nodes within the SAN fabric, which may include switch ports, HBA ports, and storage-device ports. A port or node can be a member of multiple zones concurrently. Nodes distributed across multiple switches in a switched fabric may also be grouped into the same zone. Zone sets are also referred to as *zone configurations* by some Fibre Channel switch vendors.

For redundancy and availability, VMware recommends as a best practice in a vSphere environment to adopt a *single initiator / single target* zoning strategy (see Figure 3.18). This method specifies that only one initiator and its associated target port are in a single zone. Remember that a host port and a storage-array port can be in multiple zones concurrently.

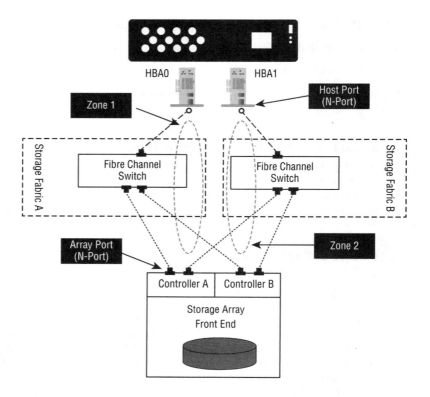

FIGURE 3.17
Fabric zoning

The zoning standards to be adopted in a typical vSphere design should include the following:

- Each host should be zoned to both the A and B fabric, to provide high availability.
- Each host should have paths to each of the storage controllers, to provide high availability.
- Each host should have at least four paths to the data for redundancy.
- A vSphere cluster of hosts should normally share the same front-end storage ports.

The key design factors associated with zoning design include these:

- Zones enable node ports within the fabric to communicate with one another.
- A zone set comprises multiple zone configurations.
- Each zone is configured with two zone members: one HBA (initiator) and one array port (target).
- Storage and host ports may exist in multiple zones concurrently.
- Zoning provides only a portion of the protection typically required in a SAN design through its access-control mechanism.

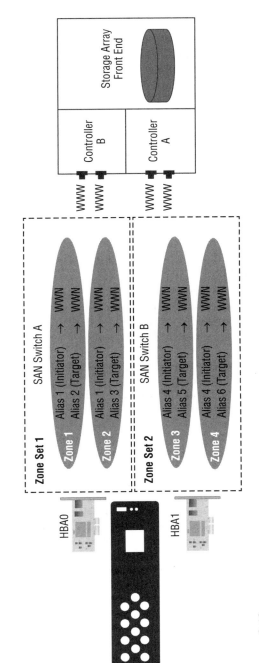

FIGURE 3.18
Zoning / zone set

LUN Masking

LUN asking provides an additional layer of data protection. Without LUN masking, any host that can access (via zoning) a storage port can see all data behind that port, and not just the LUNs that it needs to see. Therefore, typically all storage is presented via a masking view, although the implementation of this mechanism varies significantly depending on the storage system vendor.

The key design factors that are pertinent to LUN asking include the following:

- Zoning restricts server-to-storage port access, whereas masking restricts storage port-to-device access.
- If booting from SAN, a boot LUN may have to be masked first, and as LUN 0.
- If booting from SAN, each server has its own masking view, which contains only the boot LUN.
- Each ESXi host, which is a part of a vSphere cluster, is typically masked to the same single masking view, which provides visibility of that shared LUN to all of the hosts.

Virtual Fabric Design

Virtual Fabric, also known as Virtual SAN, should not to be confused with VMware's Virtual SAN, the product addressed in detail in Chapter 4. Just to clarify, some Fibre Channel switch vendors, such as Cisco, refer to this technology as Virtual SAN or vSAN. To avoid further confusion, in this book we exclusively use the term *Virtual Fabric* when referring to this logical SAN mechanism.

Virtual Fabric, a logical fabric on a Fibre Channel SAN, allows communication among a group of nodes, regardless of their physical location in the fabric. In a Virtual Fabric, a group of hosts, or storage ports, communicate with each other using a virtual topology defined on the physical SAN. Multiple Virtual Fabrics can be created on a single physical SAN, with each Virtual Fabric acting as an independent logical SAN, with its own set of fabric services, such as Name Server, and zoning. When this technology is employed as part of a storage fabric design, fabric-related configurations in one Virtual Fabric cannot impact the traffic in others.

Employing Virtual Fabric in a design improves SAN security, scalability, availability, and manageability. This technology provides enhanced security by isolating the sensitive storage data, and by restricting access to the resources located within that Virtual Fabric. As the same Fibre Channel address can be assigned to nodes in different Virtual Fabrics, scalability of the fabric is increased. In addition, events that cause traffic disruptions in one Virtual Fabric are isolated and do not affect other Virtual Fabrics within the same physical SAN.

Virtual Fabrics are an easy, more flexible, and less expensive way to manage complex Fibre Channel networks. Configuring multiple Virtual Fabrics is easier, quicker, and more cost-effective than building separate physical Fibre Channel SANs for various groups of nodes or for different customers in a multitenanted environment. For instance, in order to regroup nodes into a different Virtual Fabric, an administrator can simply change the Virtual Fabric configuration without needing to move nodes or undertake recabling in the data center (see Figure 3.19).

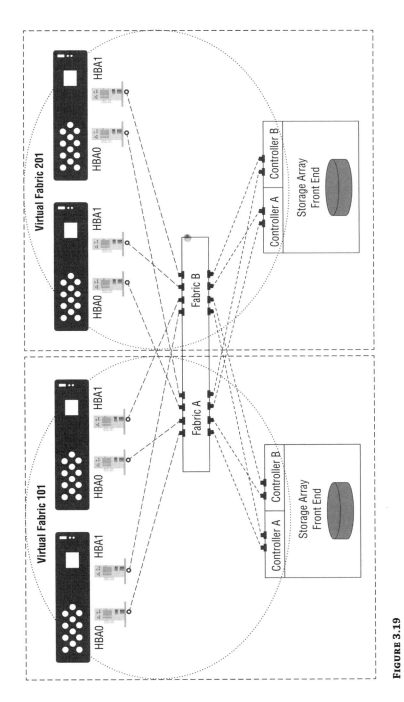

FIGURE 3.19
Virtual Fabric architecture example

The purpose of Virtual Fabrics can be compared with the use of VLAN tagging on a LAN, and like a VLAN on an Ethernet network, the Virtual Fabric has its own tagging mechanism built in.

In the example illustrated in Figure 3.20, two tenants on the same public cloud platform share the same physical SAN infrastructure. To isolate their data from one another, the service provider has created a Virtual Fabric for each customer, with different IDs used for each tenant. In the event of a misconfiguration by a member of the operational team, where the fabrics are mistakenly interconnected, the Virtual Fabrics will not merge.

Typically, trunked ISLs are created between each of the Fibre Channel switches, carrying all the Virtual Fabrics for that physical SAN. Even though it is also possible to create separate physical ISLs for each tenant's Virtual Fabric, that would consume more switch ports.

In addition, each storage port is associated with a particular Virtual Fabric, restricting which ports each tenant can access. Therefore, if the storage array system allows the administrator to control which front-end ports the individual LUNs are exposed on, the data would also be inaccessible from host servers on different Virtual Fabrics.

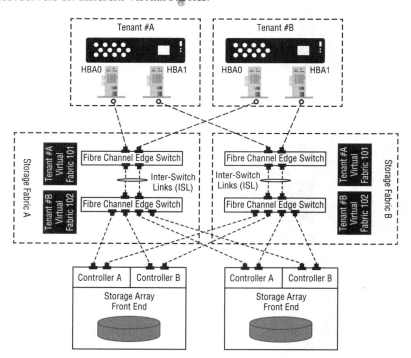

FIGURE 3.20
Virtual Fabric sample use case

FIBRE CHANNEL SAN SECURITY OPTIONS

Just as with a LAN infrastructure, various security mechanisms can be employed to protect a SAN environment, in addition to those previously addressed. Table 3.3 provides an overview of several additional SAN security options, which may form part of a fabric design.

TABLE 3.3: SAN security options

FEATURE	DESCRIPTION
Fabric Binding	Fabric Binding ensures that ISLs are formed only between authorized switches. This mechanism can prevent not only a rogue device from joining the fabric, but also unintended routing configurations, which can occur when a cable is inadvertently connected to the incorrect switch port.
Fibre Channel Security Protocol (FC-SP)	Fibre Channel Security Protocol (FC-SP) can be employed to force switches to authenticate ISL partners, as well as hosts that are attempting to register a connection with the switches.
Port Security	Port Security allows the fabric administrator to specify which WWN is allowed to connect to a specific switch port, preventing rogue devices from being connected to the fabric.
ISL Encryption	Inter-Switch Link (ISL) encryption can be employed to protect data that is traversing between switches.

N_Port Virtualization and N_Port ID Virtualization

N_Port Virtualization (NPV) and N_Port ID Virtualization (NPIV) are closely interrelated technologies. However, although NPV requires the use of NPIV, NPIV does not require the use of NPV.

N_Port Virtualization allows a Fibre Channel switch to operate without performing standard Fibre Channel switch functions, such as zoning or Name Server, and also removes the requirement for a domain ID. Essentially, N_Port Virtualization turns the switch into a hub, a simple connectivity device that aggregates connections. When a host server is connected to an NPV switch, it attempts to log in to the fabric. As the login request cannot be serviced by the NPV switch, the request is forwarded to the NPIV switch via an NP_Port, as illustrated in Figure 3.21.

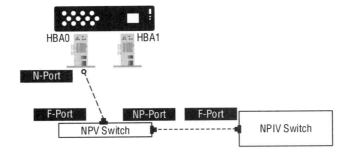

FIGURE 3.21
N_Port Virtualization (NPV) and N_Port ID Virtualization (NPIV)

The NP_Port acts as a proxy for the actual N_Port, in a similar way to an ISL attempting to log in to the fabric. Once the host server has been logged in, it is provided with an FCID from the NPIV switch.

The real advantage of these technologies lies within their ability to register multiple physical WWNs to a single switch port. Under normal circumstances, without N_Port Virtualization being enabled, each host can register only one physical WWN on a switch port. However, as shown in the previous figure, two host systems are logging into and registering their unique physical WWNs across a single NP_Port. NPIV effectively allows the end device, or in this case the NVP switch, to register multiple physical WWNs to a single switch port, which is extremely useful when employing converged blade systems.

Two typical use cases for NPV and NPIV are illustrated in Figure 3.22.

The first use case has an environment designed in the traditional way, with each blade server switch having its own domain ID. By employing NPV and NPIV, we can remove the domain ID requirement from the blade chassis switches, allowing the fabric to be significantly expanded to a much larger scale, effectively replacing the blade chassis switches with hubs.

The second use case, shown in Figure 3.22, is an interoperability issue in a mixed vendor switch environment. Using the traditional method, the two switches are interconnected in an interop mode; one switch acts as the *master* for the fabric, and the other becomes the *subordinate*. However, in this design of mixed vendor hardware, many of the switches' advanced features, which are dependent on the vendor's specific switch or firmware capabilities, are no longer available on either vendor's switch. By using NPV and NPIV, the master switch can remain in a native configuration, retaining many of the vendor's original advanced fabric switch features.

Boot from SAN

As addressed in Chapter 2, ESXi hosts can be configured to boot from a remote storage device across the SAN, in order to achieve a flexible and stateless architecture. By enabling remote boot capabilities, the ESXi host can be abstracted from the physical hardware, with no locally attached storage devices. This allows for dynamic repurposing of any node, at any given time, to alternate physical hardware and/or location. However, such a design typically requires very specific configurations from the host, fabric, and target storage device.

In the example shown in Figure 3.23, HBA0 has two paths to the target LUN through fabric A. Those two paths lead to the WWPN of the front-end storage device ports, which have masked behind it the boot device. Likewise, HBA1 has two paths through fabric B. Those two paths also lead to the same WWPN of the front-end ports, which again have masked behind them the same boot device.

The configuration of Boot from SAN (BfSAN) environments is highly dependent on the vendor's hardware and its configuration. For more information on this element of a design, you should refer to the hardware vendor's documentation.

Fibre Channel Summary

The Fibre Channel architecture forms the fundamental construct of the Fibre Channel SAN infrastructure. This fabric architecture represents true channel and network integration. It can capture many of the benefits of both channel technology and network flexibility to provide a serial data-transfer interface that operates over either optical fiber or copper wire. The Fibre Channel SAN is effectively the implementation of SCSI over a network, providing both channel speed for data transfer, with a low protocol overhead, and scalability of network technology.

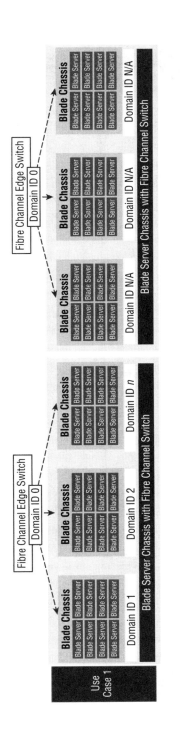

FIGURE 3.22
NPV and NPIV use cases

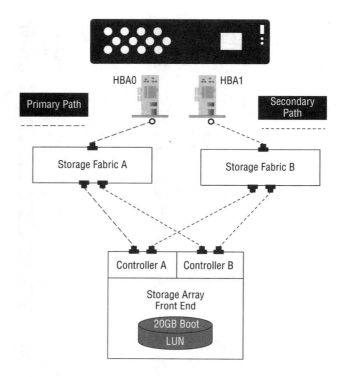

FIGURE 3.23 Boot from SAN example

In Fibre Channel Protocol architecture, all external and remote storage devices attached to the SAN appear as local devices to their host operating system. The key advantages of Fibre Channel Protocol over other storage protocols include its ability to do the following:

- Provide sustained transmission bandwidth over long distances.
- Support a larger number of addressable devices over a Fibre Channel Protocol network. Theoretically, a Fibre Channel network can support more than 15 million device addresses.
- Support for speeds of up to 20 Gb/s (20 GFC), at the time of writing.

The SAN fabric topology chosen for a specific design should first and foremost align with the design requirements. The *core-edge* model provides a simple yet scalable design that can meet the host connectivity needs for most enterprise solutions. In this model, storage is connected to the core, and if you need to add more storage than a single pair of director-class switches can support, you can scale out by adding more core switches. For the largest enterprises or cloud service provider environments, the *edge-core-edge* design allows the solution to scale up the host ports and storage ports independently of one another. In the *core-edge-core* model, the core director acts only as a connectivity layer, routing traffic between the two edge-switch tiers.

Despite which Fibre Channel topology model is employed, it is critical that the architect fully takes account of the paths of communication that are taking place between host and storage in the fabric design, and that sufficient consideration is given to the sizing of ISLs to ensure that throughput requirements can be met by the topology.

For large environments, particularly those that use blade-based hosts, utilizing NPV and NPIV will allow the design to scale even further, due to the limited number of domain IDs available to Fibre Channel switches. N-Port Virtualization removes the switching functions from a fabric switch, as well as the domain ID, therefore allowing more physical switches in the SAN, without going beyond the set size limits of the fabric.

iSCSI Storage Transport Protocol

The second storage protocol option to be addressed in this chapter is iSCSI. Like Fibre Channel, iSCSI is a well-established block protocol that can, depending on multiple design factors, be deployed using existing networking components, potentially providing a lower-cost option for a scalable storage network. However, you must also consider that performance over 1 Gb/s Ethernet is limited, and when using standard network adapters, may add overhead to the host CPU. In a typical enterprise or service provider environment, specialized 10 Gb/s adapter hardware can be employed to off-load processing from the CPU, but this can increase costs significantly, especially if a large number of hypervisors are being deployed.

The iSCSI protocol allows block-level data to be transported over IP networks, and builds on the SCSI protocol by encapsulating SCSI commands and allowing these encapsulated data blocks to be transported across an unlimited distance via TCP/IP packets, over traditional Ethernet networks, or even the Internet. With an iSCSI connection, the ESXi host system (the initiator) communicates with a remote storage device (the target) in the same way it would communicate with a local disk device. iSCSI is widely adopted for connecting servers to storage, as it can be relatively inexpensive and straightforward to implement, especially when compared with a Fibre Channel SAN infrastructure.

iSCSI Protocol Components

An iSCSI component can be either an initiator or a target. An *initiator* can be any server hosting an application, where the application makes periodic requests for data to a connected storage device. In a vSphere environment, the initiators are provided by the ESXi hosts. However, the iSCSI driver, which resides on the host, is also sometimes referred to as an initiator.

The initiators *initiate* iSCSI data-transport transactions by making an application request to send or receive data, either to or from the storage device. The virtual machine's request is immediately converted into SCSI commands, before being encapsulated into iSCSI, where a packet and header are added for transport via TCP/IP over the Ethernet network.

As illustrated in Figure 3.24, there are two types of iSCSI initiator:

iSCSI Hardware Initiators   With this adapter type, the files are accessed over the TCP/IP network via specialized hardware-based adapters.

iSCSI Software Initiators   With this adapter type, the files are accessed in the same way, over the TCP/IP network, but instead by using software-based iSCSI code provided by the hypervisor's VMkernel. This type of iSCSI initiator can be employed with any standard network adapter that can offer network connectivity to the target.

Both adapter types are examined in more detail later in this chapter.

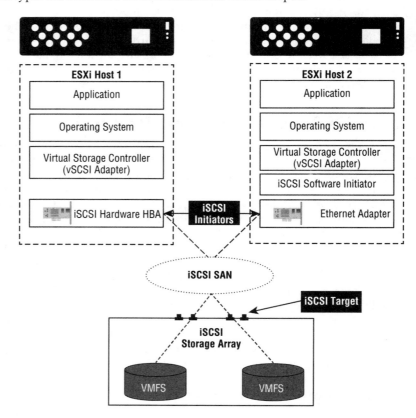

FIGURE 3.24
iSCSI protocol component architecture

An iSCSI *target* refers to the storage device, which must be located on the connected network. Targets can be any type of storage device, but in a virtual data-center environment, they are typically storage-array systems that are attached to the IP-based SAN.

Targets receive iSCSI commands from the initiators, which are then broken into their original SCSI format to allow block data to be written to or read from the storage device. The target then responds to a host's data request by sending SCSI commands back to that host. These commands are again encapsulated via iSCSI for transport over Ethernet.

Storage over Ethernet requires additional considerations over Fibre Channel. The Ethernet protocol does not guarantee delivery of data, so we use the Transmission Control Protocol (TCP) to manage session data. In general, Ethernet network traffic is deemed to be a *best effort* delivery method for data. Although every effort is made to ensure that the data transfers successfully, ultimately some packets will be lost and require resending. This approach is not a big deal for most network traffic types, but storage I/O access over Ethernet is far less tolerant of packet loss than most data types traversing the network. For storage, when data is needed to be read, typically either a user or system is waiting for it, so any latency that is added through an inefficient network design or poor equipment is considered intolerable.

In addition, iSCSI traffic should be isolated on the network. iSCSI traffic is sometimes isolated onto its own dedicated IP SAN, providing an efficient fabric for storage I/O data to reach host servers. However, in many environments, iSCSI is simply isolated to a dedicated VLAN, meaning it has to be prioritized to ensure that hosts are not waiting unnecessarily for data, and the Ethernet network does not become a bottleneck for storage performance. For this reason, even though it is not strictly required, a priority mechanism should be considered a key design factor for iSCSI traffic when crossing a shared IP network. The IEEE 802.1Qbb standard provides a method for Ethernet to perform flow control, through a mechanism know as priority flow control (PFC).

The idea behind PFC within an iSCSI network is for a specific class of service (CoS) to be assigned to this traffic type so that it has a higher priority as it traverses the network hardware. If network congestion does occur, the Ethernet switch can prioritize storage I/O traffic over communication traffic, which is considered a lower priority (for example, end users trying to watch a video of a cat having a bath on YouTube, or trying to poke fun at their friends on Facebook).

iSCSI Traffic Isolation

As already highlighted, two common approaches to traffic isolation are typically employed when designing an iSCSI storage I/O fabric solution:

- A dedicated physical IP SAN
- A dedicated VLAN

As you would expect, implementing a fully redundant and high-performance dedicated IP SAN fabric clearly has a significant cost consideration associated with its design, from both a CapEx and OpEx perspective. For this reason, identifying key requirements to establish what SLAs are in place, the budget, security obligations, and other storage fabric design factors, will provide guidance on formulating an appropriate architecture.

For reasons of cost and operational overhead, it is typically more common to see iSCSI storage I/O simply being isolated onto a dedicated VLAN, particularly when being adopted on a 10 Gb/s network platform. This approach generally has little or no CapEx implication, and only minimal operational cost, while still addressing the vast majority of iSCSI storage I/O requirements.

As a VLAN can be created without a gateway, it is possible to create nonroutable isolated networks, dedicated to iSCSI storage I/O. In addition, a VLAN creates an isolated broadcast domain, eliminating unwanted traffic, and therefore providing more-efficient data flows. It is also important to recognize that VLANs do not have dedicated bandwidth, unless configured to do so. If a network switch has a 15 Tb/s backplane capacity, which is shared among multiple configured traffic types, this is the limit of the total throughput across the switch. If the switch backplane becomes saturated with multiple traffic types, even though PFC can prioritize traffic based on 802.1Qbb tags, nothing can increase the throughput of the backplane, other than employing a different switch.

For this reason, creating a dedicated physical IP SAN exclusively for storage I/O traffic provides two key advantages. It provides you with dedicated bandwidth through the switch backplane, with no other traffic types being routed through the iSCSI network infrastructure. This additional layer of isolation also provides an enhanced fabric from a security perspective, protecting the storage I/O against potentially unwanted attention that might occur on a common network, through true traffic isolation.

Jumbo Frames

Enabling jumbo frames on an iSCSI network allows more data to be packaged into each frame, resulting in a far more efficient use of network traffic.

The default maximum transmission unit (MTU) configured on network devices is 1,500 bytes per Ethernet frame, which is fully supported and will operate without issue for iSCSI storage traffic. However, by increasing the frame size to 9,000 bytes, fewer frames are required to be sent, improving overall network efficiency. For instance, if you wished to send 27,000 bytes of data, it would take 18 regular-sized frames to complete, whereas with the MTU value configured at 9,000, it would take only 3 jumbo frames. In addition, the network adapter (initiator) will have to create and package only 3 iSCSI frames, further reducing the overhead.

Configuring jumbo frames on a network requires the entire data path to be appropriately enabled, from the vSphere host to the storage target, illustrated in Figure 3.25. There are also environments in which the MTU value for physical devices needs to be configured higher than the hypervisor's VMkernel, such as 9,216 bytes, in order to avoid fragmentation due to an additional overhead.

FIGURE 3.25
Jumbo frames data path configuration

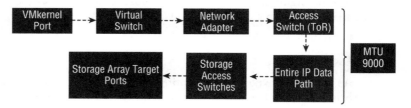

In reality, countless performance studies have shown that there is at best only a minor improvement in overall network performance when jumbo frames have been enabled. So once again, key design factors play a significant role in making a design recommendation to a customer, such as specific application requirements, network hardware, company policies, and any desire to squeeze every last piece of performance from the network infrastructure.

In a greenfield implementation, in order to obtain accurate metrics on the performance benefit of enabling jumbo frames, for a specific IP SAN fabric, customer application and use case performance testing should be carried out. Once completed, a full evaluation and comparison of the values made, in order to understand the true performance gain obtained.

However, on an existing shared network fabric, where jumbo frames are not currently being employed because of the operational overhead of implementation, which in turn may not provide any noticeable performance advantage, it is far less likely that jumbo frames will form part of the architect's design recommendations.

iSCSI Device-Naming Standards

All iSCSI devices on a network, both initiators and targets, have an assigned name. To confuse matters, there are three naming formats:

- iSCSI Qualified Name (IQN)
- Extended Unique Identifier (EUI)
- T11 Network Address Authority (NAA)

The iSCSI Qualified Name format is by far the most commonly adopted naming standard being employed by storage hardware vendors today, and you are increasingly less likely to come across instances of the other two. For this reason, this book focuses only on this naming standard.

The IQN structure shown in Figure 3.26 and described in Table 3.4 is used by both the initiator and the target to communicate with one another. The format of the IQN address provides a type field, date field, naming authority field, and unique string.

FIGURE 3.26 iSCSI Qualified Name (IQN) structure

Type: iqn
Yr/month: 1998-01
Naming Auth: com.vmware
Defined by Naming Authority: esx01-87654321

TABLE 3.4: iSCSI Qualified Name (IQN) structure

FIELD	PURPOSE	EXAMPLE
Type	This field provides the name type. All iSCSI Qualified Names begin like this, with IQN.	Iqn
Date	This field provides the date when the company (naming authority) took possession of the name, in a year-month format, as shown.	1998-01
Naming Authority	This field provides the naming authority name in reverse. This is the name of the company that produced the initiator or target device.	com.vmware
Unique String	This field provides the unique string created by the naming authority. In vSphere, it is the name of the vSphere host, with a dash and random character set, which can be manually overwritten by an administrator if required.	esx01-87654321

CHAP Security

Having already addressed the security benefits of isolating iSCSI traffic, there is still a risk that a rogue client on the network could communicate with a storage system, or host server, over the iSCSI network. To mitigate this risk, it is possible to add an additional layer of security on top of the nonroutable, isolated iSCSI network. This is done by requiring authentication, through the Challenge-Handshake Authentication Protocol (CHAP).

However, it is noteworthy to highlight that the use of CHAP does not encrypt the vSphere I/O traffic, but only acts as a mechanism to secure the connection. The CHAP mechanism can be implemented in one of two configurations: unidirectional and bidirectional. The choice between these two options will to some extent depend on the hardware choices, for both initiator and target devices, in whether they support requiring a password, referred to as the *secret*. Without the secret being exchanged, an iSCSI session cannot be established. In addition, as well as the initial session, additional hashed password exchanges occur periodically in order to prevent a replay attack.

Table 3.5 shows VMware's published guidance on support for CHAP security levels for vSphere 6.

TABLE 3.5: CHAP security levels

SECURITY LEVEL	ADAPTER SUPPORT	DESCRIPTION
None	• Software iSCSI • Dependent Hardware iSCSI • Independent Hardware iSCSI	No authentication.
Utilize unidirectional CHAP if required by target	• Software iSCSI • Dependent hardware iSCSI	The host prefers to employ a non-CHAP connection, but can use a CHAP connection if required by the target server.
Utilize unidirectional CHAP unless prohibited by target	• Software iSCSI • Dependent hardware iSCSI • Independent hardware iSCSI	The host prefers to employ a CHAP connection, but can use a non-CHAP connection if the target does not support the use of CHAP.
Utilize unidirectional CHAP	• Software iSCSI • Dependent hardware iSCSI • Independent hardware iSCSI	The host server requires a successful CHAP authentication. The connection will fail if CHAP negotiation is not possible.
Utilize bidirectional CHAP	• Software iSCSI • Dependent hardware iSCSI	Both the host server and storage target support the use of bidirectional CHAP.

In reality, CHAP implementations are rarely seen in the modern vSphere virtual data center. The simple reason for this is, if the iSCSI traffic is properly isolated with restricted physical access to the switches and other hardware, infiltrating storage I/O data is not a simple task. In addition, as well as network isolation, iSCSI storage systems typically provide some form of LUN masking, employed at the IP address or subnet level, controlling what devices can talk to the storage targets. In most iSCSI designs, this will limit the added security value that CHAP offers, and therefore its implementation is not considered viable because of the additional operational overhead associated with maintaining it.

iSCSI Network Adapters

As already been mentioned, two types of iSCSI adapter exist, both of which can act as initiators inside host servers: software-based adapters and hardware-based adapters. To add an additional design factor, hardware initiators can be provided in two varieties, each with a different price point: dependent hardware iSCSI adapters (also known as *TCP off-load engine, or TOE adapters*) and independent hardware iSCSI adapters (often referred to as *iSCSI HBAs*). Independent hardware iSCSI adapters provide significant performance and feature enhancements over the other two options, but at a higher financial cost to the design, similar to that of Fibre Channel HBAs.

The iSCSI protocol operates across layers 1 to 6 of the OSI model. SCSI commands and payload data operate at layer 6. These are then encapsulated by layer 5 into the iSCSI protocol data units (PDUs). Further encapsulation subsequently occurs lower down in layer 4, and then 3, until, finally, Ethernet frames are generated. This encapsulation process either can occur on the vSphere host, using CPU cycles, or can be off-loaded to the adapter. It is this level, at which the off-loading occurs, that defines the choice of adapter hardware for a design.

iSCSI Software Adapter

The *iSCSI software adapter* is built into the ESXi hypervisor kernel. The adapter allows iSCSI encapsulation without any requirement for specialist hardware, with the host's CPU and memory running the entire IP stack in software, in conjunction with any standard VMware-supported network adapter device. In most cases, the adapter just builds the Ethernet frame and transmits it over the network cable to its target. The physical network adapter in this case is largely immaterial.

However, as you would expect, this continuous processing of the entire IP stack does increase CPU and memory utilization on the host servers. That should be taken into account and allowances made when formulating the compute sizing design. This should be done to ensure that virtual machine workload performance is not impacted by the operation of processing the storage I/O on the hosts.

The iSCSI software adapter is simple to configure from the vSphere Web Client user interface. The adapter introduces little additional operational overhead to the design, but helps reduce costs, as there is no requirement to purchase specialist hardware. As highlighted previously, hardware-based adapters come in two forms. The choice between these two adapter types is a key design decision, as the use of iSCSI HBAs over the TOE adapter type will significantly increase the CapEx cost per host (see Figure 3.27).

Dependent Hardware iSCSI Adapter

This adapter type is commonly referred to as the TOE adapter, as stated earlier. This type acts like a hybrid option, as it is not a fully featured iSCSI HBA but does include some specialist hardware that can off-load the TCP portion of the IP stack workload from the host's CPU and memory onto the dedicated chipset circuitry found on the adapter hardware. Therefore, selecting this adapter type can be considered a compromise, or the middle option.

This adapter type is referred to as the dependent hardware iSCSI adapter because the creation of the software initiator in vSphere is still required in order to handle the creation of the iSCSI PDUs before they are passed to the TOE adapter for further encapsulation and transmission onto the iSCSI network. Note that each TOE adapter must be bound in a one-to-one relationship with a software adapter. This is referred to as a *network port binding*.

The TOE-based adapters often provide a compromise to iSCSI off-loading, a balance between performance and cost, which is so often the case when designing solutions. However, if the design requirements mandate that the entire iSCSI mechanism be off-loaded away from the host to a dedicated and specialist piece of hardware, then the independent hardware iSCSI adapter is the only design option.

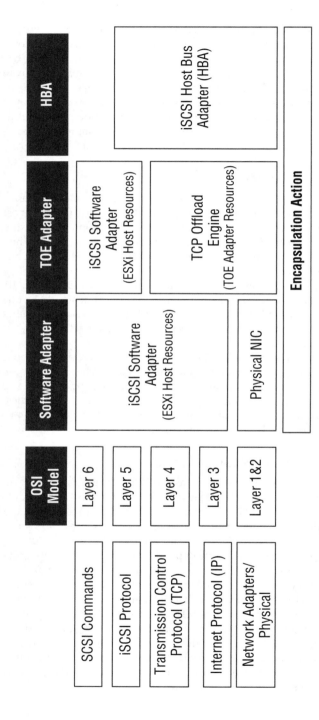

FIGURE 3.27
iSCSI off-load adapter comparison

Independent Hardware iSCSI Adapter

Typically referred to as an iSCSI HBA, this adapter type off-loads the entire iSCSI IP stack, as illustrated in Figure 3.27, to the adapter card. With iSCSI HBAs, there is no requirement to create an iSCSI software adapter, as the adapter hardware accepts raw SCSI commands, much like the Fibre Channel HBA. However, just like the iSCSI software adapter or the TOE adapter, an iSCSI HBA still requires an IP address and IQN to be configured.

When it comes to the various adapter types available on the market, it is easy to become confused. Many converged network adapters (CNAs) from various hardware manufacturers can perform multiple functions, such as standard Ethernet communications, TOE-based iSCSI off-loading, and Fibre Channel over Ethernet, but are not able to combine these protocols to operate concurrently. As many of these adapters are often proprietary technologies, particularly blade-based mezzanine cards, it's typically a good idea to liaise with the hardware vendor or supplier to ensure that you get a clearer idea of the options available for use within a design.

iSCSI Adapter High Availability

vSphere-based network adapter teaming does not provide optimal performance or load balancing between ports, and as such, is not recommended for iSCSI traffic. The reason for this recommendation is that a VMkernel port can operate on only a single interface, even if multiple adapters are configured. Therefore, unless one fails, the same single interface is always used. Even if you create two VMkernel ports for iSCSI traffic, the vSphere routing table will send traffic to only the first entry, with the second VMkernel port remaining idle until the first entry in the routing table is either reconfigured or removed. As a result, network port binding is the preferred method for iSCSI port load balancing and high availability.

Network port binding allows the use of multiple paths to process iSCSI I/O traffic to the target, so there is no reason for it not to be used in a design where adapter availability and load balancing is a requirement. Note that when Port Binding has been enabled, the routing table no longer determines the VMkernel port routing for iSCSI traffic. The path selection policy (PSP) is the mechanism now determining which adapter will send traffic. This has the added benefit of reducing the function of the vSphere network configuration, in favor of the vSphere optimized Pluggable Storage Architecture (PSA).

Virtual Switch Design

The design of virtual switching, and the decision of whether to adopt the vSphere standard switch or the vSphere distributed virtual switch, requires many factors to be taken into consideration, including virtual switch features, licensing, and use of virtualized networking technologies. A low-level discussion addressing all of these factors is beyond the scope of this book. This section covers only address-specific iSCSI traffic requirements, and cuts the design options down to either a single or multiple virtual switch architecture, the choice between which is ultimately driven by hardware, quantity of interfaces, and the layout of adapter interfaces across multiple adapter cards.

Network I/O Control

The Network I/O Control (NIOC) feature of the vSphere distributed switch provides a QoS mechanism for network traffic within the ESXi host. NIOC can help prevent *bursty* vSphere vMotion traffic from flooding the network and causing contention with other important traffic, such as virtual machine management and iSCSI communication. In addition to the use of NIOC, VMware recommends tagging traffic types for 802.1P QoS tags, and configuring the physical upstream management switches accordingly. If the QoS tagging is not implemented, the value of the NIOC configuration is limited to within the hosts themselves.

In the design example shown in Figure 3.28, two 10 Gb/s uplink interfaces, which are carrying multiple traffic flows from each host, have been configured on a single vSphere distributed switch. NIOC monitors the network, and whenever it sees congestion, it automatically shifts resources to the highest-priority traffic, as defined by the NIOC policy.

In this example, all physical network switch ports connected to the host adapters are configured as trunk ports, as recommended by the switch vendor. Figure 3.28 also shows the port groups that are being employed to segment traffic logically by VLAN, tagging traffic at the virtual switch level. Both the virtual and physical switches are configured to pass traffic specifically for VLANs employed by the network infrastructure, as opposed to trunking all VLANs.

The two 10 Gb/s network interfaces carry all ingress and egress Ethernet traffic on all configured VLANs. The user-defined NIOC resource pools should be configured port group by port group, as shown in Table 3.6.

TABLE 3.6: Sample Network I/O Control policy

PORT GROUP	VLAN ID	NO. SHARES	LIMIT (MB/s)	QOS TAG
Inside VM Network	101	150	Unlimited	2
vMotion Network	102	250	2,500	3
ESX Management Network	103	100	Unlimited	1
iSCSI Network	104	500	Unlimited	5
DMZ VM Network	105	150	Unlimited	2

In general, VMware does not recommend using limits, as they put a hard cap on usage. However, if they are to be employed as part of an NIOC design, make sure to calculate the available MB/s, and configure the limit based on the percentage of available bandwidth, as illustrated in the preceding example.

The QoS tag (802.1P), illustrated in Table 3.6, is associated with all outgoing packets. This enables the compatible upstream switches to recognize and apply the QoS tags. The default setting is None, and a value between 1 and 7 is configurable.

By segmenting vMotion, virtual machine, and iSCSI traffic onto separate VLAN-backed port groups, and employing shares and the QoS mechanism on physical hardware, as in the example shown, it should be possible to sustain a satisfactory level of iSCSI storage I/O performance, even during possible times of contention.

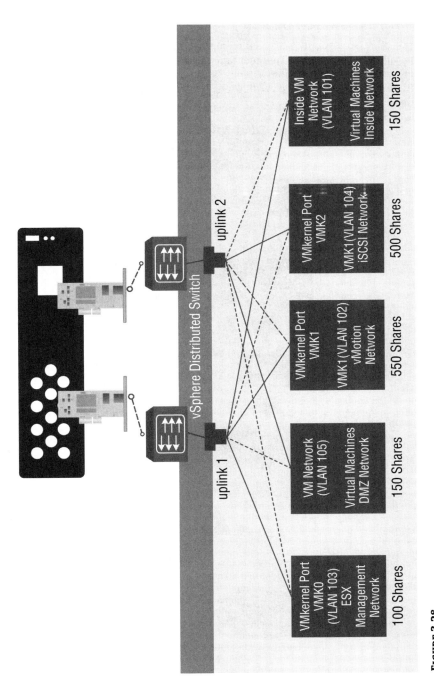

FIGURE 3.28
Network I/O Control design example

Single Virtual Switch Design

When a pair of 10 Gb/s adapters are being employed to carry not just iSCSI storage I/O traffic, but also all other host and virtual machine network communications, it is not possible, nor required, to create multiple virtual switches. This is true whether the design calls for a standard or distributed virtual switch to be configured. The standard design for this very typical use case is illustrated in Figure 3.29.

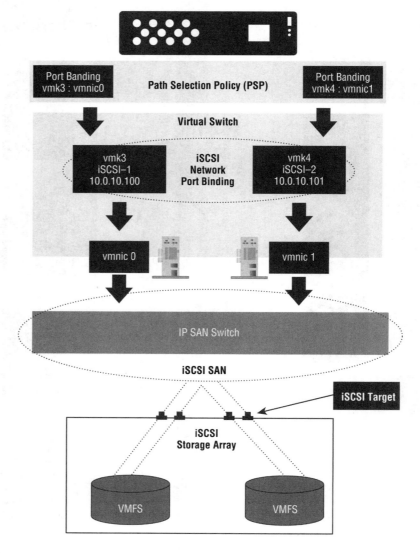

Figure 3.29
Single virtual switch iSCSI design

Note that this single vSwitch design employs an active/active failover order for the iSCSI VMkernel (VMK) ports. With this configuration, each VMkernel port is isolated to a single physical adapter. If the active adapter fails, the VMkernel will no longer process traffic, which is by design. The reason for this behavior is that the PSP will identify the dead path and move all traffic to use the surviving path. In addition, if the adapter supports round-robin (RR) I/O, both paths can be used simultaneously, which should lead to an overall improvement in data throughput.

Multiple Virtual Switch Design

The multiple virtual switch design is best suited to an environment where dedicated adapter ports are being employed, specifically and exclusively for iSCSI I/O traffic, as illustrated in Figure 3.30. In this design example, a dedicated iSCSI IP SAN fabric has been deployed to fully isolate storage traffic onto its own switching infrastructure.

Figure 3.30
Multiple virtual switch iSCSI design

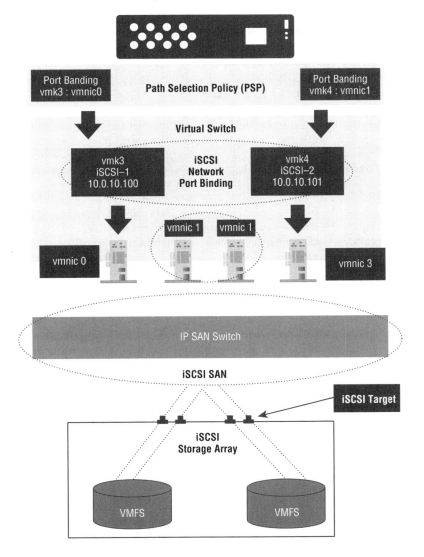

In this example, two of the four 10 Gb/s physical adapters have been configured on each virtual switch, with iSCSI traffic being isolated to its own IP SAN. In a failure scenario, where an adapter port fails, the PSP will move traffic from the failed *dead* network port onto the remaining active *live* network port binding.

iSCSI Boot from SAN

Boot from SAN in an iSCSI environment is fully supported, even with the software adapter type, as long as the adapter supports the iSCSI Boot Firmware Table (iBFT), which is a component of the ACPI 3.0b standard. The iBFT is configured with the details required to find the boot target and mount the LUN, even when the hypervisor has not yet booted.

From a design perspective, a Boot from SAN configuration widens the host failure domain to include the storage device and the SAN itself. During a complete data-center outage, hosts cannot boot until the switch fabric and target storage devices are available. These risks should be addressed as part of the design process, and where appropriate, alternative configurations should be implemented.

iSCSI Protocol Summary

When designing the IP SAN, it is true to say that you can follow many of the same guidelines as with a standard IP network. For iSCSI connectivity, it is considered an absolute design requirement to isolate the storage initiators and targets onto one or more nonrouted VLANs, or alternatively, if required, onto a dedicated physical switching network. This approach will not only avoid the performance degradation associated with routed traffic, but also protect the storage data being transmitted, which is not encrypted.

If the storage system supports it, you can use link aggregation to aggregate connections across the switch leaf, or aggregation layer, for redundancy and increased throughput (see Figure 3.31). If the vendor's storage system does not support link aggregation, it is possible to configure multiple iSCSI targets on different network adapters and manually load-balance the initiators across them.

FIGURE 3.31
Aggregated switch IP SAN design example

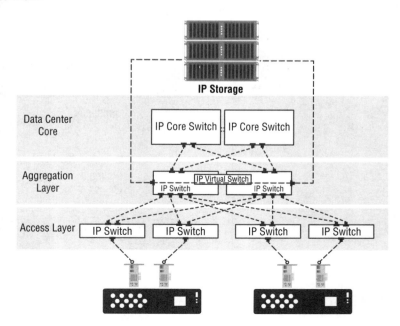

NFS Storage Transport Protocol

The second IP-based storage protocol to be addressed in this chapter is the Network File System (NFS) protocol. This protocol is also a well-established storage transport mechanism, which was previously, and continues to be, employed by Unix- and Linux-based systems. Like the iSCSI protocol, NFS can take advantage of an existing Ethernet network, while being isolated on a separate VLAN, and is also routable, allowing it to be accessed from any location. However, as with iSCSI, on a 1 Gb/s Ethernet network, NFS performance can be limited and can easily result in a storage bottleneck.

It is also important for an architect to keep in mind that a network-attached storage device employs its own, typically proprietary, operating system and integrated hardware and software components, to meet the specific needs of its file services. Its operating system is optimized for file I/O, and will therefore perform file I/O far more efficiently than a general-purpose server. However, as NFS is a file-level protocol and does not use VMFS, it will not necessarily support all of the storage features provided by the vSphere hypervisor.

Comparing NAS and SAN

When we refer to iSCSI storage, we talk in terms of the host connectivity to the storage via an IP-based storage area network, or SAN. However, when we are referring to shared storage provisioned to vSphere hosts via the NFS protocol, we refer to them as being network-attached storage, or NAS devices. So what is the difference?

The iSCSI SAN offers a block-based storage solution, whereas NAS is file based. With block-based storage, the ESXi host is consuming raw chunks of disk, and builds the VMFS datastore on top of the chunks. In a NAS environment, however, the storage array is doing all the work, by providing a preconfigured file system to the vSphere hosts, which then operates files on the remote storage device. Therefore, the primary difference between a NAS and SAN storage device is in how they process communications. A SAN employs either the Fibre Channel or iSCSI protocol, while the NAS storage connects to the network and communicates by using the shared storage device, providing a stand-alone storage solution, which can then be used for data backup or additional storage capabilities.

Despite their differences, as you would expect, many design factors that are pertinent to iSCSI architecture are equally important considerations for NFS storage. These include the following:

- Quality of Service (QoS)
- Network or VLAN isolation
- Jumbo frames
- Network I/O Control

NFS Components

Just as with the iSCSI protocol, the NFS-based storage model operates by employing various technologies and mechanisms within the host, network, and storage components.

The NAS storage devices transfer data from the storage device to a server in the form of files over the network. The NAS device, which manages its filesystems and user authentication, uses these filesystems for the storage of virtual machines and other data. In an NFS environment, because the NAS device itself, not the host, manages the storage, filesystem, and access, the vSphere host cannot manage the raw disk system, its locking, or its access mechanism.

As a result, VMFS cannot be used on NAS storage devices. This is not to say that guest operating systems can't use NAS storage just as any other network client would, but employing VMFS on a NAS device is not possible. Table 3.7 compares these technologies.

TABLE 3.7 Storage protocol comparison

Storage Technology	Protocols	Transfers	Interfaces and Transport	Performance
NAS	IP/NFS	File (no direct LUN access)	Network adapter and IP switches	Medium to high, depending on LAN integrity and storage device hardware
iSCSI	iSCSI	Block access of data/LUN	Network adapter, software initiators, TOE, iSCSI HBA, and IP switches	Medium to high, depending on LAN integrity and storage hardware
Fibre Channel	Fibre Channel/SCSI	Block access of data/LUN	Fibre Channel HBA	High, due to dedicated network and full solution stack

A NAS device can be dedicated to the vSphere virtual infrastructure or be a shared platform employed by both guest operating systems and physical non-vSphere clients, to provide shared or dedicated storage devices across a common network. For instance, NAS enables Linux, Unix, and Microsoft Windows users to share the same data seamlessly. Because NAS can provide this file-sharing mechanism in a heterogeneous environment, potentially containing many operating systems, a NAS storage device can typically serve more client types than other protocols, and therefore provide improved consolidation across the environment.

As a result, NAS devices can provide an advantage for virtual environments, by eliminating the need for multiple file servers and network infrastructures, and therefore consolidating the storage used by both the nonvirtual clients and vSphere platform onto a single system, making it more operationally efficient (see Figure 3.32).

As highlighted previously, NAS uses network and file-sharing protocols to provide access to the file data. These protocols include TCP/IP for data transfer, and the Common Internet File System (CIFS), and NFS for network file service.

vSphere 6 supports two generations of the NFS protocol. However, prior to this release of vSphere, only NFS version 3 was supported. When used in conjunction with a compatible storage device, NFS version 4.1 aims to improve the locking and performance for narrow data-sharing applications, with the following key features now supported:

- Authentication with Kerberos, which provides support to secure communication with the NFS storage device and for nonroot users to access files, in conjunction with this authentication mechanism.
- In-band, mandatory, and stateful server-side locking, in that NFS v4.1 clients use OPEN and CLOSE calls, for stateful interaction with the file server.
- Session trunking, providing near-true NFS multipathing capabilities. However, this is not Parallel NFS (pNFS). This supports the use of multiple IP addresses to access a single NFS volume, and provides improved performance and availability through load balancing and failover.
- Improved error recovery.
- Integrates the NFS, mountd Daemon, Network Lock Manager (NLM), and Network Status Manager (NSM) suite of protocols into a single protocol for ease of access across firewalls.
- Supports compound operations in order to coalesce multiple operations into a single message.
- Uses the concept of delegation to allow clients to aggressively cache file data.
- Support for IPv6 with Auth_SYS, but not with Kerberos as of yet.

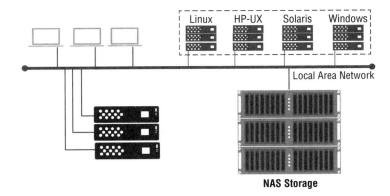

FIGURE 3.32
NAS network clients

The primary benefit of NAS is that it brings the advantages of network storage to the vSphere environment, through file-level sharing of data. Although NAS trades some performance elements for manageability and simplicity, it is by no means a second-rate technology. In addition, with 10 Gb/s Ethernet, NAS is able to scale to provide higher performance and lower levels of latency. Also, many NAS devices support multiple interfaces and can support multiple network segments at the same time.

In an enterprise or service provider vSphere environment, a NAS device is typically a dedicated, high-performance, single-purpose storage component. In this type of infrastructure, NAS devices are optimized to serve specific vSphere storage requirements, with their own operating system often integrating with a vCenter Server software plug-in. However, NAS devices are well suited to serve environments that have a mix of clients, servers, and other operations, and may be able to handle other tasks, such as web cache and proxy, local firewalling, audio-video streaming, and backups.

NAS Implementation

Network-attached storage devices can be implemented in various ways, depending on the storage vendor's hardware. For instance, unified NAS consolidates both NAS-based and SAN-based data access into a single storage platform, and provides a single pane of glass management interface for operating the mixed environment concurrently.

UNIFIED NAS SYSTEM

A *unified NAS system* provides file access and the storing of file data for both CIFS and NFS protocols, as well as facilitating access to block-level data for iSCSI and Fibre Channel storage devices.

A unified NAS storage array combines one or more *NAS heads* and storage into a single system. The NAS head is connected to the storage controllers, providing access to the disk subsystem. These storage controllers also provide connectivity to iSCSI and Fibre Channel hosts through their respective SAN infrastructures. The storage also typically consists of different disk types, such as SAS, NL-SATA, and flash devices, to meet the varying workload requirements.

Each NAS head in a unified system has front-end Ethernet ports that connect to the IP network, and back-end ports to provide connectivity to the storage controllers. The front-end ports provide connectivity to the clients and service the file I/O requests, while the back-end ports provide connectivity to the storage controllers. The iSCSI and Fibre Channel ports on the storage controllers enable hosts to access the storage directly or through a storage area network at the block level.

As a result of the consolidation of NAS-based and SAN-based access into a single storage platform, a unified NAS system can potentially reduce an organization's storage infrastructure and management costs (see Figure 3.33).

GATEWAY NAS

The second common approach to providing NAS in the enterprise or service provider environment is an architecture providing a *gateway NAS* device, which consists of one or more NAS heads but uses external and independently managed storage to provide the disk back-end.

Similar to the unified NAS architecture, the storage is shared with other applications and systems, which use block-level protocols. However, management functions in this type of solution are more complex than those in a unified NAS environment, as there are separate administrative tasks for the NAS heads and the storage array hardware. In fact, the gateway NAS device may not even be provided by the same storage manufacturer as the back-end disk system, but still use the existing or new Fibre Channel infrastructure.

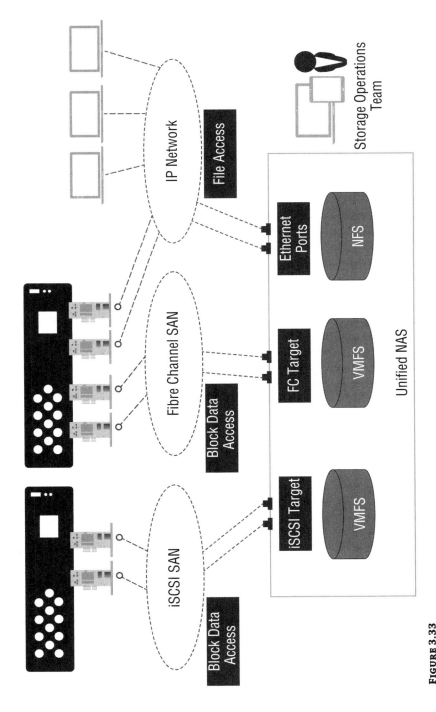

Figure 3.33
Unified NAS system architecture example

A gateway NAS architecture is typically more scalable than a unified NAS solution, as NAS heads and storage arrays can be independently scaled up, as and when required. For instance, NAS heads can be added to scale up the NAS device performance, and likewise, when the storage subsystem limit is reached, it can be scaled up independently by adding capacity to the array, regardless of the NAS heads. However, similar to a unified NAS solution, a gateway NAS can also enable extended utilization of storage capacity, by sharing it with the SAN environment and block-level protocols.

In a gateway NAS architecture, the front-end connectivity is similar to that of a unified storage solution. Communication between the gateway NAS and the storage system is achieved, as illustrated in Figure 3.34, through a traditional Fibre Channel SAN infrastructure. To architect and deploy a gateway NAS solution, key Fibre Channel design factors, such as multiple paths for data, redundant fabrics, and load distribution must also be considered.

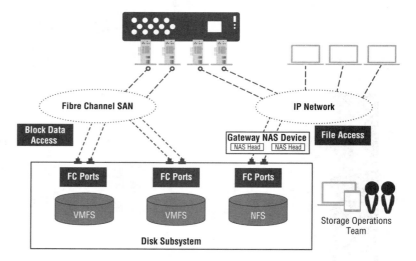

FIGURE 3.34 Gateway NAS system architecture example

NFS Exports

The NFS storage protocol has the concept of an *export*, which can be thought of as being similar to that of a target on an iSCSI storage system, as shown in Figure 3.35. An *export* is a container or directory of storage resources that is made available to systems or users to consume. A NAS storage system typically has a file called Exports within its Linux or Unix operating system, which lists the various directories that are shared, who has access to them, and what permissions they have via the access control list (ACL).

As with any logical abstraction of storage, an NFS export resides inside a volume, similar to a LUN, that is used with block storage. The volume is an abstraction of the back-end disk system, which is logically carved out. In some instances, the volume may consume the entire back-end disk device, but is more typically just a proportion of it. The back-end disk device is the raw storage, such as a pool of disks.

Figure 3.35
NFS export stack

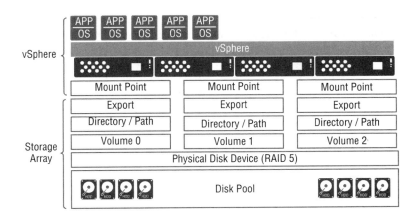

NFS Mounts

Also known as *mount points*, NFS mounts refer to the export from the vSphere host's perspective. As part of the simple process of attaching shared NFS storage to vSphere hosts, via its NFS client, the export is mounted. These storage entities are then referred to as mount points or NFS mounts.

NFS Advanced Host Configuration

Several configurable parameters are available when using NFS datastores on vSphere hosts. Table 3.8 illustrates the advanced values, which can be configured on each host in the platform, if required, in order to support a more scalable operational infrastructure. For instance, the default maximum number of NFS volumes a host server can mount is 8, although in vSphere 6, this can be increased to 256. If a design requires more than 8, you must also ensure that the Net.TcpipHeapSize is increased.

Table 3.8: NFS advanced host configuration

Parameter	Description
NFS.MaxVolumes	By default, the NFS.MaxVolumes value is 8. This sets 8 as the hard limit for the number of NFS volumes that can be mounted to an ESXi host. This can be changed if required, with vSphere 6 supporting a maximum of 256 NFS volumes mounted to an ESXi host. If you increase max NFS mounts above the default setting of 8, ensure that you increase Net.TcpipHeapSize as well.
Net.TcpipHeapSize	Net.TcpIpHeapSize is the amount of memory (in MBs) that is allocated up front by the VMkernel to the TCP/IP heap. In vSphere 6, the maximum configurable value for Net.TcpIpHeapSize is 32 MB.

TABLE 3.8: NFS advanced host configuration *(CONTINUED)*

PARAMETER	DESCRIPTION
Net.TcpipHeapMax	Net.TcpIpHeapMax is the maximum amount of memory that can be consumed by TCP/IP as heap. The maximum value for Net.TcpIpHeapMax is 128 MB. If changes are made to the default NFS.MaxVolumes as discussed previously, you must adjust the heap space settings for TCP/IP accordingly. In vSphere 6, the maximum configurable value for Net.TcpIpHeapMax is 1,536 MB.

NOTE Changing Net.TcpipHeapSize and/or Net.TcpipHeapMax requires a host reboot for the changes to take effect.

Access Control Lists with NFS

vSphere requires both read (ro) and write (rw) access to exports in order for them to host virtual machines. However, specific use cases, such as a shared ISO library, may require only read access to the export.

The specific permission configuration on the storage device varies from vendor to vendor, although typically manufacturer guidance will include ensuring that the export is assigned read-write permissions, and also configuring the export with the no_root_squash parameter. This is required, as vSphere prefers to mount NFS storage devices with root permissions, although by default, the NFS array's software is likely to *squash* root access to the exports unless the no_root_squash parameter is implemented.

The NFS ACLs are also used to limit access via IP address or subnet. For instance, say you have two business units, R&D and HR. R&D exports need to be accessed by hosts on the 10.10.20.0/24 subnet, while HR hosts are on the 10.10.30.0/24 subnet. To configure the correct access to the path of the filesystem, the *Exports* file on the storage array would appear similar to the following:

```
/RnD 10.10.20.0/255.255.255.0 (rw)
/HR 10.10.30.0/255.255.255.0 (rw)
```

One of the enhancements available with vSphere 6 supporting NFS version 4.1 is security, and the ability to use Kerberos to allow nonroot user authentication. As highlighted previously, with NFS version 3, files were typically accessed by hosts with root permissions, and NAS storage devices had to be configured with the no_root_squash option to allow this root access to files; this is known as the AUTH_SYS mechanism.

The AUTH_SYS method of access is still supported with NFS version 4.1. However, employing Kerberos is a much more secure and operationally efficient mechanism. When employing Kerberos, the same NFS user is defined on each host, which is accessing the storage, using the esxcfg-nas -U -v 4.1 command. This user is then used for remote file access to the storage device. The same user account must be employed on all hosts accessing the storage device, or it is likely that vMotion operations will fail. In addition, for this mechanism to be employed, a common Active Directory must be used, with each vSphere host configured on the same Active Directory domain.

Network Adapters for NFS

Unlike iSCSI-based storage, NFS has no specific design requirements and no adapter types that are specifically designed for NAS devices. Having said that, there is an advantage in employing a network adapter type that includes a TOE. This card type off-loads the TCP/IP stack from the vSphere VMkernel onto the network adapter engine, therefore freeing up CPU resources on the hypervisor. Depending on the hardware vendor, most adapters include TOE functionality by default, so the choice is likely to be moot. Furthermore, in most environments, the CPU resource is not in constraint, so there may be limited value in forcing TOE-enabled devices into a design where it is not already naturally occurring on the vendor's hardware.

NFS Virtual Switch Design

As with iSCSI, various options are available for virtual switch design in an NFS environment. Many design factors relate specifically to requirements set out in the virtual and physical networking design, such as the use of the vSphere distributed switch versus the vSphere standard switch, feature requirements, and scalability. However, some NFS storage design factors are also relevant and should be considered as part of the overall network design.

NFS storage is probably the most challenging protocol to design for when attempting to achieve high availability and consistent, high performance. This stems from vSphere traditionally supporting only NFS version 3, which provided no native support for multipathing. VMware added support for NFS version 4.1 capabilities in vSphere 6, which when used in conjunction with a compatible array, and with several caveats associated with support for some vSphere advanced storage features, can provide support for load balancing and multipathing through the use of session trunking. The limitations are addressed in more detail later in this chapter.

As NFS version 3 has no support for multipathing, NFS can maintain only one active path for I/O for each export. This does not mean that the design must accept the risk of a single point of failure. It is perfectly possible to design NFS version 3 storage for high availability with a passive path, although only one path can be active at any one time, as illustrated in Figure 3.36. However, while NFS version 4.1 introduced better performance and availability through load balancing and multipathing with session trunking support, at the time of writing, vSphere does not support the use of Parallel NFS (pNFS).

Session trunking is similar to multipath I/O (MPIO) in an iSCSI or Fibre Channel environment. With session trunking, you can create multiple paths or sessions to the NAS device, and distribute load across each of those sessions. It is the storage array's software and physical active/active or active/standby architecture that determines whether this can be achieved across NAS controllers or is confined to a single point on the physical architecture.

Next we present virtual switch designs employing NFS version 3. Each of these example designs offer multiple options that allow you to provision multiple exports in order to facilitate additional active paths for storage I/O.

Single Virtual Switch / Single Network Design

The *single virtual switch design* outlined here is the simplest overall configuration and is best suited to an environment that employs a 10 Gb/s storage network. In this example, a dedicated VLAN has been configured to isolate network storage traffic. As shown in Figure 3.36, 10.0.10.0/24 is the network segment used on VLAN 250.

In this example design, VLAN 250 is fully isolated and dedicated to NFS traffic, which is nonroutable. The VMkernel port has been placed onto a vSphere standard switch, with two 10 Gb/s vmnic uplinks to the physical infrastructure. The interfaces have been teamed with the failover policy configured as active/active. However, as a VMkernel port can maintain only a one-to-one relationship with a vmnic, traffic can traverse only one route at any given time. In this example, the host has placed the VMkernel port on vmnic0.

FIGURE 3.36
Single virtual switch / single network design example

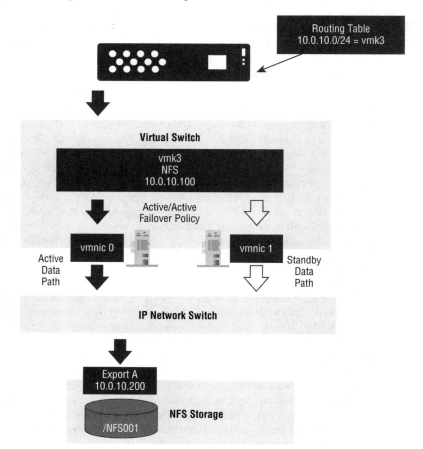

In this design, if vmnic0 fails, the VMkernel is failed over by the vSphere host to vmnic1. As both vmnic ports are marked as active in the failover policy, the link will not fail back to vmnic0 when it is brought back online. This behavior is, by design, to avoid unnecessary *flapping* of vmnic interfaces in an unstable environment. As a result, the fail-back task would require manual operational team intervention during a soft maintenance window.

It is noteworthy that if this design instead employed the vSphere distributed switch with a Link Aggregation Control Protocol (LACP) load-balancing policy by way of a link aggregation group (LAG) on the two vmnics, it would be unlikely to improve traffic throughput, as there is a single source IP address (the VMkernel port) and a single target IP address (the export). As a result, the LAG-hashing algorithm would always use the same uplink port, so there would be no additional performance or availability gain, only an increase in operational overhead and complexity, in maintaining the LACP configuration on the physical infrastructure.

The key design factors associated with this single virtual switch / single network solution include the following:

- In this example, there is a single export on the storage array, although multiple are supported and can be load-balanced across both vmnic ports.

- A single NFS network and isolated VLAN are used in this example, but designing for multiple instances is a common configuration.

- A single target IP address on the storage array is used, although support for multiple addresses is common.

- There is a single active path to the NFS export.

Single Virtual Switch / Multiple Network Design

One of the limitations of NFS version 3 is that each export can use only a single active I/O path. One design option for implementing multiple storage paths, in order to load-balance traffic across all available interfaces, is to use multiple exports on different networks.

As illustrated in Figure 3.37, this design extends the previous example by adding a second VMkernel port on a new network, 10.0.11.0/24. This is, however, likely to require additional configurations on the network as well as the target storage array. First, the array must be able to support multiple target IP addresses. This is typically achieved by storage vendors through either virtual IPs (VIPs) or logical IP addresses (LIPs), by associating multiple IP addresses to a single interface. This could also be achieved, if necessary, by adding additional I/O cards to provide more target ports to the array's system, and subsequently assigning the required IP addresses and configuration to the new physical interfaces. This approach has the added benefit of increasing the throughput capability of the NAS device's front-end ports.

This configuration is illustrated in Figure 3.37. Each export has been assigned a target IP address on a unique network segment.

If this design were employing only two physical vmnic interfaces, it would require an active/unused failover policy in order to avoid the risk of both VMkernel ports utilizing, or trying to use, the same vmnic during an outage of one port. However, such a design would result in the vmnic being a single point of failure. The failover policy for the architecture shown in Figure 3.37 is clarified in Table 3.9, which as you can see, results in no single point of failure within the architecture.

TABLE 3.9: Design example vmnic configuration

vmk4	NFS-01	vmnic0	Active
		vmnic1	Active
vmk5	NFS-02	vmnic2	Active
		vmnic3	Active

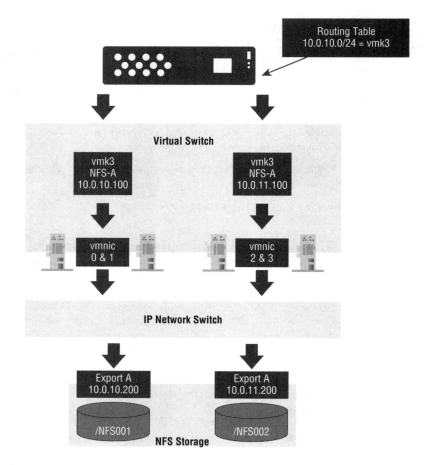

FIGURE 3.37
Single virtual switch / multiple network design example

In a typical use case for this design scenario, a host has a large number of 1 Gb/s interfaces, and the design calls for high availability and load balancing across the multiple available uplinks. This configuration helps reduce saturation of the lower-bandwidth interfaces, while still providing active/passive availability for a range of failure scenarios, including adapter failure, switch port failure, and physical switch failure, where a redundant pair of physical devices has been employed. In a 10 Gb/s network design, this multiple network solution offers little advantage over a single network configuration.

The key design factors for a single virtual switch / multiple network solution architecture include the following:

- There are multiple storage paths to the array, one per export.
- There are multiple exports employed on the storage array system.
- There are multiple IP addresses configured on the storage array system.
- There are multiple NFS networks to be configured, and multiple NFS isolated VLANs to be maintained.

vSphere 6 NFS Version 4.1 Limitations

At the time of writing, limitations exist when employing NFS version 4.1 datastores and vSphere core features. However, as releases move forward, some (if not all) of these considerations should be addressed. Therefore, it is advised to consult with the latest official VMware documentation to confirm support if it is a factor in a design. In vSphere 6.0, Storage DRS, Storage I/O Control, Site Recovery Manager, vSphere Storage APIs for Array Integration (VAAI), and Virtual Volumes are not supported with NFS version 4.1. In addition, only symmetric multiprocessing fault tolerance (SMP-FT) virtual machines are supported. The legacy FT mechanism of previous vSphere releases is also not supported with NFS version 4.1. Other limitations that may prove pertinent to a design include the following:

- NFS version 3 and version 4.1 datastores can coexist on the same host.
- You can mount an NFS volume as a version 3 or 4.1 volume, but not both, as they use different locking mechanisms. NFS version 3 uses proprietary client-side cooperative locking, while NFS version 4.1 uses server-side locking. Using both concurrently can lead to data corruption.
- The upgrade path to version 4.1 requires the use of SvMotion to migrate from NFS version 3 to a 4.1 datastore, in a nondisruptive manner. A direct in-place upgrade is not possible.
- NFS 4.1 does not support hardware acceleration (VAAI); as a result, you cannot create thick virtual disks on an NFS version 4.1 datastore (thin only), or use any of the VAAI-NAS primitives, such as Fast File Clone.

NFS Protocol Summary

The NFS protocol, while a good choice for specific use cases, does have several drawbacks, as we have addressed in this chapter. Even though it is arguably the simplest protocol to support operationally, scalability in an enterprise or service provider environment has to be considered as part of an initial design, in order to facilitate the required growth. For instance, 10 Gb/s switching and distributing storage I/O traffic to exports across multiple network interfaces are key scalability requirements. In addition, functionality and feature limitations may play a deciding factor in its ability to meet customer prerequisites.

Fibre Channel over Ethernet Protocol

Fibre Channel over Ethernet (FCoE) is the last and newest of the storage protocols to be addressed in this chapter. FCoE brings the flexibility of Ethernet and the lossless reliability of Fibre Channel together into a single converged network infrastructure protocol that provides several distinct advantages over running separate isolated environments for each traffic type.

As you have seen, traditional LAN infrastructure can provide a platform for IP storage networks to function effectively. Storage protocols, such as iSCSI and NFS, all use TCP and/or UDP over IP-based networks. It was this flexibility that made IP SAN an attractive option when looking to deploy a lower-cost storage solution, as compared to a native Fibre Channel fabric. However, performance for these IP-based storage protocols was always raised as a concern, as they were often traditionally limited to 1Gb/s, in comparison to the enhanced performance offered by 4, 8, 16, or even 20 Gb/s Fibre Channel environments. For this reason, most enterprise

IT organizations and service provider data centers typically maintained separate networks to handle each traffic type.

However, in order to support these multiple network protocols, servers in a data center must be equipped with multiple redundant physical network interfaces, such as multiple Ethernet and Fibre Channel adapter cards. In addition, to enable the required communication, different types of networking switch and physical cabling infrastructure must be implemented, which increases the overall cost and complexity of data center operations.

As time went on, and 10-Gigabit Ethernet grew in popularity, these IP-based storage protocol technologies became more attractive to enterprise and service provider customers, as these performance limitations were no longer of concern. However, companies that had already invested large amounts of capital into Fibre Channel networks were hesitant to discard their proven solutions in favor of adopting a very different technology with a completely different operational model.

Data center bridging (DCB), also referred to as *converged enhanced Ethernet* (CEE), helps to bridge this gap between the traditional Fibre Channel model and Ethernet-based IP SAN networks, by providing additional enhanced functionality required to handle Fibre Channel traffic. DCB employs a 10 Gigabit Ethernet infrastructure with all the traditional functionality, including support for traditional IP SAN technologies, DCB also can support Fibre Channel over Ethernet across the same physical infrastructure, thereby combining the functionality and reliability of Fibre Channel with the flexibility of Ethernet (see Figure 3.38).

FIGURE 3.38
Fibre Channel over Ethernet converged protocol

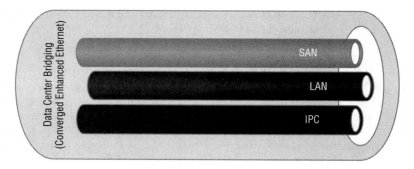

Fibre Channel over Ethernet helps IT organizations and service providers address the challenge of having multiple discrete network infrastructures through the use of DCB technology. By employing DCB and using the same physical link to send Fibre Channel frames over Ethernet, the number of I/O cards, cables, and physical switches can be reduced in the data center by up to 50 percent.

While the cost savings are obvious, the required switching hardware is not standard Ethernet or Fibre Channel, meaning there is typically no opportunity to reuse existing network infrastructure equipment. In addition, distance limitations for Fibre Channel over Ethernet can be a constraint, as shown in Table 3.10.

TABLE 3.10: Fibre Channel over Ethernet distance limitations

CABLE TYPE	DISTANCE
Twinaxial copper	10 meters
Multimode fiber	300 meters
Single-mode fiber	40 kilometers

Fibre Channel over Ethernet Protocol

The Fibre Channel over Ethernet protocol encapsulates the entire Fibre Channel frame, including the start of frame (SOF) and end of frame (EOF) into a single FCoE wrapper and eventually into an Ethernet frame.

Each Fibre Channel over Ethernet frame is created approximately 2,500 bytes in size, to accommodate its 2,112-byte payload and all the additional header metadata. Unlike other Ethernet frame types, the FCoE frame cannot be fragmented while traversing the network. Therefore, if any switch in the data path does not support the frame size, it will be dropped. Also, unlike traditional jumbo frames, where the frame size can be increased up to 9,000 bytes so larger payload data can be carried, the FCoE frames cannot be increased beyond the 2,500-byte size; the frame size is always fixed. In addition, the Fibre Channel over Ethernet protocol has stringent requirements for the composition of the frame itself, as illustrated in Figure 3.39.

This industry-standard enhancement to Ethernet divides the bandwidth of the Ethernet pipe into eight virtual lanes, with DCB functions classifying different data types into different classes of service (CoS). The Fibre Channel over Ethernet network can be configured with up to eight CoS values, one for each virtual lane. This mechanism then divides the bandwidth of the Ethernet pipe into the eight virtual lanes accordingly, allowing DCB to determine how the bandwidth in these virtual lanes is allocated across the entire network.

Each CoS value reserves a specific segment of bandwidth for a particular traffic type, providing a level of data-flow management even in an oversubscribed system. Each CoS can have different attributes associated with it, in order to control how the Fibre Channel over Ethernet switches and end devices handle that specific traffic type. For instance, *Best Effort* is a predefined system class (CoS value) that sets the quality of service for the lane reserved for basic Ethernet traffic. Some properties of this system class are preset and cannot be modified, such as its drop policy, which allows it to drop data packets if required. Likewise, the *Fibre Channel* system class, which sets the quality of service for the lane reserved for Fibre Channel traffic, also has some preset properties that cannot be modified, such as its no-drop policy, which guarantees it will never drop Fibre Channel data packets.

DCB is composed of various subprotocols. Table 3.11 briefly describes its component parts.

FIGURE 3.39
Fibre Channel over Ethernet frame

TABLE 3.11: Data center bridging attributes

NAME	IEEE STANDARD	DESCRIPTION
Priority Flow Control (PFC)	IEEE 802.1Qbb	Provides class-of-service flow control by enabling PAUSE functionality on IEEE 802.1P lanes. In doing so, provides controls on how congestion is handled between two devices.
Enhanced Transmission Selection (ETS)	IEEE 802.1Qaz	Manages bandwidth and assigns priorities to groups of IEEE 802.1P lanes, based on class of traffic, as there can be several types of traffic competing for a fixed amount of bandwidth. This is critical to ensure that applications obtain the bandwidth they require to meet service levels. ETS allows operational teams to determine the amount of bandwidth that a CoS value is guaranteed to have at any one time. In addition, ETS provides a mechanism to reallocate unused bandwidth to other CoS values, if they are in need.
Data Center Bridging Capabilities Exchange (DCBX)	IEEE 802.1Qaz	Provides autonegotiation of enhanced Ethernet capabilities. When a host is initialized, its CNA employs DCBX to learn what functionality is supported by the FCoE switch. In addition, many DCB parameters, such as the CoS configuration, will be automatically transferred to the host, eliminating the need for manual configuration, which can also occur across switch-to-switch links.
Congestion Notification	IEEE 802.1Qau	Provides quantized congestion notification across network links. It also provides a network-wide congestion management solution, which operates with similar capabilities to TCP. Whereas PFC pauses traffic between devices, Congestion Notification attempts to throttle traffic, in order to avoid having to employ PFC.

Fibre Channel over Ethernet Physical Components

A Fibre Channel over Ethernet infrastructure requires various components to operate, many of which are typically proprietary, and therefore can be used only in conjunction with other equipment from that vendor.

CONVERGED NETWORK ADAPTER

The *converged network adapter* (CNA) combines the functionality of both the standard Ethernet network adapter and Fibre Channel HBA into a single hardware device that consolidates both traffic types. This adapter eliminates the need to deploy separate hardware and cables for each type of traffic, and therefore reduces the number of server slots required in the hosts and the number of switch ports required on the access switches.

As with the TOE adapters addressed previously in this chapter, the CNA off-loads the FCoE protocol-processing tasks from the host server onto its own chipset, thereby freeing the server

CPU and memory resources for application processing. The CNA chipset contains separate modules for 10 Gigabit Ethernet, Fibre Channel, and FCoE application-specific integrated circuits (ASICs). The FCoE ASIC encapsulates the Fibre Channel frames into Ethernet frames. One end of this ASIC is connected to the 10GbE and Fibre Channel ASICs for server connectivity, while the other end provides a 10GbE interface to connect upstream to the FCoE switch, illustrated in Figure 3.40.

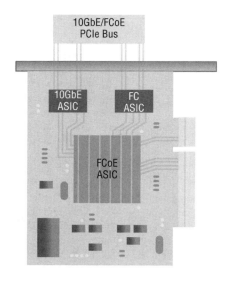

FIGURE 3.40
Converged network adapter (CNA)

FIBRE CHANNEL OVER ETHERNET SWITCH

The *Fibre Channel over Ethernet switch* has both Ethernet and Fibre Channel switch functionalities. In addition, the FCoE switch has a Fibre Channel forwarder (FCF), Ethernet bridge, set of Ethernet ports, and optional Fibre Channel ports. The function of the FCF is to encapsulate the Fibre Channel frames received from the port into the FCoE frames, and also to de-encapsulate the FCoE frames received from the Ethernet bridge to the Fibre Channel frames, as illustrated in Figure 3.41.

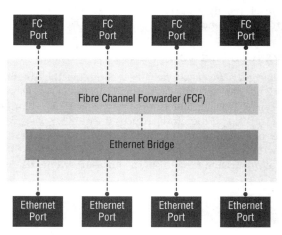

FIGURE 3.41
Fibre Channel over Ethernet switch architecture

The Ethertype is used to indicate which protocol is encapsulated in the payload of an Ethernet frame. The Ethertype of a Fibre Channel over Ethernet frame is always 8096, which allows the switch to identify it has an FCoE frame and act accordingly. Upon receiving the incoming traffic, the FCoE switch inspects the Ethertype of the incoming frames and uses that to determine the destination. If the Ethertype of the frame is FCoE (8096), the switch recognizes that the frame contains a Fibre Channel payload and forwards it to the Fibre Channel forwarder. From there, the Fibre Channel is extracted from the FCoE frame and transmitted to the target, either to the Fibre Channel SAN over the Fibre Channel ports or directly to the storage device, if in an end-to-end Fibre Channel over Ethernet design. If the Ethertype is not FCoE, the switch handles the traffic as usual Ethernet traffic and forwards it over the Ethernet ports.

Fibre Channel over Ethernet Infrastructure

Figure 3.42 illustrates the I/O consolidation in a Fibre Channel over Ethernet environment, which is utilizing FCoE switches and CNAs. As we have addressed, this reduces the requirements for multiple network adapters in the host server to connect to different networks, and reduces the requirement in the infrastructure for multiple cable types and switches. It can also reduce the cost of operational management of the environment, as multiple network infrastructures are consolidated, and therefore fewer hardware devices are required.

Fibre Channel over Ethernet Design Options

Fibre Channel over Ethernet design employs two basic approaches:

- Edge Fibre Channel over Ethernet design
- End-to-end Fibre Channel over Ethernet design

EDGE FIBRE CHANNEL OVER ETHERNET DESIGN

In the *edge Fibre Channel over Ethernet* model, the access layer switches must support FCoE, Fibre Channel, and Ethernet protocols. The single connection type, although typically redundantly configured, provides the converged FCoE connection from the host to the access layer switch. The connectivity between the access layer switches and the aggregation layer needs to support only traditional Ethernet and native Fibre Channel protocols. Therefore, in this design model, the aggregation layer switches do not need to be converged.

As illustrated in Figure 3.43, the connectivity from the access layer to the SAN is provided using native Fibre Channel ISLs from the converged access layer to the Fibre Channel SAN switches, with each access layer switch connected to only a single Fibre Channel fabric, A or B. Connecting them to both fabrics, as in a meshed approach, is likely to cause the fabrics to merge if not configured correctly, and does not provide any benefit because of the typical zoning and masking configuration found in such an architecture.

As illustrated in Figure 3.42, one example of a vendor technology that employs this model is Cisco's UCS Blade system. This model is ideal for environments that want to migrate to the Fibre Channel over Ethernet protocol over time. It provides a hybrid environment that supports both protocols, allowing for significant flexibility in the design.

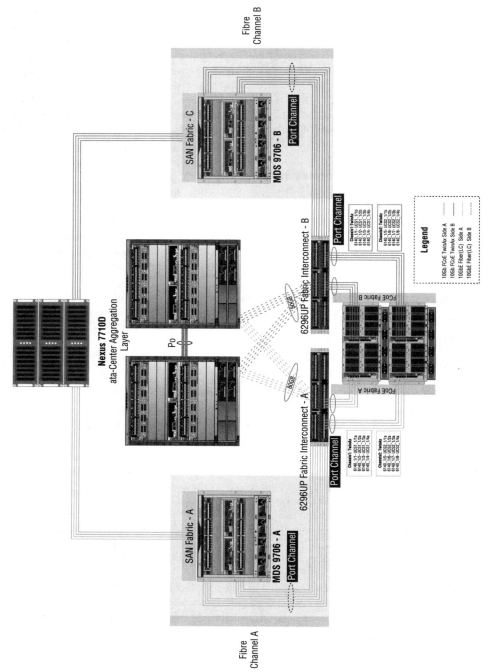

Figure 3.42
FCoE infrastructure example (Cisco UCS Blade system)

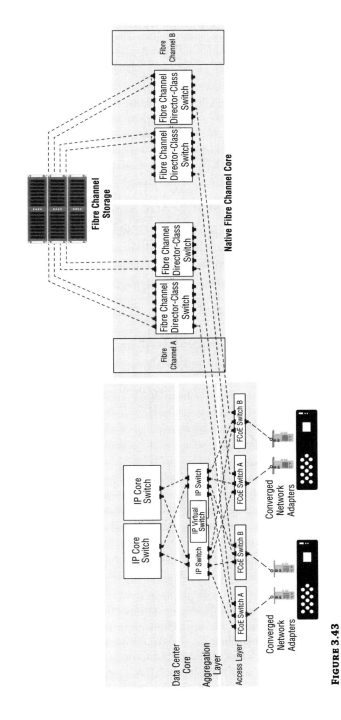

FIGURE 3.43
Edge Fibre Channel over Ethernet design

END-TO-END FIBRE CHANNEL OVER ETHERNET DESIGN

In the end-to-end Fibre Channel over Ethernet design, the converged network is extended further northbound to the aggregation layer. However, each FCoE switch is a member of only a single FCoE fabric, as again, you do not want to cross-connect as you would typically do for an Ethernet environment. This design sees connections only between each access layer switch, or multiple switches in the same fabric, to the aggregation layer switch.

As you can see in Figure 3.44, in this model the storage is connected directly to the aggregation layer and not to the core network. Because there is typically no layer 2 connectivity between the aggregation layer switches and the core, FCoE traffic cannot traverse the link, since it is not routable. This design model typically best suits new greenfield deployments, as it is generally less flexible from a coexistence perspective.

Fibre Channel over Ethernet Protocol Summary

Key design factors associated with deploying converged Fibre Channel over Ethernet environments include, but are not limited to, the following:

- FCoE is a protocol that transports Fibre Channel data over Ethernet networks.
- FCoE can be positioned as a storage networking design option for the following reasons:
  - It enables the consolidation of Fibre Channel SAN traffic and Ethernet traffic onto a common merged network infrastructure.
  - It reduces the number of adapters, switch ports, and cables required.
  - It can decrease capital expenditure costs and simplify data-center management.
  - It can reduce power and cooling costs, and also decreases floor space requirements in the data center.

Multipathing Module

In any business-critical environment, a vSphere host must be able to access a volume on a storage array through more than one path, in order to facilitate failover, and in some instances, traffic load-balancing. Having more than one path from a host to a volume is called *multipathing*.

The vSphere host supports multipathing in order to maintain a constant connection between the server and the storage device, in case of failure that results in an outage of an HBA), fabric switch, storage controller, or Fibre Channel cable. Multipathing support does not require specific failover drivers. However, in order to support path switching, the server does require two or more HBAs, from which the storage array can be reached by using one or, typically, multiple fabric switches.

As illustrated in Figure 3.45, each host connects to multiple paths in order to provide redundant access to the Fibre Channel storage device. In this Fibre Channel multipathing example, if HBA0 or the link between HBA0 and the Fibre Channel switch fails, HBA1 takes over and provides the connection between the host and the fabric. The process of one HBA taking over from another is referred to as an *HBA failover*. Similarly, if storage controller A fails, or the links between storage controller A and the fabric switch stop working, storage controller B takes over and provides the connection between the switch and the storage device. This process is called *storage controller failover*. The multipathing capability of the vSphere stack supports both HBA and controller failover.

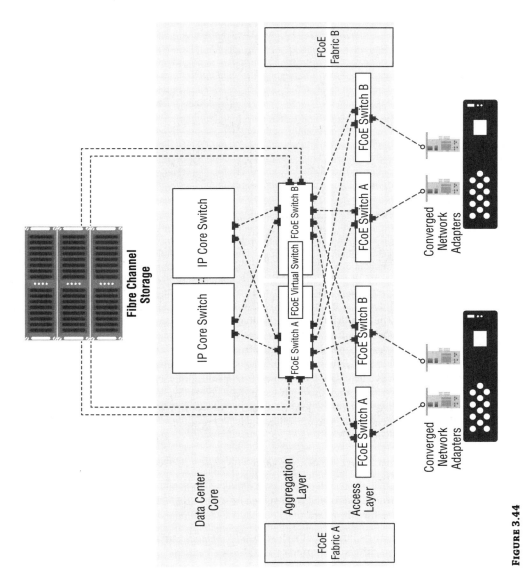

FIGURE 3.44
End-to-End Fibre Channel over Ethernet design

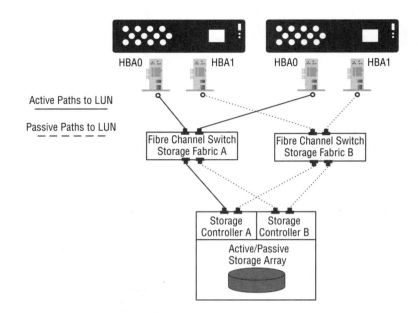

FIGURE 3.45
Fibre Channel multipathing example configuration

Before addressing the VMware multipathing module, we must also take into account the difference between active/active and active/passive disk arrays. An active/active disk array allows access to the volumes simultaneously, through all the available storage controllers. In an active/active array, all the paths are active at all times, unless a path failure occurs. In contrast, in an active/passive disk array, one storage controller is actively servicing a given volume, and the other storage controller acts as backup for the volume and may be actively servicing other volumes' I/O. In an active/passive array, I/O is sent to only the active processor, and if the active storage controller fails, the secondary storage controller becomes active for its volumes, typically automatically, and without administrator intervention.

In Figure 3.46, one storage controller provides the active array, and the other a passive array to a set of volumes, with data arriving through the active array only. Using active/passive arrays, like the one shown here, coupled with a Fixed path policy, can potentially lead to path thrashing. *Path-thrashing* occurs when two host servers access the same volume on a storage array via different storage controllers, and as a result, the volume experiences low throughput because of the length of time taken to complete each I/O request.

To avoid path-thrashing:

- Ensure that all hosts sharing the same set of volumes on active/passive arrays access the same storage controller.

- Correct any cabling inconsistencies between hosts and SAN targets, so that all HBAs see the same targets in the same order.

- Ensure that the path policy is set to Most Recently Used, the default for active/passive storage devices, unless otherwise stated by the storage vendor.

Figure 3.46
Active/passive disk arrays

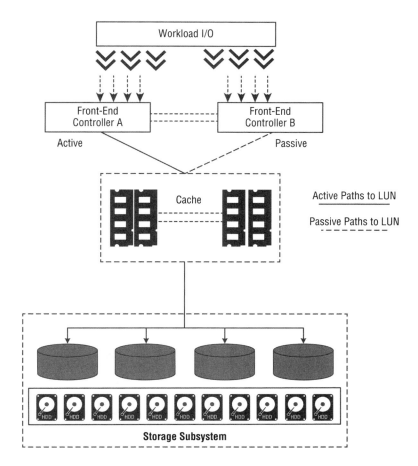

Another approach to resolving the path-thrashing problem has been Asymmetric Logical Unit Access (ALUA), illustrated in Figure 3.47. ALUA is a technology that allows a storage array to use the interconnect between the controllers to service I/O. In an ALUA-capable array, a volume can be accessed via both storage controllers, but still only one of the controllers owns the volume. The primary path is the optimized path, the one with the most direct path to the owning storage controller. However, the nonoptimized path also has a connection with the volume, although still doesn't own it, but instead has an indirect path to the storage controller via the internal interconnected bus. This technology can also prevent failover events in certain types of failure scenario, but the downside of this approach is that this proxying via the internal interconnected bus adds additional workload on the controllers and can increase latency.

One additional consideration relating to active/active, active/passive arrays and ALUA is that these capabilities are often further confused by storage vendors who try to ensure that their product stands out from the crowd. You will often see storage vendors' marketing collateral referring to active/passive ALUA-capable arrays as active/active, as the nonowning controller is able to accept I/O for volumes it does not own. This of course does not make them active/active arrays, but can often cause confusion when evaluating and comparing different vendors' storage products.

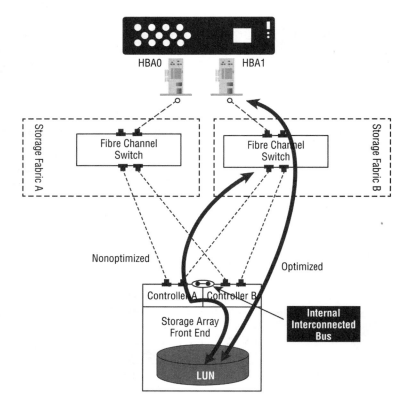

FIGURE 3.47
ALUA-capable array path

Pluggable Storage Architecture

In vSphere 4, VMware introduced a full rearchitecture of its storage layer and introduced the *Pluggable Storage Architecture* (PSA), which provided third-parties the ability to plug customized multipathing software into the vSphere storage stack, to allow optimized load-balancing, failover, and performance for vendor-specific storage devices. This opened up a more flexible modular framework and provided storage vendors with the opportunity to integrate their multipathing software solutions with the vSphere host platform. For instance, EMC and Dell are both storage vendors that have taken advantage of this functionality and produced their own multipathing plug-in (MPP).

Hosts configured without any third-party plug-in use the Native Multipathing Plug-in (NMP), where the vSphere host's own software dictates the multipathing functionality. The NMP is split into two modules: the Storage Array Type Plug-in (SATP) for path failover, and the Path Selection Plug-ins (PSPs) for path load-balancing and path selection. As illustrated in Figure 3.48, by drilling into the VMkernel components, you can see how the vSphere host's storage stack provides this default NMP module, without any requirement to install additional software or any intervention from operational teams.

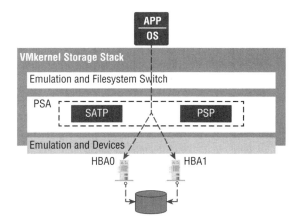

FIGURE 3.48
vSphere Pluggable Storage Architecture

The SATP is used by the host to identify the storage device, determine the failover type for a volume, and handle path failover for a given storage array. SATPs run in conjunction with the NMP and are responsible for array-specific storage operations. The vSphere host offers an SATP for every type of array that VMware supports, including generic SATPs for nonspecified storage arrays and a local SATP for direct-attached storage. Each SATP provides characteristics of a certain class of storage and can perform the array-specific operations required to detect path state and to activate an inactive path. This is done by confirming the array's details against a set of claim rules defined in the host's /etc/vmware/esx.conf file, which lists all of the certified hardware. From this, the PSA determines which multipathing module should be loaded to claim the paths to a particular device, and become responsible for managing that device. Based on the information available, the host sets the pathing policy for each presented volume, and it is this classification that determines whether the array is classed as active/active or active/passive.

As a result of its flexibility, the NMP module can work with multiple storage arrays without having to be aware of the specific storage device. After the NMP determines which SATP to call for a specific volume, it associates the SATP with the physical paths to the storage array. The SATP then goes on to monitor the health of each physical path and reports changes in its state to the NMP. The SATP also performs array-specific actions necessary to allow for storage failover. For instance, for active/passive devices, it can activate passive paths.

The PSP also runs in conjunction with the NMP module and is responsible for choosing the physical path for I/O requests. The NMP assigns a default PSP for every logical device (volume) based on the SATP associated with the physical paths for that storage target. The PSP controls three types of pathing policy:

Fixed   This is the default policy for active/active storage devices. When the Fixed pathing policy is selected, the vSphere host always uses the preferred path to the disk, if that path is available. If it cannot access the disk through the preferred path, it tries the alternative paths.

Most Recently Used   MRU is the default policy for active/passive storage, and is typically required for those devices. When the Most Recently Used pathing policy is selected, the vSphere host uses the most recently known path to the disk, until that path becomes unavailable. With this policy enabled, the host does not automatically revert to the preferred path after a failed path once again becomes available.

Round Robin The RR policy rotates through all available optimized paths, providing a rudimentary load-balancing mechanism. When the Round Robin pathing policy is selected, the vSphere host uses an automatic path selection, which rotates through all available paths. In addition to path failover, RR supports basic load balancing across all available paths.

While the most appropriate policy is determined by the SATP and is selected automatically, this can be manually overwritten. When determining the most appropriate multipathing configuration for a design, this policy and its configuration will typically be based on clear recommendations from the storage vendor.

NOTE It is noteworthy, however, that vSphere hosts, by default, can utilize only one path per I/O, regardless of the number of available paths. In environments that use active/active arrays, the path is selected on a volume-by-volume basis. In contrast, in an active/passive array, the hosts determine the active path themselves by employing the Most Recently Used path selection policy.

However, if the design requires specific storage array characteristics that the host can use, then depending on the hardware vendor, there may be an option to use a third-party SATP. The SATP would be provided by the storage vendor or by a partnering software company specializing in optimizing the use of that vendor's storage array. This option may allow the design to take advantage of more-complex I/O load-balancing algorithms, which can be installed onto the third-party PSP. When installed, the third-party sub-plug-ins are coordinated by the NMP, and can run alongside and be simultaneously used with the vSphere native sub-plug-ins (see Figure 3.49).

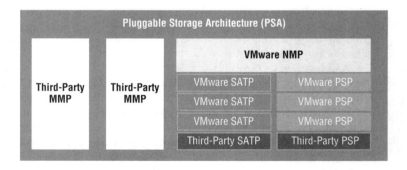

FIGURE 3.49
Native and third-party multipathing plug-ins

By including third-party MPPs in a design for specified storage arrays, the behavior of the vSphere host's NMP can be completely replaced, and full control over path failover and load-balancing operations entirely handed over to the third-party software. This also means that third-party storage vendors can add support for new arrays into the ESXi storage stack, without having to provide internal or intellectual information about the array to VMware.

The third-party multipathing plug-ins fall into one of three categories:

- Third-party SATPs are typically developed by the storage hardware manufacturers, which of course have expert insight into their storage devices. This can lead to optimized plug-ins that accommodate specific characteristics of their storage arrays. Storage vendors may also need to provide specific SATPs when the behavior of their arrays does not match the behavior of any existing SATPs that the PSA offers. When installed, the third-party SATPs are coordinated by the NMP, and can run alongside and simultaneously with the VMware SATPs.
- Third-party PSPs can provide more-complex I/O load-balancing algorithms than are supported natively. In general, these plug-ins are developed by third-party software companies and can help achieve higher throughput across multiple paths. When installed, the third-party PSPs are coordinated by the NMP and can run alongside and be used simultaneously with the VMware PSPs.
- Third-party MPPs can also define entirely new fault-tolerant and performance-optimized behavior. Third-party MPPs run in parallel with VMware's NMP, and for certain specified arrays, replace the behavior of the NMP by fully taking control over the path failover and load-balancing operations.

These three categories are summarized in Table 3.12.

TABLE 3.12: Pluggable Storage Architecture (PSA) third-party plug-in categories

PLUG-IN	SCOPE	DETAILS
SATP (Storage Array Type Plug-in)	Failover	• Lightweight plug-in to VMware's NMP module • Essential for achieving failover check mark on VMware SAN HCL
PSP (Path Selection Plug-in)	Load-balancing	• Lightweight plug-in to VMware's NMP module • For each I/O, selects which physical path to use
MPP (Multipathing Plug-in)	Failover & load-balancing	• Larger level of effort • Entails implementing an alternative to VMware's NMP • Requires taking an I/O and sending it to the storage driver (making all policy decisions) • Handles all I/Os for a given LUN

iSCSI Multipathing

In an iSCSI storage environment, both the hardware and software initiator types are supported for multipathing. In iSCSI multipathing, if one port goes down or is overloaded, traffic is moved to another port. The iSCSI target can force this behavior by telling the initiator to log out, and new logins are then pointed to other ports. However, some iSCSI storage systems use only one target, which switches to an alternative during a failover event.

In designs using hardware initiators, multipathing works by employing the same mechanism as with Fibre Channel environments. The host identifies the iSCSI HBAs as storage adapters, and employs the NMP with the SATP to identify the storage device, determine the failover type, and handle path failover for a given storage array and path selection policy. Then through dynamic discovery, iSCSI HBAs obtain a list of target addresses that the initiators can use as multiple paths to iSCSI target LUNs for failover purposes.

The software iSCSI initiator requires additional configuration, over the hardware HBA approach, in order to take advantage of vSphere's storage multipath I/O capabilities. By default, with the software-initiated iSCSI adapter, network adapter teaming is employed to provide failover, but as the initiator provides a single endpoint, no load-balancing capabilities are available through this mechanism. Therefore, as addressed earlier in this chapter, this is not considered good design practice and does not produce optimum performance of the failover capabilities provided by the software initiator.

To take advantage of the vSphere storage NMP and enable load-balancing functionality across software initiators, you must use the technique addressed earlier in this chapter known as port binding (see Figure 3.50).

NAS Multipathing

Multipathing for NAS-based storage is fundamentally different from that of Fibre Channel or iSCSI storage devices, as vSphere multipathing I/O components, such as SATPs and PSPs, are not available for NAS-based designs. As NFS version 3 provided no native support for multipathing, traditionally, in order to provide a highly available design, the only option for an architect was to use IP-based redundancy and routing, which relied entirely on the physical network stack.

However, as previously highlighted, as of the release of vSphere 6, NFS version 4.1 was supported, which provides the ability to support multipathing for servers through session trunking. By employing session trunking technology, vSphere hosts can use multiple IP addresses to access a single NFS volume, which in turn can optimize performance and availability through load balancing and failover capabilities.

Traditionally, with NFS version 3, for each mounted NFS export, only a single physical interface could be used, despite any link-aggregation technologies being employed to bind multiple interfaces together. While network adapter teaming could provide failover, for redundancy purposes, it could not provide load balancing to and from an export. However, as addressed earlier in this chapter, by creating multiple exports alongside multiple connections on different network segments, an architect could design a solution that makes it possible to statically spread the load of NFS datastore traffic across multiple interfaces.

As you can see, with NFS version 3, it was challenging for a design to provide a good NFS load-balancing mechanism, and in reality, NFS storage design was often limited to providing failover capabilities only. In the past, this was one of the key design factors when employing NFS storage and addressing how it would scale in a successful design. However, with vSphere 6 supporting session trunking, you can now provide multiple IP addresses to access the same NFS export, which can fundamentally help a storage architecture in terms of both performance and availability.

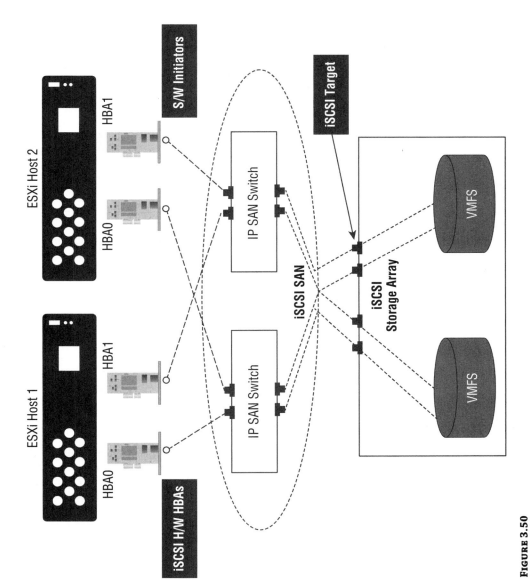

Figure 3.50
iSCSI storage multipathing failover and load balancing

As illustrated in Figure 3.51, vSphere 5.5 and earlier, with NFS version 3, simply provides an active/passive configuration; each target has a maximum of one session, despite what the fabric design provides. In an NFS version 3 design, all other connections are merely mitigating hardware failures, such as network interfaces and switches. However, what vSphere 6 with NFS version 4.1 brings to the design is the ability to add another session to the same target. As shown in Figure 3.52, by deploying this type of architecture, you can provide an increase in bandwidth and a decrease in latency, while still maintaining the ability to provide failover should a hardware outage occur.

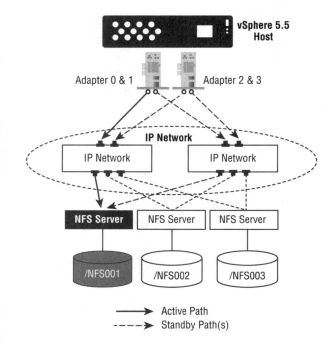

Figure 3.51
NFS version 3 configuration example

Direct-Attached Storage

Direct-attached storage, or DAS, is the most fundamental level of storage provided by storage vendors. Disk devices are either integrated internally, as part of the host server hardware, or directly connected to an external storage device. As a result of this architecture, hosts that want to access the storage device must first access the directly connected server in order to facilitate I/O.

Historically, DAS-based products were widely adopted, and still in many data centers make up a large share of the installed storage systems. DAS remains a viable option for some very small environments, as it is simple and inexpensive to deploy, and in most cases requires only basic configuration. However, data availability can be an issue in DAS environments, as any host requiring access to storage that is not directly attached depends on the availability of the host that is directly attached to the device. This limitation fundamentally rules out DAS from providing shared storage to vSphere clusters employing a traditional architecture.

The DAS model is ideal for localized file sharing, and in environments with a single server or a few servers (see Figure 3.53). In addition, from an economic perspective, the initial investment in DAS is smaller than for NAS or SAN. However, its limited scalability makes it inappropriate for most enterprise or service provider IT organizations that anticipate rapid data growth or want to deploy shared storage to vSphere host clusters. From both a cost-efficiency and administration perspective, the NAS or SAN model is far better suited to highly scalable business storage requirements.

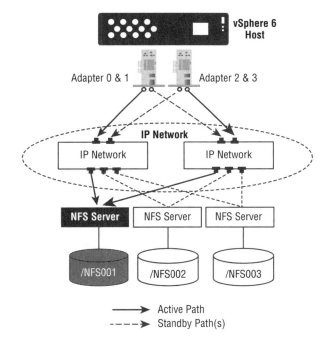

FIGURE 3.52
NFS version 4.1 configuration example

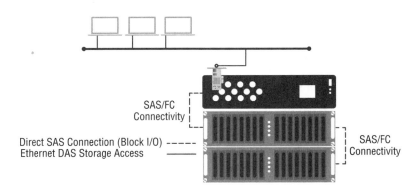

FIGURE 3.53
Direct-attached storage model at ROBO site

Despite these limitations, DAS can be a viable option for specific use cases. For instance, as illustrated in Figure 3.53, if you are deploying a single vSphere host at a remote office / branch office (ROBO) site, introducing an additional shared storage device might introduce another single point of failure and be cost prohibitive. In addition, in this design scenario, shared storage would probably provide lower levels of performance and offer no additional redundancy.

Another use case is to provision specific hardware with high-density DAS to provide external storage to blade servers with few or no local disks, in order to deploy a blade-only direct-attached JBOD solution in a Virtual SAN environment. This approach enables Virtual SAN to scale on blade servers by adding more storage through high-density DAS. Examples of hardware providing this type of functionality for Virtual SAN use cases such as these include Lenovo's Flex SEN with x240 Blade Series compute nodes and Dell's FX2 with 12G controllers.

Figure 3.54 illustrates Lenovo's Flex SEN solution, with x240Series Blades, providing high-density DAS for Virtual SAN disk group consumption, through support for direct-attached JBODs. This architecture allows blade server designs to scale Virtual SAN clusters to meet much larger capacity requirements. VMware's Virtual SAN solution is covered in depth in the following chapter.

FIGURE 3.54
Lenovo's Flex SEN with x240 Blade Series

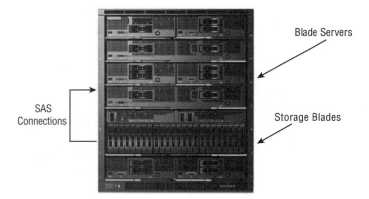

Evaluating Switch Design Characteristics

When evaluating SAN and LAN switch options, many features and considerations must be taken into account before it is possible to select a vendor and appropriate switch model that meets the customer's design requirements. Enterprise-class switches typically come in three form factors:

- Modular switches
- Fixed switches
- Hybrid switches

Modular switches are larger and more flexible by design; they allow you to install different types of line cards to perform different tasks, or simply to add ports. *Fixed switches* have a *fixed* number of ports, which cannot be increased and therefore cannot support as many features or functions as modular switches. *Hybrid switches* have fixed ports but also provide support for installable modules that are similar to, but typically smaller than, line cards. Hybrid switches provide additional flexibility over fixed switches, but without the high cost associated with larger modular hardware.

Evaluating and comparing hardware from various vendors can be challenging, particularly when the vendor's technical terminology describing their features is inconsistent. However, some key common design factors can be evaluated globally, across switch types and vendors. These include, but are not limited to, the following design factors:

Port Density   Port density should be one of the first considerations when evaluating a switch. A smaller number of larger switches might reduce operational overhead, but will also increase the failure domain, so oversubscription may need to be factored into the design. Port density for converged switches is more complex, as they might have different port types specifically for Fibre Channel uplinks. Other switch port types might support all three protocols, providing additional flexibility in the numbers of each device type that can be connected.

Throughput   Throughput is the amount of data that the switch can forward at any one time. This is typically measured in gigabits or terabits per second. If the aggregate bandwidth for the front-end ports exceeds the total switch throughput by a significant margin, the switch might become a bottleneck, as it will not be able to manage all inbound and outbound traffic flows.

Oversubscription   Oversubscription is the ratio of ingress traffic to egress traffic. Depending on the switch architecture, the frames may need to be forwarded to a backplane or arbiter mechanism. Because of space constraints, or sometimes to reduce costs, some switches have more front-end bandwidth than back-end bandwidth, which results in oversubscription. Oversubscription is not necessarily bad. It's typically not common for hosts to sustain the full rate of a switch port continuously during normal operations, although it is not uncommon for ISLs and ports that connect directly to storage devices to do exactly that. As a result, if multiple hosts have simultaneous bursts of network activity, the environment could experience performance degradation as the hosts compete for available back-end bandwidth.

Inter-Switch Links   Inter-Switch Links (ISLs) are critical to the success of a distributed network design. To scale a network beyond that of a single switch, a design needs to use links between the layers of switches. ISLs typically carry traffic for multiple VLANs or virtual fabrics, and are typically built from multiple bonded physical links to create a single virtual link. It is important to understand as part of the storage fabric design which protocols are supported for ISLs, such as the Link Aggregation Control Protocol (LACP), Port Aggregation Protocol (PAGP), EtherChannel, and port channels, as not all switches, vendors, or even switches from the same vendor necessarily support the same options. It is also important to understand how many ports can be aggregated into a single channel, so that the design can allow for the number of switch ports required by links being aggregated into a single virtual bonded connection, for inter-switch connectivity, and therefore the total capacity of the ISLs. In addition, while you can have multiple ISLs in a Fibre Channel fabric, you cannot in a standard Ethernet environment, as it will create loops in the topology. In converged environments, ISL scalability is of particular concern in an edge Fibre Channel over Ethernet topology. This results from the discrete Fibre Channel ports, which connect northbound to a native Fibre Channel SAN, and the ratio between the converged ports and

Fibre Channel ports on the converged switch, which could potentially constitute a source of oversubscription.

Switch Aggregation Switch aggregation allows two or more switches to be merged into a single physical or logical entity, and can be employed by both Ethernet and converged switches. Fibre Channel switches do not need to use this technology, as the Fibre Channel Protocol supports multiple paths between switches.

Finally, you should determine whether the switch supports any additional or advanced features. For instance, with Fibre Channel switches, this could include Fibre Channel over IP (FCIP), Inter-VSAN Routing (IVR), encryption, or other technologies. In an Ethernet environment, examples of advanced features could include layer 3 routing, layer 4 packet filtering, or other similar enhanced technologies.

Fabric Connectivity and Storage I/O Architecture Summary

Having now addressed each of the protocol choices available in a vSphere environment, how do you map the most appropriate storage transport to a specific design or use case, and what type of shared storage is best in a vSphere environment?

As you would expect, no single answer can align with every specific design use case. Each type of shared storage protocol has advantages and drawbacks, and some environments require a combination of several storage protocol types in order to meet complex customer requirements. In fact, vSphere's ability to use a mix of storage protocol types within one infrastructure is one of the competitive advantages that the product offers.

Figure 3.55 illustrates the most common design factors associated with each of the protocols addressed in this chapter.

In addition to these key factors, the following customer-specific design elements can influence your choice of an appropriate protocol:

Familiarity and Manageability Most IT organizations typically favor sticking with or extending what they already know and understand. This is likely to reduce CapEx and OpEx costs, and avoids a complete rip-and-replace of existing infrastructure. In addition, it's much easier to extend the existing technical skills of operational teams (if a good level of expertise is already in place) than to take a whole new approach. If a design does lead an IT organization to use an infrastructure with an unfamiliar protocol, for which preexisting knowledge and experience are not in place, this should be considered a project risk.

Performance As always, infrastructure performance is a key design factor. For almost all use cases, Fibre Channel, FCoE or 10 Gb/s iSCSI is likely to be more than sufficient to meet customer needs. For most environments, assuming it is appropriately designed, the storage transport protocol is not the bottleneck. The storage array system, its front-end cache and back-end disk system are far more likely to provide latency to the environment than the transport protocol itself. However, if performance is a key design factor for low-latency virtualized business-critical applications, such as real-time or rational databases, then Fibre Channel or Fibre Channel over Ethernet is likely to be the preferred choice.

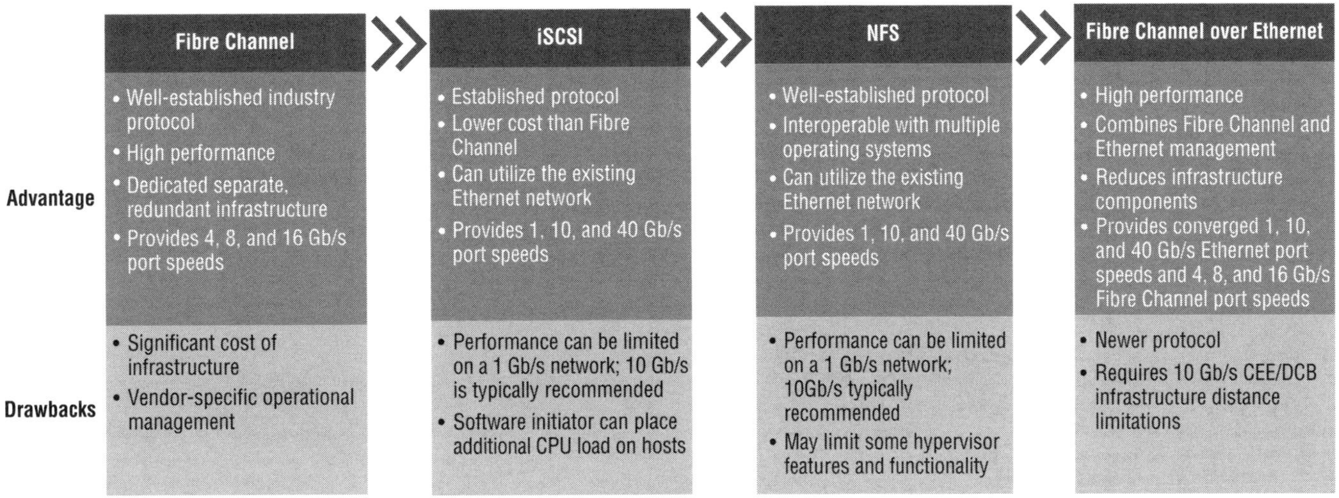

	Fibre Channel	iSCSI	NFS	Fibre Channel over Ethernet
Advantage	• Well-established industry protocol • High performance • Dedicated, separate, redundant infrastructure • Provides 4, 8, and 16 Gb/s port speeds	• Established protocol • Lower cost than Fibre Channel • Can utilize the existing Ethernet network • Provides 1, 10, and 40 Gb/s port speeds	• Well-established protocol • Interoperable with multiple operating systems • Can utilize the existing Ethernet network • Provides 1, 10, and 40 Gb/s port speeds	• High performance • Combines Fibre Channel and Ethernet management • Reduces infrastructure components • Provides converged 1, 10, and 40 Gb/s Ethernet port speeds and 4, 8, and 16 Gb/s Fibre Channel port speeds
Drawbacks	• Significant cost of infrastructure • Vendor-specific operational management	• Performance can be limited on a 1 Gb/s network; 10 Gb/s is typically recommended • Software initiator can place additional CPU load on hosts	• Performance can be limited on a 1 Gb/s network; 10Gb/s typically recommended • May limit some hypervisor features and functionality	• Newer protocol • Requires 10 Gb/s CEE/DCB infrastructure distance limitations

FIGURE 3.55

Storage protocol design factors

Cost   Cost can also be key to making an architectural decision on a storage protocol. NAS-based systems are typically considered the lower-cost option, but many storage arrays provide flexible I/O connection choices, allowing multiple protocols to be employed. Fibre Channel was traditionally considered the costliest option in the data center. However, a dedicated 10 Gb/s iSCSI design or an FCoE converged infrastructure implementation are also likely to require a similar CapEx investment for a greenfield deployment.

Design factors that might also be pertinent to the design relate to the fact that some vSphere features are not available with certain types of shared storage or storage protocol. In addition, keep in mind when addressing the network requirements for each of the protocols discussed that it is critical to understand how they will impact the customer's existing and future network design requirements. From this perspective, storage resources cannot be taken out of context of other functions, such as networking and compute, within the overall data-center architecture.

Finally, as previously mentioned, vSphere does not require a design to use the same storage type throughout the environment. For instance, lower-cost iSCSI or NAS might be the best design choice for storing templates or archived virtual machines, whereas Fibre Channel may be required for low-latency business-critical application use cases. By reviewing the virtual machine workloads and applications as part of the design process, an appropriate storage solution can be targeted in a more granular manner, within a large and complex multiprotocol environment. In addition, as business requirements typically grow and change over time, a multiprotocol approach provides flexibility to accommodate these shifting business needs.

Chapter 4

Policy-Driven Storage Design with Virtual SAN

Enterprise storage system development has been one of the more stagnant, non-innovative, and conservative components of the data center for many years. Many storage vendors have reinvented old technologies, often with a new name, year after year, to keep the IT industry interested and maintain their market share.

This approach clearly does not align well with the software-defined data center (SDDC) model, or an organization that wants to become a truly software-defined enterprise (SDE). The SDE must be agile in its approach and able to react quickly to ever-changing requirements. In the SDE, storage must be consumed and provisioned as a service and also be aware of both the perceived and actual limitations of the applications and services being hosted. In addition, the storage services being delivered in the SDE must be automated, policy driven, integrated, and aligned from the top down to the needs of the application. As a result, a flexible and efficient virtual machine–centric storage system is required for the SDE to thrive. VMware has introduced Virtual SAN and Virtual Volumes to meet this need (see Figure 4.1).

The vision for the software-defined storage (SDS) component of the software-defined enterprise is to simplify operations through policy-driven automation, which in turn enables a more agile approach to storage consumption. This SDS approach must not only simplify the delivery of storage services throughout the organization, but also deliver service levels to individual applications by providing a finer level of control over hardware resources and allowing for dynamic adjustments in real time, as the applications require it. In a nutshell, SDS uses virtualization and automation to remove the unnecessary complexities associated with storage provisioning, and aims to put the application back in charge of its disk-based requirements.

Challenges with Legacy Storage

As addressed in Chapter 2, "Classic Storage Models and Constructs," storage administrators traditionally provisioned and managed storage through the use of large groups of LUNs. When LUNs are provisioned and presented to the vSphere environment, those datastores can be tagged with just one level of service through the use of storage policy–based management. As a result, any virtual disk deployed or migrated to that datastore will receive the same back-end data services that were preallocated by the storage administrator, regardless of the workload requirements of the application.

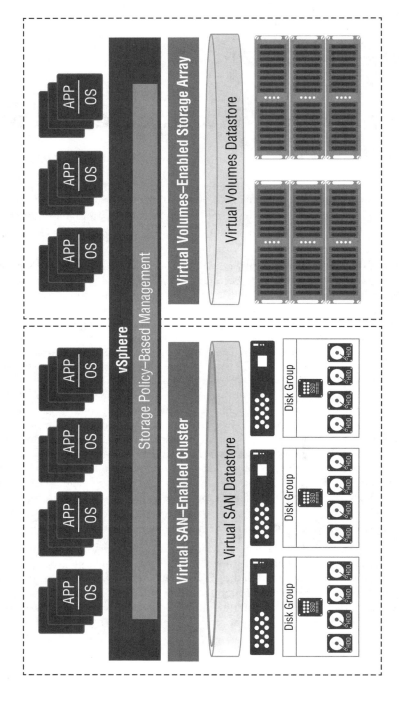

FIGURE 4.1
Software-defined enterprise storage

One of the key principles of the SDS model is that application owners and vSphere administrators are given the tools to select the required class of storage dynamically. To achieve this, VMware has introduced a control plane that can access a single policy catalog, which enables you to publish the various service levels offered by the environment. VMware refers to this feature as *storage policy–based management*, or SPBM. The aim of SPBM is to allow vSphere storage administrators to create multiple storage policies, each providing a different service level with assigned capabilities. This is then combined with the virtual machine–centric storage management platforms provided by Virtual SAN and Virtual Volumes, allowing the administration team to offer differentiated services to each virtual machine or even virtual disk.

The aim of these new storage platforms is to relieve vSphere storage administrators of the daily challenges previously seen when using the classic storage model, examined in Chapter 2. Before we progress, let's first address some of the storage-related challenges faced by enterprise IT organizations and service providers, which traditionally exacerbated these pain points and led us toward these newer technologies.

When using the classic storage model, it can

- Be difficult to align storage capabilities with application requirements
- Require specialized storage hardware, which has often been inefficiently used and operationally complex to maintain
- Be unreliable, and sometimes have unpredictable performance
- Be storage device–centric, creating vendor storage silos, which has led to a lack of granular control over the environment
- Lack end-to-end management visibility through a single pane of glass
- Result in manual workload provisioning
- Require complex processes, which are slow and lack the ability to be automated in a meaningful and helpful way
- Require specialized hardware, operational teams, and costly vendor support

The next question to ask is, how does SDS allow organizations and service providers to be more efficient in managing their storage infrastructure and bring an end to these challenges outlined?

First, the SDS vision is divided into two components: the *control plane*, which provides policy-driven automation and dynamic control across all storage subsystems, and the *data plane*, which is built using virtualized hardware in a distributed, resilient architecture.

This new control plane uses the SPBM mechanism for seamless automation and orchestration. The data plane abstracts and pools physical storage resources into a flexible virtual datastore, while at the same time the data plane dynamically makes adjustments to underlying storage pools, ensuring that policies are compliant and service-level agreements (SLAs) can be met. These virtual datastores can either be provisioned through the Virtual Volumes model, on a shared SAN or NAS storage system, or be hypervisor-converged, enabled by Virtual SAN.

The second component to SDS is providing virtual machine–level data services, such as replication, snapshots, and caching. The final vision is to enable an application-centric approach, using a common policy-based control plane. This is facilitated through SPBM: the storage requirements for each virtual machine are captured in a simple and intuitive policy, which then follows that virtual machine throughout its life cycle, on any supported vSphere infrastructure.

Another key driver for SDS is the level of integration and interoperability between VMware's software layer and its storage ecosystem. Integrating storage through APIs , in order to extend capabilities on top of the core vSphere infrastructure, such as through a cloud management platform, is key to automation and scalability for both service providers and large enterprise IT organizations.

With this approach, the SDS concept is able to provide an operational that is built on the integration of policy-driven automation, virtualization of external hardware, and the abstraction of virtual machine–centric storage pools. The final aim of this model is to streamline the storage deployment and provisioning process to allow for more-flexible on-demand consumption of storage resources, provide application-specific SLAs, and ultimately, to lower the total cost of ownership of storage infrastructure for enterprise IT organizations and service providers.

Policy-Driven Storage Overview

As you have already seen, storage in the vSphere infrastructure is changing and has been doing so for some time. Traditional LUN-based storage mechanisms applied storage capability at the datastore level, as the SCSI LUN was presented to the host or multiple hosts. The underlying storage array had no knowledge of (or at best, limited integration with) the hypervisor, its filesystem, guest operating systems, or workloads. With this classic storage approach, it was left up to the hypervisor and vCenter Server, or other management tools, to map the various files, such as VMDKs, to the corresponding extents, pages, and logical block address (LBA) understood by the storage system. In addition, when a NAS solution was used, an additional layer of abstraction was placed over the underlying block storage, to handle the file management mechanism and the associated file-to-LBA mapping activity.

This new policy-based storage-provisioning mechanism, provided by Virtual SAN, often also referred to as VSAN and Virtual Volumes, which can often be seen represented as VVOLs in some documentation, moves us forward significantly in the journey toward the SDDC by providing storage-driven policy management. Instead of a storage system that simply presents a SCSI LUN or NFS mount point and has limited (VASA 1) or no visibility into how the underlying storage array is presenting its disks, this shift toward a more intelligent relationship between the hypervisor and storage gives us our next generation of storage architecture, which is far more aware of the virtual machines and their workload requirements than ever before.

This policy-driven storage mechanism is one of the foundational concepts of VMware's new generation of storage offerings. As already highlighted, the primary use case for employing policy automation for storage placement is the ability to deploy virtual disks onto the correct tier of storage, based on their application requirements and service levels, which is a key deliverable for the software-defined enterprise. This capability is provided by the SPBM mechanism, which will allow operational teams to become more efficient and automated in their approach to the initial placement of virtual machines, ongoing maintenance, and changing requirements of workloads.

In the software-defined enterprise, fully automated placement of virtual machines through SPBM means that no one will ever need to be looking for the correct cluster to place an application on, or be trying to establish whether it resides on the right set of datastores. VMware's SPBM mechanism comprises the following key components:

- A common policy framework across all virtual machines, whether they reside on Virtual Volumes, Virtual SAN, or classic VMFS storage volumes
- A common API layer for cloud management platforms (CMPs) such as vRealize Automation, OpenStack, vCloud Director for Service Providers, scripting users (PowerShell, JavaScript, Python, and so on), and orchestration platforms including vRealize Orchestrator
- Represents application and virtual machine service-level requirements
- Consumes capabilities published via VASA 1 and 2

With SPBM, vSphere operational teams can create profiles and present them to the cloud and automation layers of the management platform, thus providing storage consumers with service levels instead of technical details. Under the covers, SPBM ensures that the storage system conforms to the policies embedded in the storage profile, as well as doing the following:

- Enabling a stable, robust, repeatable, and standardized storage platform
- Providing intelligent placement and control of services and capabilities at the virtual machine or virtual disk level
- Shielding the automation and orchestration changes from operational teams, by abstracting them into the underlying storage

The SPBM framework allows vSphere administrators to start by creating a storage policy, which could be a nontechnical storage class, such as *tier-1-business-critical* or *test-noncritical*. These storage policies can not only define performance levels, but also provide far more granular features, based on the capabilities of the underlying storage hardware or software, such as the *percentage of cache reservation* or *replication RPO threshold*.

Availability capabilities, including recovery point objective (RPO) and recovery time objective (RTO), encryption, and retention, can also be delivered through the SPBM framework, allowing storage administrators to create policies that are aligned with application and business requirements for data protection. In addition, SPBM ensures that the disk subsystem complies with the assigned workload service levels, which are embedded into the storage policy's capabilities.

Later in this chapter, we cover the capabilities of SPBM and how it relates specifically to Virtual SAN. Chapter 8, "Policy-Driven Storage Design with Virtual Volumes," shows how this framework is used by Virtual Volumes in order to take advantage of third-party storage capabilities.

VMware Object Storage Overview

VMware's next-generation policy-driven storage offerings are no longer based around classic VMFS storage volumes, but are based on an object storage system model. However, that doesn't mean that a Virtual SAN or Virtual Volumes datastore is like other object-based storage systems, such as Amazon's S3.

This new generation of virtual datastores are not LUNs, like the regular block storage systems we examined in Chapter 2. Likewise, a virtual datastore or distributed datastore, which is covered in more detail later in this chapter, is not accessed like any of the various object storage solutions available, such as Amazon's S3. Instead, this new generation of datastore is a VMware-specific implementation of object-based storage, and while Virtual SAN and Virtual Volumes both provide this new form of object storage access, not all object storage is the same. A multitude of object storage access types and architectures are available, which all differ from the traditional block LUNs and NAS volumes addressed previously in Chapter 2. Even though VMware uses terms such as *object* and *object storage*, in the context of Virtual SAN and Virtual Volumes these are not the same as other object storage solutions.

This newer object store approach from VMware is completely software defined, which allows directly attached commodity disks inside host servers to be used to provide cluster-wide persistent shared storage to objects, in addition to storage arrays. This approach also provides solid enterprise features to IT organizations and service providers, allowing them to transform their legacy storage platforms into a more service-oriented infrastructure by delivering the following:

- Commodity-based object storage resources
- RESTful API access to objects
- Enterprise-class levels of availability
- Bucket and object versioning and object metadata

Virtual SAN Overview

Virtual SAN has a relatively short history. VMware first introduced its Virtual SAN product with the launch of vSphere 5.5 Update 1, but development of the platform has been aggressive, with a number of releases since then.

At the time of writing, four major releases of Virtual SAN have been launched, as outlined in Table 4.1.

TABLE 4.1: Virtual SAN major releases

GENERATION	VIRTUAL SAN RELEASE	vSPHERE RELEASE	FIRST AVAILABLE
First generation	Virtual SAN 1.0	vSphere 5.5 Update 1	March 2014
Second generation	Virtual SAN 6.0	vSphere 6.0	March 2015
Third generation	Virtual SAN 6.1	vSphere 6.0 Update 1	September 2015
Fourth generation	VMware Virtual SAN 6.2	vSphere 6.0 Update 2	March 2016

To establish a baseline, this book focuses on Virtual SAN 6.2 with vSphere 6.0 Update 2. However, when pertinent or of interest, previous releases are also referenced.

VMware Virtual SAN is a SDS feature built into the vSphere hypervisor, enabling locally attached storage to be configured to create a cluster-wide distributed datastore. While vSphere abstracts and aggregates CPU and memory into logical pools of compute resources, Virtual SAN, embedded into the hypervisor's VMkernel, pools together server-attached disk devices to create a high-performance, cluster-wide distributed datastore on which virtual machine workloads can reside with cloud agility and efficiency.

To be able to leverage Virtual SAN in a vSphere environment, you must associate at least one flash device and one capacity disk in the hosts that are participating in the cluster. The flash device does not contribute to the storage capacity, but instead acts as a read-caching and write-buffering mechanism. The aggregation of disk devices from each host server in the Virtual SAN cluster forms a distributed Virtual SAN datastore. If designed correctly, Virtual SAN has no single point of failure, as at the core of the technology is a distributed object-based redundant array of independent nodes (RAIN) architecture.

This approach can easily meet the shared storage requirements of the most demanding infrastructure at a lower cost than classic SAN or NAS storage devices. The Virtual SAN hyper-converged infrastructure architecture has the following key features:

- The system is inherently fault tolerant. If designed correctly, there are no single points of failure, and failures can be handled without downtime.
- The entire system is tightly integrated with, and automated by, vCenter Server.
- Locally attached storage is aggregated from the ESXi hosts in the cluster to create a distributed datastore.
- Virtual SAN is flash optimized. The sole purpose of the solid-state flash devices is to provide I/O acceleration.
- The solution is based on virtual machine–centric data operations and policy-driven management principles.
- Virtual SAN is fully integrated with vSphere at the VMkernel layer.

Virtual SAN, as a converged-hypervisor product, delivers virtual machine storage and compute through the same $x86$ server platform running the vSphere hypervisor. One of the key differentiators for Virtual SAN, over most other third-party HCI products, is that it was built from the ground up to be a fully integrated component of the ESXi VMkernel, specifically designed to host virtual machine storage resources. As a result of this tight integration, Virtual SAN seamlessly interoperates with the entire vSphere stack, including features such as vMotion, High Availability (HA), Symmetric Multiprocessing Fault Tolerance (SMP-FT) and Distributed Resource Scheduler (DRS). In addition to this high level of integration, Virtual SAN is also firmly focused on delivering high performance, enterprise-class features including availability and scalability, without compromising on the stability required to deliver virtualized business-critical applications.

The remainder of this chapter focuses on the components pertinent to a Virtual SAN storage design and examines the technology itself, its architecture, and implementation considerations. Any design recommendations made in this section are based on VMware best or common practices when used in the context of the SDS model.

Virtual SAN Architecture

The Virtual SAN storage platform operates through various easy-to-understand components and technologies, all of which need to be designed and considered both independently and in conjunction with one another. This approach to design allows each Virtual SAN component technology to interact seamlessly together, in order to provide the high levels of performance, continuous availability, and the low levels of latency customers require from their entire storage platform. The first of those components to be addressed are disk groups.

Virtual SAN Disk Groups

Virtual SAN uses *disk groups*, shown in Figure 4.2, to pool together flash drives employed as caching devices, and mechanical or flash drives employed as capacity devices, as a single management construct.

The exact configuration of a disk group can vary based on several design factors. For instance, a Virtual SAN hybrid disk group is composed of one write-buffering and read-caching solid-state *flash* device (SSD) and at least one mechanical disk capacity device (also referred to as a high-density disk, or HDD), magnetic disk, or spindle to provide persistent storage. In a hybrid disk group, the solid-state flash device is used purely to provide a performance layer, and acts as a read cache and write buffer, whereas the capacity devices (mechanical disks in the hybrid model) are used to provide persistent storage capacity. Each vSphere host can support up to five disk groups, with each disk group comprising a single caching device, and between one and seven capacity devices, as illustrated in Figure 4.3.

Virtual SAN 6, the second generation of VMware's hyper-converged storage platform, introduced an *all-flash* disk group configuration in addition to the original mixed-disk *hybrid* option. This all-flash architecture enables a Virtual SAN datastore to be carved entirely out of solid-state flash devices. Flash drives are employed for both tiers: write buffering at the endurance flash layer, and persistent data storage at the capacity flash layer.

The Virtual SAN all-flash disk group configuration follows the same model as the hybrid offering, in that it is made up of two tiers. As illustrated in Figure 4.4, the key difference with the all-flash configuration is that flash devices are used on both tiers of the two-tier setup, providing endurance-based write buffering and capacity. The capacity flash devices are used for persistent data storage, similar to magnetic disks in the hybrid model. Like the hybrid model, all-flash disk groups cannot be created without an endurance flash device defined during the disk group configuration.

As part of designing a Virtual SAN, you are required to make appropriate design decisions relating to the type and capacity of disks in each disk group, the number of disk groups per host, and the number of disks required in each disk group. By providing these key design metrics, you can dictate the ratio of flash to magnetic disk, or endurance flash to capacity flash, which is required for the design's final configuration.

The more capacity disks that are configured in a disk group for persistent storage, the higher the SSD requirement for the caching and buffering mechanism, in order for the environment to maintain an acceptable level of read and write performance. Having a low ratio between the two storage tiers typically leads to a higher cost to deliver the design, because of the caching device per disk group limit of one. Such a design results in more endurance flash devices being required, as each disk group requires one flash device for caching and at least one device for capacity. However, as we will go on to address, the lower the ratio used, the higher the expected performance characteristics of the storage platform, as a direct result of higher levels of data being maintained on the flash devices.

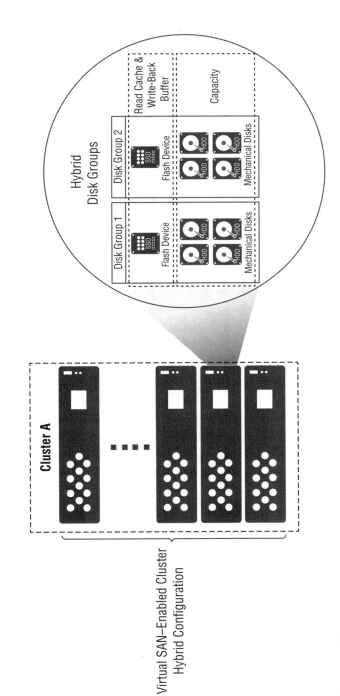

FIGURE 4.2
Disk group configuration

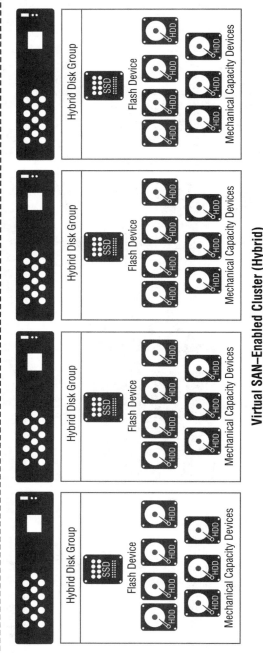

FIGURE 4.3
Virtual SAN hybrid disk group configuration

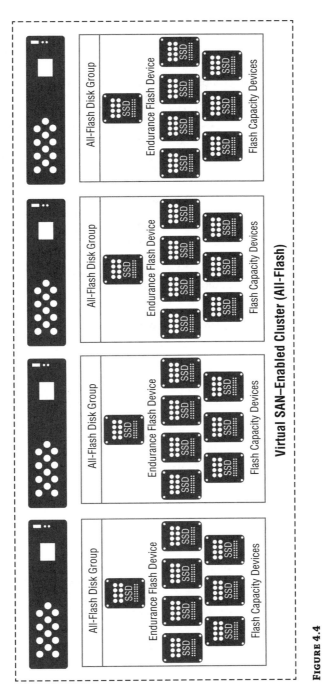

FIGURE 4.4
Virtual SAN all-flash disk group configuration

When determining the number of disk groups per host and the capacity-to-flash ratio, you must also consider the amount of storage capacity required on the Virtual SAN datastore. This in turn requires availability design factors, such as defining the Number of Failures to Tolerate storage policy, which is addressed in detail later in this chapter.

The optimum number of disk groups per host and the correct ratio between endurance flash devices and capacity drives can be established after the customer's requirements for performance and capacity have been identified. The key design factors driving these decisions are based on finding the appropriate balance between performance, availability, capacity, and cost, while still achieving all of the customer's other requirements.

By adding more capacity disks per disk group, the design will not only increase the total space available on the Virtual SAN datastore, but also improve virtual machine availability, by providing additional options for object and component placement across the host. Alternatively, employing a larger number of smaller disk groups in each host will likely significantly improve performance, as a result of a higher percentage of working data being cached. But the cost of this methodology is likely to quickly become prohibitive, as a result of the increased number of expensive endurance flash devices required for the design.

Depending on the number of capacity disks configured in each disk group, and the number of disk groups configured per host, the total amount of storage space in the configuration can be quite significant. As a starting point for a typical workload design, configure the disk groups based on the following:

- The total storage capacity required on the datastore.
- The tolerance for failure required in the design. This will include the Number of Failures to Tolerate (FTT), fault domain (FD) configuration, and stretched Virtual SAN cluster use cases.
- The overall percentage of flash capacity per disk group.

In the absence of any specific customer requirements, a good starting point is to use two disk groups per host, each composed of three capacity disks and one caching device. This approach provides each host with two failure domains and increased IOPS. Each host would contain two flash devices for caching, and six flash or mechanical disks for capacity, as illustrated in Figure 4.5. However, you can use more or fewer disks per disk group, as defined by the customer's workload and capacity requirements. As the architect, it is your role to ensure that the customer's sizing and workload prerequisites for the Virtual SAN platform are properly met by the design.

The ratio between capacity disks and the caching device should typically depend on specific customer use cases and workload characteristics. However, the best-practice guideline from VMware is to use a total caching and buffering capacity that is a minimum of 10 percent of the total consumed persistent capacity in each disk group, in both the hybrid and all-flash configuration. This is to be calculated before taking into account the replica copies of object components configured by the FTT storage policy. Datastore sizing calculations and the caching and buffering formulas for endurance flash devices are covered in more detail later in this chapter.

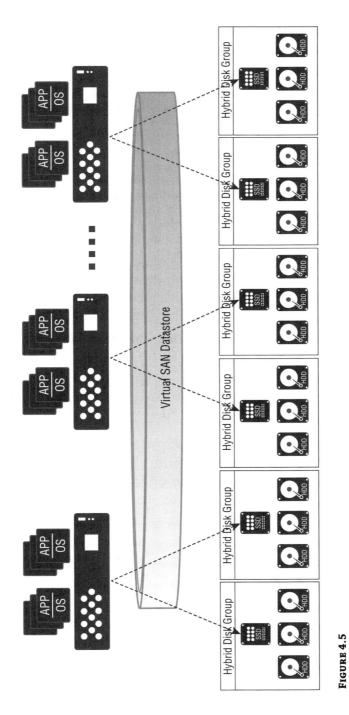

FIGURE 4.5
Disk group configuration example

Comparing Virtual SAN Hybrid and All-Flash Models

As highlighted previously, the first VMware release of Virtual SAN, with vSphere 5.5 Update 1, had only one disk group configuration option, which we now refer to as the *hybrid* model. However, with the release of Virtual SAN 6, support for two disk group configuration options was made available: the original hybrid configuration, which leverages both flash-based devices and mechanical disks, and the new all-flash model, which uses two tiers of flash-based storage.

At the time of writing, the hybrid configuration is the most common approach used in Virtual SAN design. Hybrid solutions use a single flash-based device to provide the caching tier, and one or more mechanical disks to provide capacity and persistent data storage. In an all-flash configuration, however, flash-based devices are used for both write buffering and capacity, although the type of device for each function will differ. The cost of implementing an all-flash solution can be significant, with 10 Gb/s networking and Advanced Virtual SAN licensing being mandatory, in addition to the flash storage devices themselves. For this reason, at the time of writing at least, this model appears to be deployed for only very specific workload use cases, but expect this to change in the not-too-distant future.

However, for this additional cost, the all-flash Virtual SAN solution brings with it another level of improved, highly predictable, and uniform performance, regardless of workload type, when compared with the hybrid model. Both hybrid and all-flash Virtual SAN clusters include the same VMware recommendation of a minimum of 10 percent of consumed capacity for the caching layer. However, the way in which cache is used across the two Virtual SAN disk group models is different.

In hybrid designs, the caching algorithm maximizes both read and write performance by allocating 70 percent of the cache for storing frequently read disk blocks, which aims to minimize access to the slower magnetic disks. The remaining 30 percent of available flash is used for write buffering (see Figure 4.6). Even though this is a configurable parameter in Virtual SAN, it is strongly recommended that the ratio between read caching and write buffering is not modified, unless explicitly recommended to do so by VMware support, based on specific workload characteristics.

FIGURE 4.6
Anatomy of a hybrid solution read, write, and destaging operation

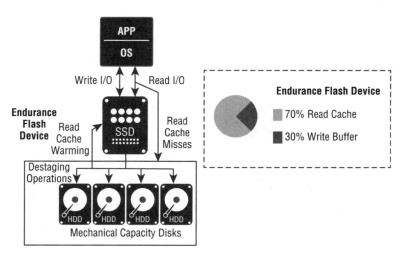

In addition to the read-caching mechanism offered by Virtual SAN, the Virtual SAN caching algorithm also maintains metadata in host memory, in that a small amount of host RAM is used to provide an in-memory read cache to the most recently accessed cache lines from the SSD's read cache. This in-memory cache is assigned dynamically, based on the memory available in the host system. In Virtual SAN 6.2, 0.4 percent of the host's available memory is used in this way, up to a maximum of 1 GB, with the memory blocks being used on the host where the virtual machine is running.

In all-flash solutions, two types of flash are used: a very fast and durable write cache, referred to as the *endurance flash device*, and larger more cost-effective capacity flash, referred to as *performance-class devices*.

In a Virtual SAN hybrid configuration, the primary purpose of the flash drive's caching and buffering mechanism is to enhance performance. However, in an all-flash configuration, this purpose changes to providing enhanced endurance. Of course, performance is also typically enhanced by implementing a higher-functioning drive to deliver the write cache mechanism. However, endurance remains the primary goal. By providing a single high-endurance drive as part of each disk group, none of the heavy I/O-intensive write workload touches the capacity flash disks directly. A lower grade of disk can be used, therefore lowering the overall cost of the solution.

In contrast to the hybrid solution, in an all-flash configuration, 100 percent of the capacity offered by the endurance device is allocated for write buffering. The flash-based endurance caching tier is intelligently used as a write-buffer only, while the second tier of capacity SSDs form the persistence storage layer. As illustrated in Figure 4.7, in the all-flash architecture, read requests are primarily handled directly by the capacity flash devices. The read speeds provided by these SSDs should be almost as quick as if the data was being read from the endurance tier, making the algorithm more than sufficient to handle even the most demanding of read-intensive workloads. However, if data is freshly written and is still *hot* (it has not yet been destaged), the read request will have to be served from the write cache tier.

This significant difference in architecture between the hybrid model and all-flash configuration results in no read cache misses. In the hybrid model, this can cause performance degradation if the ratio of flash to mechanical disk is not optimal for the workload. This difference in architecture also means that there is no requirement to move the blocks from the capacity layer up to the caching layer, to prestage the read buffer with data, which is what happens in a hybrid configuration.

FIGURE 4.7
Anatomy of an all-flash solution read, write, and destaging operation

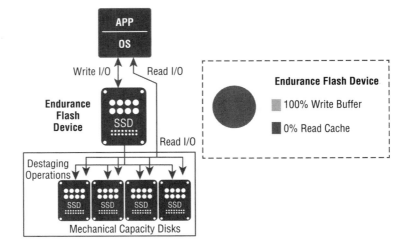

In summary, the all-flash architecture allows the tiering of SSDs, a write-intensive high-endurance caching tier for the writes, and a read-intensive cost-effective SSD tier for data persistence, which results in reducing the overall cost of the architecture. By using a Virtual SAN all-flash architecture, in which flash devices are intelligently used for both caching and data persistence, it becomes possible to deliver high performance through predictably low latencies, with a sub-millisecond response time, while also achieving an extremely high I/O per second, of typically up to 90K per host. As a result, this makes an all-flash Virtual SAN architecture ideal for performance-intensive workloads or tier 1 business-critical applications with specific performance requirements that must be achieved through the design.

In addition, the following key design factors must be taken into account when considering an all-flash solution:

- At least vSphere 6 must be used with Virtual SAN Advanced licensing.
- 10 Gb/s networking is a strict requirement.
- The maximum number of host nodes in an all-flash configuration is 64.
- The Flash Read Cache Reservation policy is not used in an all-flash solution.
- All drives need to be tagged/marked as flash.
- Endurance drive specifications are an important consideration for the caching tier.

All-Flash Deduplication and Compression

Deduplication and compression for all-flash solutions were introduced with the Virtual SAN 6.2 release. This technology provides software-based deduplication and compression to help optimize all-flash storage capacity by lowering the total cost of ownership through the reduction of the storage footprint.

Solid-state flash devices are, at the time of writing, still expensive in terms of dollars per gigabyte, but more cost-effective when you consider the I/O per second. For instance, say you purchase a 200 GB SSD for $400, which is capable of serving 45K IOPS (equal to $0.004 per I/O operation). Then compare that with acquiring a single 1 TB mechanical disk for half of the same cost, which can deliver 100 IOPS (equal to $1.00 per I/O operation). As you can see, the most cost-effective way to use an all-flash configuration is to maximize I/O per second, and not capacity.

In addition, flash devices have a finite life expectancy. This longevity is measured by the number of writes the device can tolerate over its expected lifespan. Technologies that reduce the amount of data written, such as deduplication and compression, can extend hardware life expectancy and therefore lower the total cost of ownership per I/O operation. Using this technology can help meet these endurance device design challenges, by providing up to 7× data reduction in highly duplicated virtual desktop infrastructure (VDI) environments with minimal impact on the host CPU and memory consumption.

Deduplication and compression operations occur during the destaging of data, from the endurance write cache tier to the capacity performance tier of flash devices, when you enable Space Efficiency at the cluster level, on a Virtual SAN–enabled cluster.

Deduplication and compression are disabled by default. You can enable them, without downtime, as a single cluster-wide configurable option by modifying the parameters shown in Figure 4.8. However, it is important to recognize that doing so initiates a rolling reformat of each disk group within the cluster. This might not only take a significant period of time, but also affect current virtual machine I/O operations.

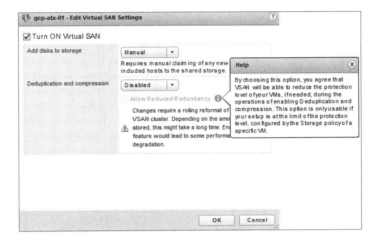

FIGURE 4.8
Deduplication and compression web client configuration

The scope of deduplication and compression exists only within each disk group. Therefore, multiple copies of a data block within the same disk group are reduced to a single copy, with multiple copies across disk groups not being deduplicated. As a result, larger disk groups typically result in a higher ratio of deduplicated data.

Maintaining the deduplication domain within the boundaries of a single disk group avoids the architectural requirement to implement a global lookup table across the entire cluster, at a significant resource overhead, and therefore frees up compute resources for other operations. In addition, *write hot data*, which is data currently only in write cache, is not deduplicated or compressed, in order to further avoid wasting compute resources.

After the data blocks are deduplicated, they are compressed by the LZ4 lossless data-compression algorithm. Deduplication occurs at 4 KB block sizes. Compression is then performed for each unique 4 KB block. If the resulting block is less than or equal to 2 KB, the compressed block will be saved in place of the 4 KB block. However, if the block output after compression is greater than 2 KB, the data block will be written to disk uncompressed and tracked accordingly. Virtual SAN will commit compressed data to disk only if the deduplicated 4 KB block can further be reduced to 2 KB or less. If not, the block is written uncompressed, in its original unique 4 KB state. This mechanism serves to avoid block-alignment issues across the capacity drives, as well as reduce the CPU overhead for data decompression, which is higher when also compressing data with low compression ratios (see Figure 4.9).

The additional CPU cycles required to compress and uncompress data adds performance overhead to the host that is typically expected to be less than 5 percent (which in most designs is negligible). You also have to take into account that the decompression operation typically occurs in the latency-sensitive I/O read-request path, which might result in some overall guest performance degradation.

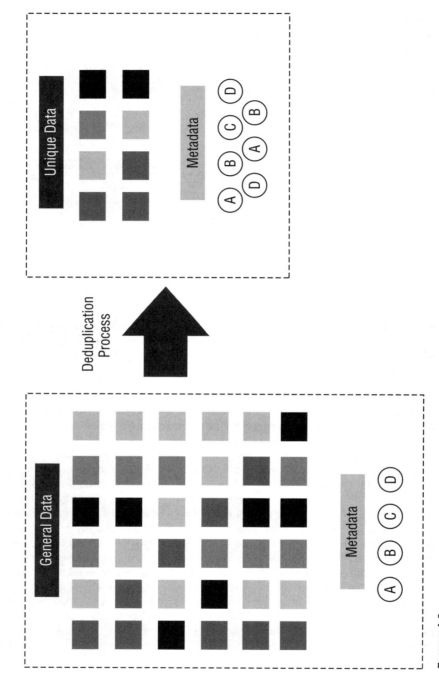

FIGURE 4.9
Deduplication mechanism

The use of deduplication and compression also has an implication on SPBM, Object Space Reservation capability. Any value that has been configured other than the default of 0 percent or maximum of 100 percent (any other setting from 1 percent to 99 percent) is not supported when deduplication has been enabled on the cluster. In addition, if an object has been assigned an Object Space Reservation of 100 percent, that object will be analyzed for deduplication (as the entire capacity for that object has been reserved), but no capacity reduction will occur.

When deduplication and compression are enabled as part of a design, capacity flash requirements can be reduced by between 5× and 7×. However, this is completely dependent on the running workloads. The actual savings in capacity tier flash will vary significantly, depending on the types of data and how much this data can be compressed, as well as the number of duplicated blocks and the distribution of these blocks across disk groups. For this reason, the best results are typically seen in VDI or similar environments, where high levels of block duplication often exist.

Data Locality and Caching Algorithms

Virtual SAN uses a distributed persistent cache mechanism on the caching devices across the cluster, in front of the capacity disks where the data replicas live. This distributed caching mechanism results in better overall utilization of the caching device, which is the most valuable storage resource in the cluster.

The Virtual SAN architecture does not implement a persistent client-side read cache, which is only local to the host on which the virtual machine is running, as many other similar third-party solutions do. There are several design- and engineering-based reasons that this is the case:

- Poor balancing and utilization of flash devices, for both capacity and performance, across the cluster.

- After a vMotion operation, the virtual machine would incur a significant performance impact, while the cache rewarmed on the new host.

- Only limited benefits in performance and reduced latency can be gained by implementing a local-only, client-side read cache.

Virtual SAN uses a distributed read cache mechanism whereby reads and writes are distributed to all hosts holding replicas. This way, if one host is busy, the other hosts holding replicas can still service I/O requests.

In addition, with vMotion or DRS, when virtual machines are migrated across hosts in the cluster in order for them to utilize their compute resources more effectively, the caching mechanism used by Virtual SAN means there is typically no requirement to migrate gigabytes of data across hosts in the cluster. As a result, little or no impact is caused to I/O flow, and therefore no additional performance overhead is incurred. Effectively, this means not having to rewarm caches every time a virtual machine vMotion operation occurs, relocating a virtual machine to a new host.

The reason this mechanism works so well in a Virtual SAN environment is the nature of the distributed storage system. Virtual SAN locates data in two or more locations across the distributed Virtual SAN datastore, in order to withstand host or disk failures. With data locality operating in this way, I/O can come from any of the data replicas across the cluster, helping to mitigate potential host or disk bottlenecks and allowing Virtual SAN to run more efficiently, while still maintaining data availability and optimum performance.

The next question you might ask is, what about the latency the network introduces, if data is being accessed from a host, other than where the virtual machine resides? Typical latencies in 10GbE networks are in the range of 5 to 50 microseconds, and the specifications of endurance flash devices typically indicate latencies in the range of 50 to 100 microseconds. When you operate endurance flash devices at thousands of I/O per second, adding approximately 10 microseconds on top of a 1 millisecond latency makes no significant difference in terms of virtual machine performance.

However, it is noteworthy that the metadata stored in host memory (highlighted earlier in this chapter), which provides a small in-memory read cache to the most recently accessed cache lines from the SSD's read cache, is used on the host where the virtual machine is running, making this mechanism a *client-side* cache operation.

Virtual SAN Destaging Mechanism

Data from committed writes from different virtual machine disks accumulate quickly on the endurance flash device's write buffer space, which is 30 percent by default. As this space fills up, the newly written data residing on the cache device must be destaged to the capacity layer, which may be flash or mechanical disks, depending on the disk group model used in the design. The mechanism for destaging differs between the two Virtual SAN models, hybrid and all-flash; mechanical disks are typically good at handling sequential write workloads, so Virtual SAN uses this to make the process more efficient.

In the hybrid model, an elevator algorithm runs independently on each disk group and decides, locally, whether to move any data to its capacity disks, and if so, when. This algorithm uses multiple criteria and batches together larger chunks of data that are physically proximal on a mechanical disk, and destages them together asynchronously. This mechanism writes to the disk sequentially for improved performance. However, the destaging mechanism is also conservative: it will not rush to move data if the space in the write buffer is not constringed. In addition, as data that is written tends to be overwritten quickly within a short period of time, this approach avoids writing the same blocks of data multiple times to the mechanical disks. Also note that the write buffers of the capacity layer disks are flushed onto the persistent storage devices before writes are discarded from the caching device.

In the all-flash model, Virtual SAN uses 100 percent of the available capacity on the endurance flash device as a write buffer. In all-flash configurations, essentially the same mechanism is in place as that in the hybrid model. However, Virtual SAN does not take into account the proximal algorithm, making it a more efficient mechanism for destaging to capacity flash devices. Also, in the all-flash model, changes to the elevator algorithm allow the destaging of cold data from the write cache to the capacity tier, based on their data blocks' relative hotness or coldness. In addition, data blocks that are overwritten stay in the caching tier longer, which results in reducing the overall wear on the capacity tier flash devices, increasing their life expectancy.

Virtual SAN Distributed Datastore

The *Virtual SAN datastore* is an object store that is presented to vSphere hosts as a filesystem. This object store mounts the volumes from all the hosts in the Virtual SAN–enabled cluster, and presents them as a single shared datastore that is distributed and visible across all nodes. This is an important point and worthy of repeating: Virtual SAN simplifies the storage configuration, as there is only a single datastore for virtual machines. This distributed datastore uses storage from each vSphere host in the Virtual SAN cluster, which is configured with a disk group and stores all of the virtual machine object files across this single storage entity.

The Object Storage File System (OSFS) enables VMFS volumes from each vSphere host to be mounted as a single datastore. There are no directories in the OSFS. Data on a Virtual SAN datastore is stored in the form of data containers, referred to as *objects*, which are logical volumes that have their data distributed across hosts in the Virtual SAN cluster. An object can be a VMDK file, VM swap, snapshot, or virtual machine namespace. For each virtual machine on a Virtual SAN datastore, an object is created for each of its virtual disks. A container, called the *virtual machine namespace object*, is also created that maintains a VMFS volume and stores the virtual machine metadata files. I address objects in more detail shortly in this chapter.

For all intents and purposes, a Virtual SAN datastore appears the same as any traditional block-based VMFS datastore in most views in the vCenter Web Client, which is the only interface which allows the configuration, management, and monitoring of the Virtual SAN environment. The classic Windows C# client is not supported for Virtual SAN operations.

All hosts in the cluster, and *only* hosts in that cluster, can view the distributed Virtual SAN datastore. It is not required that all hosts contribute to the storage, but it is typically recommended. The recommended configuration is typically to maintain a consistent host configuration, in terms of disk groups, CPU, memory, storage controller, and network adapters, across all nodes in the Virtual SAN cluster (see Figure 4.10).

Each Virtual SAN cluster results in only a single datastore entity in vCenter Server. But vCenter Server can manage multiple vSphere clusters with Virtual SAN datastores, each to meet different customer requirements, with different performance capabilities and characteristics, and with different default storage profiles assigned, as illustrated in Figure 4.11.

Objects, Components, and Witnesses

Virtual SAN is built on the concept of objects and components for the storage of virtual machines. An object is composed of one or more components, which are distributed across the Virtual SAN datastore, based on the assigned storage policy for that object. In a Virtual SAN–enabled cluster, each virtual machine can be made up of four object types, as described in Table 4.2.

TABLE 4.2: Virtual SAN object types

OBJECT	DESCRIPTION
Virtual machine home Namespace object	Location for the virtual machine configuration and log files.
Virtual machine swap object	Created for the virtual machine swap file. This object type is created only when the virtual machine is powered on.
VMDK object	Stores the data that is on a virtual disk.
Snapshot delta VMDK objects	Created when a virtual machine has a snapshot created on it. A memory object is also created when the snapshot memory option is selected, when creating or suspending a virtual machine.

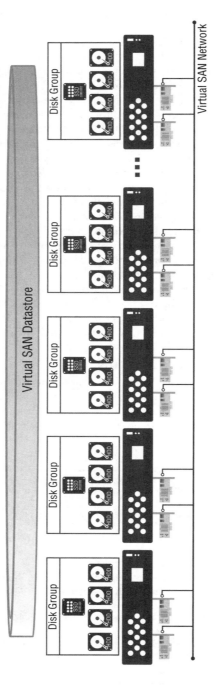

Figure 4.10
Virtual SAN distributed datastore

FIGURE 4.11
Multiple virtual SAN datastore design

Each component can be a maximum of 255 GB. If an object exceeds this size, it is split into multiple components and is therefore, by virtue of size, striped. In addition, objects that are larger than the capacity of a single physical disk are automatically divided into multiple components. This might be particularly pertinent in an all-flash configuration. For instance, if a disk group is made up of 200 GB capacity SSDs, and one of the resident virtual machines includes a virtual disk that exceeds this 200 GB limit, then this virtual disk, also by virtue of size, will be required to be striped across multiple capacity flash devices. Objects are also split into multiple components, based on their performance and availability requirements, as defined in their assigned storage policy.

An object's component parts are automatically written across multiple disk groups and cluster nodes by utilizing Virtual SAN's distributed RAIN architecture, which uses two techniques: striping (RAID 0) and mirroring (RAID 1), as illustrated in Figure 4.12. These two techniques allow a Virtual SAN–enabled cluster to tolerate failures, accommodate large objects, and meet performance requirements. The number of component replicas and stripes created is based on the factors outlined previously, and on the object policy definition set out through the SPBM framework, which is covered in more detail later in this chapter.

In Virtual SAN 6.0, 6.1, and 6.2, the maximum component count per host is 9,000, with more components required for larger and more highly redundant virtual machines. This can be a design constraint that restricts how far a single host can be scaled up, in terms of the number of disk groups per host and the number of large mechanical drives per disk group. For instance, at a maximum of five disk groups per host, and seven capacity drives per disk group, using 4 TB mechanical disks could result in almost 140 TB of raw storage on a single host. Although this is an extreme example, the maximum number of components per host should be factored in as part of a typical design, but is unlikely to be a constraint when using the VirstoFS on-disk format, which is addressed further in the following section. Nevertheless, sizing an environment appropriately to meet the customer's needs is critical if the maximum configuration limits on the Virtual SAN–enabled cluster are to be avoided.

To calculate the likely number of components consumed by a typical customer workload, create an estimate based on the current state analysis of the environment and any new requirements gathered from the customer. For instance, a single virtual machine with a single 500 GB disk and no snapshots will always consume the following components:

- 2 for the virtual machine's *VM namespace* object (where the FTT = 1)
- 2 for the virtual machine's *VM swap object* (assuming that there is less than 255 GB of RAM assigned to the virtual machine)
- 4 for the virtual machine's 500 GB *VMDK objects* (assuming no additional stripes and FTT = 1)
- 1 witness component

Witnesses are zero-length components, containing just metadata. The purpose of the witness is to ensure that only one network partition can access an object at any one time. For instance, consider a failure scenario in which two vSphere hosts communicate over a Virtual SAN network. The VMDK object has been configured with a replica, so there is a copy of the VMDK on a second vSphere host. If the Virtual SAN network goes down, the two hosts are no longer able to communicate with one another, even though both hosts are still up and running and accessing other networks successfully. From which partition does the virtual machine access the data?

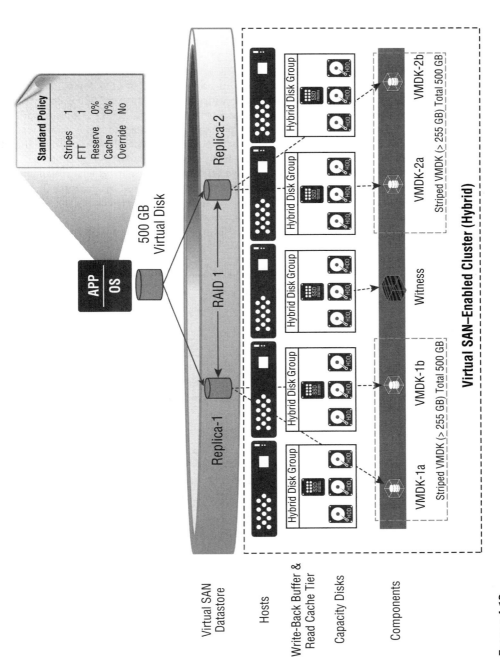

FIGURE 4.12
Virtual SAN disk components

As illustrated in Figure 4.13, in this failure scenario, the witness metadata is used to provide the deciding vote, in order to avoid a split-brain scenario. This ensures that only one network partition can access the object.

On-Disk Formats

Virtual SAN 6 introduced a new on-disk format. This new version 2 on-disk format leverages VirstoFS and significantly increases the supported component count from 3,000 in the vSphere 5.5 U1 release to 9,000 in version 6. The new on-disk format also enables higher-performance characteristics and efficient and scalable high-performance snapshots and clones. The new VirstoFS mounts the volumes from all hosts in the cluster and presents them as a single shared datastore, in the same way as the VMFS-L on-disk format did previously.

New installations of vSphere 6 automatically use the VirstoFS format, which is available only with Virtual SAN 6 or later. These improved scalability features offered by VirstoFS allow IT organizations to deploy far more virtual machines per host then was previously possible, with multiple disk objects and without reaching the component maximums.

Table 4.3 provides an overview of supported configurations. Other configurations, such as Virtual SAN 1 with VirstoFS on-disk versioning, are unsupported.

TABLE 4.3: On-disk file format version history and support configuration

Virtual SAN Version	Format Type	On-Disk Version	Overhead	Supported Component Count (per Host)
1	VMFS-L	v1	750 MB per disk	3,000
6	VMFS-L	v1	750 MB per disk	3,000
6	VirstoFS	v2	1% of physical disk capacity	9,000
6.1	VirstoFS	v2	1% of physical disk capacity	9,000
6.2	VirstoFS	v2	1% of physical disk capacity	9,000

From a design perspective, the recommendation should be to always use VirstoFS in a greenfield vSphere 6.x deployment. The only design factor that could affect this is a customer constraint stipulating that backward compatibility with vSphere 5.5.x hosts is required for the cluster, which would not be a recommended configuration.

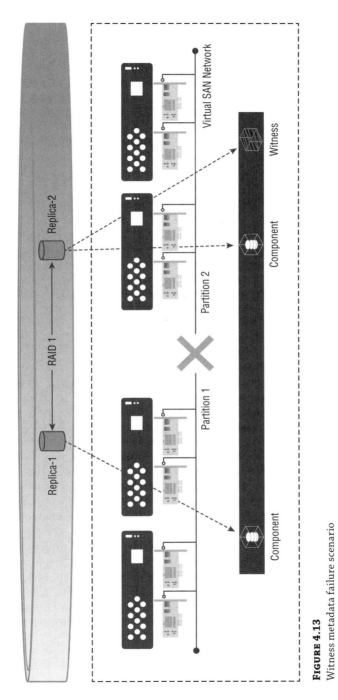

FIGURE 4.13
Witness metadata failure scenario

Upgrades from version 1 of the on-disk format to version 2 can be carried out via a rolling upgrade process, via a built-in utility. This upgrade, however, is optional when transitioning an environment to vSphere 6, as the VMFS-L on-disk format for Virtual SAN 1 continues to be supported. Upgrading the on-disk format to VirstoFS is recommended when updating the environment to vSphere and Virtual SAN 6 or later.

In addition to the new on-disk format introduced in Virtual SAN 6, this release also saw the introduction of a new virtual machine snapshot mechanism. This mechanism provides comparable performance to the legacy system, but takes advantage of the new *always sparse* on-disk format.

Virtual SAN 1 snapshots were based on vmfsSparse (redo logs). However, the new mechanism in Virtual SAN 6 or later uses redo logs in a vsanSparse format. The new vsanSparse format takes advantage of the new Virtual SAN on-disk format, VirstoFS, by writing and extending in-memory caching capabilities to deliver a much more efficient level of performance. All disks in a vsanSparse disk-chain will be vsanSparse, except the base disk.

The vsanSparse snapshot still consumes storage, with each disk snapshot adding a Virtual SAN object. The new snapshot format might consume more storage on a short disk-chain, but level off with old snapshot consumption over time (hours, not days).

Swap Efficiency / Sparse Swap

Swap efficiency, also referred to as *sparse swap*, was introduced in Virtual SAN 6.2, in order to reclaim disk space being used by memory swap-file objects.

A virtual machine creates a swap file when it is powered on. The swap file is created based on the amount of allocated memory assigned to the virtual machine, minus the amount of RAM that has been reserved. For instance, for a virtual machine assigned 8 GB of allocated memory with a 4 GB reservation, vSphere will create a 4 GB virtual swap file. When deployed on a Virtual SAN–enabled cluster, assuming that the FTT is equal to 1, the same virtual machine will result in 8 GB of disk space being consumed. If the operations team then goes on to deploy 800 virtual machines with the same memory configuration, over 6 TB of capacity will be consumed on the datastore.

Unlike other object types, swap objects are not managed by the SPBM mechanism, but are instead managed by the VMkernel and are by default provisioned thickly, despite any SPBM policy assigned to that virtual machine.

As of the Virtual SAN 6.2 release, it is possible to provide efficient capacity utilization to swap files by using the advanced host-level option `SwapThickProvisionedDisabled`. By enabling this advanced host setting, the VMkernel creates swap files on the Virtual SAN datastore as sparse objects. The swap file will consume capacity on the datastore only as the blocks are accessed, which in turn provides additional space savings. This can result in significant savings in consumed capacity, particularly in designs that utilize linked clone VDI workloads. This disk savings does, of course, assume that the virtual machines are not being overcommitted, which requires the use of the swap file. For example, when physical host memory is exhausted, the swap file is used in place of actual physical memory for the virtual machines.

VIRTUAL SAN DISTRIBUTED RAID

Prior to the release of Virtual SAN 6.2, components were distributed across disk groups and cluster nodes via just two techniques: striping (RAID 0) and mirroring (RAID 1). Virtual SAN 6.2 introduced a new set of capabilities referred to as *erasure coding*. This is addressed in detail later in this chapter, alongside the FTT capability.

Striping is controlled by and referred to as `NumberofDiskStripesperObject`. This is the rule set assigned to the virtual machine or virtual disk, via a storage policy, on which you want to use this characteristic. Striping splits the object's data into component chunks or segments, so that those components of data can be accessed simultaneously. This can be done to improve read performance or may occur as a result of an object's size.

Mirroring is controlled by and referred to as `NumberofFailurestoTolerate`, which controls the number of replica copies created of an object, in order to provide increased availability and data protection against hardware failure.

Mirroring and striping can be used in combination, providing both redundancy and read performance benefits. This is achieved by combining the `NumberofDiskStripesperObject` and `NumberofFailurestoTolerate` rule sets into a single storage policy. We cover this in more detail later in this chapter.

Software Checksum

Introduced in Virtual SAN 6.2, the *software checksum* feature, also known as *disk scrubbing*, provides data integrity across the Virtual SAN datastore by performing an end-to-end checksum of the data, allowing this feature to automatically detect and resolve disk errors (see Figure 4.14).

This feature is enabled cluster-wide by default, but can also be modified object by object through SPBM. The aim of this feature is to use a cyclic redundancy check (CRC32) algorithm to detect data corruption that might have occurred on either hardware or software during read or write operations.

FIGURE 4.14
Software checksum web client configuration

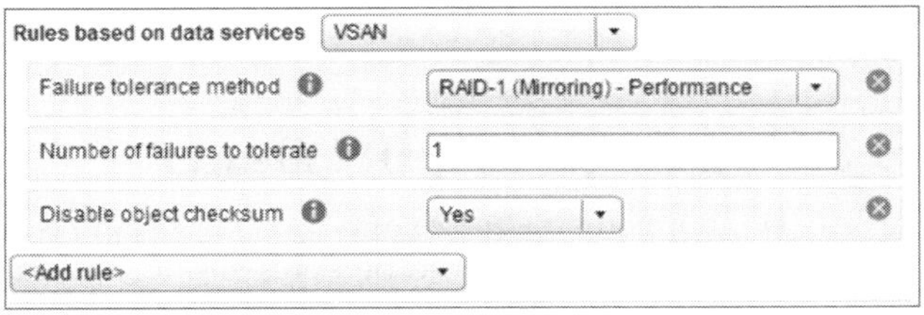

In the case of hard drives, two types of corruption can occur:

◆ Latent sector errors

◆ Silent corruption or silent disk errors

Latent sector errors typically occur as a result of physical disk drive malfunction, while *silent corruption* can take place without warning and may result in the loss of data, corrupted data, or even an outage. The software checksum provides an effective means of detecting such corruption through end-to-end integrity checking during the read and write operations, ensuring the validity of the data based on the checksum. If the checksum operation detects that data is not valid, it will take the steps required to either correct the data or report it to the operations team for action. Typical corrective actions taken by the software checksum mechanism include the following:

◆ Retrieving data from a different replica copy, or in the case of erasure coding RAID 5 and 6 (to be addressed later in this chapter), from parity data. This is referred to as *recoverable data*.

◆ In a failure scenario where no valid copy of the data is available, an error will be returned; this is referred to as a *nonrecoverable error*.

Typical reporting actions taken by the software checksum mechanism include the following:

- When an error occurs, the issue is reported in both the logs and vSphere Web Client. This error includes the impacted data blocks and impacted virtual machines, allowing operational teams to identify workloads that are hit by nonrecoverable errors.

- The operational team also can see historical and trending errors on each drive, helping them to proactively identify hardware issues.

As noted previously, the software checksum mechanism enhances data integrity by using a CRC32 algorithm with CPU off-load support to reduce the overall overhead, to scrub data in the background at two levels: component-level scrubbing and object-level scrubbing. In *component-level scrubbing*, every block of each component is checked, and if a checksum is mismatched, the scrubber tries to repair the block by reading the data from other components. In *object-level scrubbing*, each block of data that makes up the object of each replica copy or parity (in an erasure coding RAID 5 or 6 configuration) is read and checked for inconstant data. When corruption is detected, all data is marked as bad, and the repair operation can be initiated either by the Distributed Object Manager (DOM) owner or by the scrubber during normal I/O operations.

It is noteworthy that the repair mechanism for a RAID 1 mirror differs from that used for an erasure coding RAID 5 or 6 configuration. When the software checksum operation results in a verification error, the DOM owner or the scrubber will read the data from a different copy, or in the case of a RAID 5 or 6 configuration, it will read data from the same stripe, before rebuilding correct data over the corrupted copy, based on blocks of 4 KB. This end-to-end checksum mechanism helps to ensure data integrity, which could result from either a silent disk error or latent sector error.

Virtual SAN Design Requirements

In a Virtual SAN environment, the choice of physical hardware is critically important to the final design and its configuration. This section presents not only hardware requirements to be considered as part of any Virtual SAN design, but also key configuration considerations that need to be addressed as part of any meaningful production architecture.

Host Form Factor

Suitable Virtual SAN vSphere host servers are available in several form factors, from stand-alone rackmount servers, which can typically range from 1U to 4U in size; blade system servers, which in themselves can come in multiple form factors; and multinode, single-chassis servers, which are often favored for hyper-converged infrastructure architectures. Each of these form factors can be used for a Virtual SAN environment, but each has a different set of characteristics that can significantly affect the design.

Rackmount servers, particularly in their 2U form factor, are ideally suited as the foundation for a Virtual SAN infrastructure. They are typically constructed with 16 or 24 available disk slots, which enables the configuration of multiple disk groups, facilitating a scale-up approach.

Even though blade servers are supported, their limited disk capacity (with two disk slots on half-height and half-width blades, and four slots on full-height and full-width blades) and their dependency on chassis and other internal components, typically makes them unsuitable for this type of Virtual SAN– enabled deployment. However, Virtual SAN does support blade-only direct-attached JBODs, which provide a form of high-density, direct-attached storage to blade-based hosts. This enables Virtual SAN to scale disks sufficiently with this type of blade system, allowing designs that are based on this type of supported hardware to fulfill design requirements. Examples of this type of scalable blade system, which could potentially meet customers hyper-converged infrastructure prerequisites include, Lenovo's Flex SEN with x240 blades, and Dell's FX2 with 12G controllers.

While technically possible, and supported, the approach of providing compute-only nodes via blade servers and compute and storage nodes via rackmount servers is not recommended for most designs. This type of architecture and configuration complicates operational support and requires strictly adhered-to sizing from both a compute and storage perspective in order to not waste resources.

The newer form factor, used for many hyper-converged solutions, merges a blade-sized compute node with the flexibility of a rackmount server chassis, delivering a viable solution for many designs. This form factor provides the scalability of 2U or 4U rackmount servers, with 24 disk slots or more, with the compute provided by dense blade type sized systems. Depending on the specific design requirements, this new-generation form factor can be an ideal hardware platform for a Virtual SAN–enabled solution.

Host Boot Architecture

The ESXi hypervisor binaries can be installed to several locations on the host. Traditionally, VMware ESX or ESXi was deployed to local hard drives, often configured as a RAID 1 mirrored pair in order to provide improved host availability. However, in recent years, and for various reasons such as cost, power, and simplification, there has been a wider adoption of the use of internally mounted USB flash and SD cards. This trend has continued to grow with the increased use of Virtual SAN. However, these nonvolatile memory locations are considered removable storage by the ESXi installer and are therefore subject to the caveats and considerations for log file retention.

In a typical Virtual SAN environment, hosts are configured to boot from USB, SD, or other nonpersistent storage, to maximize the number of persistent disk slots available on the hardware, thereby increasing the potential for virtual machine storage. In a Virtual SAN environment, the ability to leverage the maximum number of disk slots on the host for virtual machine datastore capacity is seen as a good practice. However, before you decide whether the removable media option is suitable for a design, consider the following design factors:

Scratch Partitions   ESXi does not create scratch partitions on USB flash drives or SD cards during installation, even if they have the capacity, as the potentially heavy disk I/O from the user-world swap could damage them.

Media Quality   The design should specify industrial-grade SD cards or USB devices for the ESXi hypervisor installation, redundantly configured in RAID 1 if supported by the hardware.

Support   This configuration is not supported for vSphere hosts with 512 GB of memory or greater.

Log Files

The physical location where logs are written will depend on the device used during the ESXi installation. When the ESXi installation device is an SD card, USB flash device, or remote Boot from SAN environment, a local scratch partition is not automatically created on the installation media during the deployment. Despite its capacity, ESXi always sees this type of installation as remote, and as such, logs, including Virtual SAN logs, are stored in RAM disk, a disk drive made up of a block of volatile memory that is lost when the host is rebooted. Removable flash memory, such as USB and SD devices, is sensitive to high amounts of I/O, so by design, the installer will not place the scratch partition on this type of device.

During the host installation, the ESXi installer first scans for a local 4 GB VFAT partition. If it is unable to find one, it will then scan for a local VMFS volume to use to create a scratch directory. If no local VFAT partition or VMFS volume is found, the last resort is to put the scratch partition in the /tmp/scratch location on the local RAM disk. After this type of installation, you will see a warning on the ESXi hosts in vCenter, indicating that their log files are stored on nonpersistent storage. When this is the case, scratch space should be manually configured on the ESXi host using the VMware vSphere Web Client, CLI, or as part of a scripted installation procedure.

When log messages are stored on RAM disk, they are not retained after a reboot, so troubleshooting information contained within the logs and core files will be lost. If a persistent scratch location on the host is not configured properly, you might experience intermittent issues due to lack of space for temporary files, and the log files will not be updated. This can be problematic in hosts with low memory, but is not typically a critical issue for other ESXi operations.

If the installation device is considered local during the host's deployment, the ESXi host does not usually need to be manually configured with a scratch partition. The ESXi installer creates a 4 GB FAT16 partition on the target device during the installation if there is sufficient space to do so. If persistent scratch space is configured, these logs (see the following table) are located in the /var/log/ directory, and the scratch volume contains symlinks (symbolic links) to the persistent storage location.

NOTE A symlink is a special type of file that contains a reference to another file in the form of an absolute or relative path.

Many log files are generated automatically by ESXi components and services. Table 4.4 provides a list of Virtual SAN logs, their locations, and descriptions.

TABLE 4.4: Virtual SAN logs and descriptions

Log File and Persistent Location	Description
/var/log/clomd.log	Cluster Level Object Manager (CLOM) logs, for the CLOM daemon
/var/log/osfsd.log	Object Storage File System (OSFS) logs, for the OSFSD daemon
/var/log/vsanvpd.log	Virtual SAN vendor-provider logs

The next design consideration is the type of persistent storage that should be used for the log files and Virtual SAN trace files, if the only available datastore is the Virtual SAN datastore itself. If persistent storage is not considered as part of the design, a circle of dependency can result as operational teams try to access Virtual SAN logs and trace files to troubleshoot a problem. If the files they require are located on a Virtual SAN datastore, the only persistent storage in the environment, those files could be unavailable during any significant Virtual SAN outage.

If the design results in this circle of dependency—where ESXi has been installed on either a USB or SD device, and all local storage has been allocated to Virtual SAN disk groups, with no local disks or external datastores available for persistent logging—it is critical that the design includes the configuration of a central syslog collector and the VMware Dump Collector.

Configuring a syslog collector, such as vRealize Log Insight, to direct system logs over the network as opposed to, or in addition to, a Virtual SAN datastore provides access to these critical troubleshooting files by operational teams when they are needed the most. In addition, by configuring hosts to use a central logging server, aggregate analysis and searches of Virtual SAN and other logs becomes possible, providing visibility into events that can impact multiple hosts in the cluster. This centralization of logs also provides operational teams and auditors with increased administration and security investigation capabilities. However, it is also key to appreciate that trace files, to be addressed next, do not get forwarded to the host syslog daemon.

Likewise, configuring ESXi memory dumps across the network provides a copy of the state of the host's working memory at point of failure. The VMware vSphere ESXi Dump Collector allows you to keep these core memory dumps on a centralized server for use during the debugging process and, if necessary, for long-term retention. If available, a core dump from the VMkernel includes everything seen on the physical console when the Purple Screen of Death (PSoD) occurred. A core dump can expedite the resolution of hardware issues, but the analysis can be performed only by members of VMware technical support staff.

VIRTUAL SAN TRACE FILES

Virtual SAN generates additional logs, referred to as *trace files*, or `vsantraces`, that need to be written persistently in order to facilitate troubleshooting activities.

Virtual SAN trace files allow VMware Global Support Services and engineering teams to gain insight into what is occurring within Virtual SAN's internal mechanisms in order to troubleshoot problems. Virtual SAN trace files are not forwarded to the host's syslog daemon, because of the bandwidth overhead that would be required. As a result, if the design uses a centralized syslog server or servers to capture logs, Virtual SAN trace files will not be available as part of this operational monitoring solution. Virtual SAN trace files can grow quickly, up to approximately 500 MB, so their persistency and retention (specifically, when `/scratch` is not configured on the boot storage device) should be addressed as part of the design.

As noted previously, when the host operating system is installed to an SD or USB storage device, the Virtual SAN trace files exist, by default, only on RAM disk, although they also get copied to `/locker` for retention purposes via `/etc/init.d/vsantraced` when the host reboots. However, because the `/locker` location is small, the Virtual SAN trace files will not typically fit in their entirety. Therefore, they are copied in order so that the most recent, and therefore the most significant, trace information is captured first. This automated copy mechanism may not occur during a host failure or PSoD event. It is for this reason that trace file retention and availability should be addressed during the planning stage of a Virtual SAN design project.

Because of the lack of bandwidth and I/O limitations associated with removable media, it is not practical to persist the Virtual SAN trace files to an SD or USB device. The number of writes generated by `vsantraces` can be significant and could cause the boot device to fail. By avoiding the placement of Virtual SAN trace files on removable media, this default Virtual SAN behavior also helps to preserve the life expectancy of the SD or USB storage device and reduces the risk of burnout.

The location of vsantraces depends on the boot device being used in the design. Virtual SAN trace files can be located on the scratch partition, on the local boot HDD if one exists, on a scratch location defined on an available VMFS-formatted device, or on RAM disk. Table 4.5 defines the locations where the Virtual SAN trace files are stored, based on the boot device being used.

TABLE 4.5: Virtual SAN trace file location

Host Description	Boot Device	Virtual SAN Trace File Location
ESXi host with 512 GB of memory or less	SD card or USB	For designs that include hosts with up to 512 GB of memory, VMware supports booting the hypervisor from SD card or USB device, with no persistent storage required. A RAM disk is utilized to store VMkernel logs. A RAM disk is also configured by the system to store Virtual SAN trace files, although these are copied to /locker for persistency when the host reboots.
		The core dump partition is also configured on RAM disk in a compressed format. If a core dump event occurs, 2.2 GB of the USB device will be utilized to store it, although the vSphere Dump Collector can also be employed in conjunction with this configuration. The minimum supported SD card or USB device capacity is 4 GB, although 8 GB or larger is recommended. Virtual SAN trace files that exist in RAM disk may not all be persisted during a reboot or host-failure event. These are significant design factors when considering a boot device strategy.
ESXi host with 512 GB of memory or greater	HDD or LUN	For hosts configured with greater than 512 GB of memory, the use of an SD card or USB boot device is not supported. A physical disk device with persistent storage is required. As a result, /scratch is provided on persistent storage, typically on a local disk or disks, and therefore VMkernel logs are persisted in the /tmp partition.
		Rather than copying the Virtual SAN trace files from RAM disk to /locker upon host reboot, the trace files are copied to the persistent storage device.
		If the design installs the ESXi operating system on local physical magnetic disks, enough space is typically available for a local VMFS partition, partition 3, to be created during the installation. The /scratch partition is available by default on the local VMFS volume, which does not require manual configuration. The default location is /vmfs/volumes/vmfs-datastore/.locker/.

TABLE 4.5: Virtual SAN trace file location *(CONTINUED)*

Host Description	Boot Device	Virtual SAN Trace File Location
ESXi host with SATADOM boot device	SATADOM	SATADOM, in effect, is a flash SSD device that looks like a USB thumb drive but connects directly to a SATA connector on the host server's motherboard. Therefore, this device is indistinguishable from the persistent disk configuration described in the previous table entry. As SATA has significantly higher data-transfer rates than USB, and the storage device is far less likely to burn out, trace files are located in the /scratch partition created during the host operating system's installation. Other aspects of log-file behavior also align with those when employing a local HDD boot device.

This differentiation between hosts that are configured with above or below 512 GB of memory relates to the PSoD core dump size.

vSphere releases prior to 5.5 employed a VMcore partition of only 100 MB. However, as it has become far more common for hosts to support much higher capacities of physical memory, even multiple terabytes, this partition has become too small to support core dumps from these hosts. With the release of vSphere 5.5, a new 2.2 GB VMKdiagnostic partition was introduced to allow the capture of core files on ESXi hosts with higher memory capacities.

As previously noted, VMware supports a minimum SD or USB device capacity of 4 GB for a boot device. In a design that uses hosts with less than 512 GB of memory, 2.2 GB of the SD/USB device is set aside for the core dump. However, if the memory capacity of the host is above 512 GB, there may not be enough capacity to capture the core dump on an SD or USB device, and therefore the operation will fail, resulting in lost troubleshooting data.

The vSphere Dump Collector can also be used to redirect core dumps to an external, centralized location. In a design that uses the network dump collector, the core dump is sent over a UDP connection after a PSoD event occurs. However, design considerations associated with employing the core dump collector include the unreliable UDP transport mechanism, and the fact that if a transmission failure were to occur, it would result in a failed core dump capture, and no data would be available for VMware support engineers to provide a root cause analysis.

An additional design option for Virtual SAN trace files is to redirect them to an available NFS datastore, if one is available, which can be used as a centralized maintenance location. The Virtual SAN trace files can be redirected by employing the esxcli vsan trace set command. For instance:

```
vsantraces -> /vmfs/volumes/NFS-Extent/
```

Boot Devices

Virtual SAN, by design, uses storage configured in the local slots of the host server. It is natural, and a common practice, to therefore want to maximize the available disk slots in a host for virtual machine storage. The selection of the boot device to be used in a Virtual SAN–enabled cluster is a key design decision.

Virtual SAN supports booting from removable flash memory devices, such as USB and SD cards, or traditional mechanical or flash drives. Booting from SAN, and as of vSphere 6, SATADOM devices, is also supported.

When using removable flash devices for boot media, the quality of that device should be considered carefully. The design should specify industrial wide-temperature SD cards or USB devices for the ESXi installation, configured redundantly in a RAID 1 set if supported by the hardware. Media quality is important, because many consumer-focused low-quality devices are available on the market. Although the minimum supported SD card or USB boot device size is 4 GB, an 8 GB industrial-grade, wide-temperature USB/SD flash drive should be considered the minimum specification for a business-critical environment.

As indicated previously, Virtual SAN trace logs are stored within either the scratch partition on the local boot HDD, if available, or a scratch partition manually configured on an available VMFS-formatted LUN (often referred to as a *maintenance LUN*). If no persistent storage is available, the trace logs are stored only in RAM disk, and are therefore not persisted during reboot or host failure. Consider this key design factor, outlined previously, when selecting a host boot architecture.

SATADOM Boot Devices

The release of Virtual SAN 6 added support for SATADOM as a boot device. SATADOM, or Serial ATA Disk on Module, is similar to a flash SSD device, but looks like a USB thumb drive and connects directly to a SATA connector on the host server's motherboard. However, SATADOM has significantly higher data-transfer rates than USB, and the storage device is far less likely to burn out during write-intensive operations.

When a SATADOM device is used to boot a vSphere host, the design must specify that a single-level cell (SLC) device be used to ensure that a higher level of endurance is provided. This helps prevent burnout of the boot device when Virtual SAN trace files are being written. As previously noted, this is specifically required for this boot storage type, because when SATADOM flash storage is used, the Virtual SAN trace files are written directly to the SATADOM device, significantly increasing the level of operational write I/O.

Virtual SAN Hardware Requirements

All flash-based devices and storage controllers *must* be listed on the VMware Compatibility Guide for Virtual SAN, for the design to be supported and therefore successful. The key hardware design considerations are as follows:

- A minimum of three hosts is required in a cluster configuration. All three hosts must contribute storage to the datastore. The maximum number of hosts per Virtual SAN cluster aligns with the vSphere 6 maximum of 64. Previous releases of Virtual SAN and stretched cluster configurations support fewer hosts.

- It is strongly recommended that all hosts in the Virtual SAN cluster are configured with the same or similar hardware configurations.

- Supported storage controllers, with an appropriate queue depth, must be used in all designs.

- 1 GB/s Ethernet or 10 GB/s Ethernet is required for the Virtual SAN Network, although 10 GB/s is preferred and highly recommended in a production environment. However, 10 GB Ethernet is mandatory for an all-flash configuration.

Virtual SAN Host Memory Requirements

The memory requirement for a vSphere host operating as part of a Virtual SAN cluster is dictated by the number of disk groups and the number of disks that the hypervisor manages. To support the maximum number of five disk groups, a minimum of 32 GB of RAM is required in each host.

The memory sizing of hosts, as part of a full vSphere design, must take into account numerous factors. To fully use the processing capabilities that the compute nodes offer, host systems must be configured with sufficient memory. In recent years, memory costs have been reduced and the memory density of hardware has increased. With this new extended memory capability, at the time of writing, as much as 768 GB of memory is becoming common on a single half-size blade, and rackmount servers are being introduced into the data center that offer in excess of 1 TB of memory. However, when designing clusters that incorporate Virtual SAN, you also have to take into account the maximum configurable storage capacity available on a host, and balance that out to find the *sweet spot*, or most optimal and balanced configuration between CPU, memory, and storage.

It is considered a best practice when architecting a robust technical platform to configure a consistent amount of memory in each host in the cluster. If a disparity exists in the amounts of RAM in available hosts, consider designing a solution with multiple Virtual SAN clusters.

Additional design considerations which should be included when calculating the memory requirements for a Virtual SAN enabled cluster include:

- Virtual SAN reduces its memory consumption when hosts are configured with less than 32GB memory.
- Virtual SAN will increase its memory consumption in designs which have clusters with greater than 32 nodes.
- An all-flash disk group configuration utilizes additional memory resources, when compared to a hybrid configuration.

Host CPU Overhead

VMware does not set out specific CPU requirements for Virtual SAN. However, you should be aware that Virtual SAN introduces a small amount of overhead to CPU utilization, although this should not be greater than 10 percent. This CPU overhead affects available resources on hosts, and therefore should be considered a design factor that you take into account as part of the compute sizing exercise that calculates projected CPU resources for the environment. This will ensure that virtual machine performance will not be impacted because of a lack of resources or overcommitment issues.

Special consideration and allowances may also be required when evaluating the impact on available resources in highly consolidated environments, such as cloud service provider platforms, or in other special use cases, which could include CPU-intensive virtualized business-critical applications.

Storage Controllers

Storage controllers are a critical piece of component hardware when designing for Virtual SAN–enabled environments. The choice of hardware and its configuration play an important part in the performance of the overall solution, and is therefore a key component that must be carefully considered for the design to be successful.

Virtual SAN supports SAS, SATA, and SCSI adapters, in either their pass-through or RAID 0 modes. These are the only two modes that are supported by Virtual SAN, with many storage adapter devices supporting both, as well as others. The next critical point is that performance is highly dependent on the I/O controller chosen for the design. The following are key design factors that you should consider when evaluating I/O controller hardware:

Modes Supported by the Device   RAID 0, pass-through mode, or both must be supported by the controller. In RAID 0 mode, SSD performance should be a factor reviewed. The choice of RAID 0 mode also comes with operational overhead, which can impact scalability and performance, in that manual interaction with the storage-controller interface may be required to add disks to a disk group. For these reasons, pass-through mode is typically considered the optimum design choice.

Storage Controller Interface Speed   Confirm with the device vendor for PCI-e device performance when behind their controller.

Number of Disks Supported for the Controller   Confirm with the controller vendor the number of supported disks for the controller type.

Number of Controllers to Be Used   Multiple controllers in each host can reduce the failure domain and improve performance, but as a consequence, will also increase the required hardware budget.

Controller Queue Depth   Controller queue depth is key to performance. A queue depth of 256 or higher should be selected. This is a key determining factor over the end performance of the Virtual SAN–enabled cluster, with a higher storage controller queue depth improving overall performance.

Regardless of the hardware and configuration choices made, Virtual SAN requires complete control of the disks. With the controller in pass-through mode, the vSphere hypervisor provides direct access to the underlying drives, giving Virtual SAN complete control over the host's flash and mechanical drives via the storage controller. The performance impact in choosing pass-through mode or RAID 0 is typically negligible for most storage I/O devices. However, when using RAID 0 mode, the storage controller write cache and read-ahead cache should be disabled, in order to ensure that it does not conflict with the flash drive cache controlled by Virtual SAN's internal mechanisms. In addition, direct I/O should be enabled. These hardware configurations should be investigated when evaluating hardware components, as the storage controller cache is a configurable value on some storage controller devices, but not all.

The other main consideration when using RAID 0 on storage I/O controllers is the potential increase in operational overhead. As highlighted previously, RAID 0 mode requires manual interaction with the storage controller software, typically as part of a system BIOS interruption, or alternatively requires interaction with the storage controller firmware through vendor-specific tools, in order to manage the addition and removal of disks from a Virtual SAN disk group. With this mode type, each drive must be configured as an individual RAID 0 array, so as to present each drive to Virtual SAN individually, rather than as a single array of disks, with the single endurance flash device being tagged via an `esxcli` command. The use of RAID 0 mode devices might add a significant layer of operational overhead to the design, as Virtual SAN performance and stability will almost certainly be compromised if configuration changes are not managed correctly. In addition, when RAID 0 mode is being used in the design, Virtual SAN is unable to manage the hot-plug capabilities of disks; instead, hot-plug is managed by the storage controller firmware.

The following are key design factors that should be taken into account when evaluating storage controllers best suited for the design of the environment:

- The available disk modes—pass-through mode is considered optimal for most designs.
- The make and model of the flash devices, to ensure support.
- The make and model of the mechanical disks, to ensure support.
- The queue depth of the controller—minimum supported queue depth is 256.
- The number of disks and corresponding disk groups being configured on the hosts in the design.

STORAGE DEVICES

It goes without saying that storage devices are the core components in a Virtual SAN environment. As already highlighted, a Virtual SAN cluster uses both mechanical and solid-state flash media in a variety of configurations to enhance performance and/or endurance, through a nonvolatile write buffer and a read cache, and provides capacity in an all-flash configuration.

The choice and configuration of these disk storage components is critical for Virtual SAN to operate successfully. This section addresses the performance, endurance, and capacity characteristics of these devices in far more depth, in order to help you understand the options available across the various Virtual SAN configurations.

Solid-State Flash Drives

Solid-state flash drives are the most popular storage medium used in modern computers for storing and accessing data, and for performance-intensive applications. Flash drives support rapid access to random data locations, which means data can be written or retrieved quickly for a large number of simultaneous users or applications. Mechanical disks, however, typically have a much larger capacity than flash devices, although this is slowly changing. Virtual SAN can be configured with both disk types to provide a flexible balance between enhanced performance, write endurance, and increased capacity.

Flash drives, or SSDs, as they are most commonly known, use NAND modules and provide high performance and low power consumption in comparison to traditional mechanical disks. Flash drives do not have rotating media or mechanical positioning, so the response is near instantaneous. In addition, enterprise flash drives can process nearly 30 times the number of I/O per second and maintain straight-line response as the workload increases. Therefore, a single flash drive can process the same number of IOPS as up to 30 traditional 15,000 RPM mechanical drives. However, unlike a magnetic disk, data stored on NAND flash needs to be erased before new data can be written or *programmed*; this is known as the *program-erase cycle* (PEC).

Another key aspect of drive performance is queuing. Any disk drive, whether it be flash or mechanical, can perform only a single operation at one time. User or system I/O requests are sent to the drive at electronic speeds, but if more than one request is sent to a drive, only one is processed at a time, and the others must wait. Intelligent drive algorithms optimize the order in which requests are processed, but only one request at a time is handled, and the rest are queued, and response time for queued requests increases exponentially. Flash drives can minimize the performance impact caused by queuing, as the faster you can service an I/O request, the shorter the queue typically is.

Solid-state drives can use either SATA, SAS, or PCIe interfaces. Each of these common interfaces provides different performance characteristics, relating to its maximum throughput, as outlined in Table 4.6.

TABLE 4.6: Interfaces supporting solid-state drives

INTERFACE	INTERFACE SPECIFICATION	THROUGHPUT
SAS drives	SAS-1	3 Gb/s
	SAS-2	6 Gb/s
	SAS-3	12 Gb/s
SATA drives[1]	SATA 1	1.5 Gb/s
	SATA 2	3 Gb/s
	SATA 3	6 Gb/s
PCIe drives	1 to 32 lanes	Drive throughput is dependent on the PCIe vendor, and the generation of the device. ◆ Gen v1.x: 250 MB/s (2.5 GT/s) ◆ Gen v2.x: 500 MB/s (5 GT/s) ◆ Gen v3: 985 MB/s (8 GT/s) ◆ Gen v4: 1,969 MB/s (16 GT/s) For instance, a PCIe 2nd Generation with 8 lanes can provide a maximum of 4 Gb/s per lane, producing a total of 32 Gb/s of throughput.

[1] SATA 3 Gb/s or 6 Gb/s drives can be connected to a SAS interface. However, SAS drives cannot be connected to a SATA interface.

Even though PCIe SSD devices typically outperform SAS and SATA SSDs, in terms of interface throughput, interface performance is only one factor in choosing an SSD device; I/O performance is also critical.

The flash device used in a Virtual SAN configuration is vital, as it largely dictates the performance that can be achieved. In a Virtual SAN hybrid configuration, the use of the flash device is split between a nonvolatile write buffer, which takes up approximately 30 percent of the total drive capacity, and a read cache, which uses the remaining 70 percent of the drive's capacity. While this ratio is a configurable parameter, it should not be modified for most workloads and configurations unless specifically directed to do so by VMware support services. The level of endurance, and the number of I/O operations per second that can be achieved and sustained by the flash device are the critical factors in delivering the performance required for mixed, I/O-intensive, virtualized workloads.

In contrast to the hybrid model, in an all-flash configuration, the nonvolatile write cache value is configured at 100 percent, as read performance is not a factor in an all-flash disk group. This parameter is fixed and cannot be modified by vSphere storage administrators.

In addition to performance, endurance is another key factor that should dictate the choice of flash device used in a design. Endurance of the flash device is defined in standard industry write metrics, which are the primary measurements used by the IT industry to gauge the reliability of the drive. VMware recommends selecting an appropriate endurance-class device, based on the customer's requirements for the environment.

SSDs are categorized by VMware into five classes, based on write performance. The class of SSD used in a design can significantly impact the performance of the Virtual SAN–enabled cluster. The following table lists the various classes of SSD and recommendations for the endurance-class device required for the write buffer and read cache mechanism, based on workload. For optimal performance of an all-flash Virtual SAN solution, select a higher-performance class of SSD, as highlighted.

Table 4.7 lists the designated SSD classes, as specified within the VMware hardware compatibility guide.

Table 4.7: SSD endurance classes and Virtual SAN tier classes

Category	SSD Class	Writes per Second	Virtual SAN Model/Tier
Performance-class devices	Class A	2,500–5,000	All-flash model / capacity tier only
	Class B	5,000–10,000	All-flash model / capacity tier only and hybrid model / caching & buffering tier
Endurance-class devices	Class C	10,000–20,000	All-flash model / caching & buffering tier (medium workload)
	Class D	20,000–30,000	All-flash model / caching & buffering tier (high workload)
	Class E	30,000–100,000	All-flash model / caching & buffering tier (write-intensive workload)
	Class F	100,000+	All-flash model / caching & buffering tier (extreme write-intensive workload)

Write performance is used as the primary means by VMware to categorize SSD devices. In flash-based storage, writes are a far more limiting factor, and therefore a better overall gauge on performance than either random or sequential reads. Other factors that can impact SSD performance characteristics, and therefore affect workload performance, include the devices' queue depth and maximum drive latency. In addition, flash storage devices have a finite level of endurance, and therefore an expected lifespan, that is measured in the number of PECs that the device can withstand. Each time data is written to a device cell, the device must perform a PEC on the cells being used.

SSD manufacturers categorize their drives as single-level cell (SLC), multilevel cell (MLC), or enterprise multilevel cell (eMLC) NAND, with each type using different storage mechanisms that affect both performance and endurance characteristics.

An SLC device stores a single bit of data (0 or 1) in each cell, whereas in an MLC device, the NAND flash modules can use multiple levels per cell, typically 4 bits, allowing more bits to be stored. SLC flash provides the highest level of performance, which typically comes at the highest cost. SLC also has the highest endurance, at around 100,000 PECs per cell.

An MLC is a lower-grade device, which typically costs less than an SLC device. However, the NAND modules are expected to have a shorter life expectancy. An MLC device provides endurance typically at a level of 10,000 to 30,000 PECs per cell. Therefore, an MLC is not typically recommended for applications that perform many writes, which often makes them unsuitable as an endurance flash device in an all-flash disk group configuration. In addition, as an MLC device uses the same number of transistors as an SLC device, there is also an increased risk of errors occurring within each NAND module.

An eMLC NAND device provides a balance between cost and life expectancy, and is therefore a good compromise option between SLC and MLC modules. An eMLC device can typically store 2 bits per cell. Because this media type uses a higher PEC rate than consumer MLC, it can also benefit from higher endurance, at around 20,000 to 30,000 PECs per cell, and can therefore tolerate the increased endurance required by enterprise application workloads. However, as SSD vendors use different internal architectures in their devices to improve the reliability of the drive, this does not necessarily mean that an SLC device is more reliable than an eMLC, as NAND modules can have their longevity enhanced by software features within the SSD's controller.

Other technologies, such as triple-level cell (TLC), provide a higher maximum density but poor endurance at 1,000 to 5,000 PECs, making them suitable for only consumer-grade electronics. In addition, 3D NAND TLC, a newer type of flash storage device, provides similar performance and endurance as an MLC device but at a lower cost, and has yet to see any significant market penetration.

As a result of this variability in performance and endurance between device types, VMware does not differentiate between SLC, MLC, and eMLC types in its Virtual SAN hardware compatibility list (HCL). From a Virtual SAN design perspective, the key consideration is to ensure that the flash devices used meet the minimum performance and reliability metrics defined by VMware, within each of the given performance classes outlined previously in Table 4.7. These metrics should be considered, regardless of the vendor hardware architecture, in terms of NAND module type versus controller features used to enhance endurance, in order to achieve the device's published maximum values.

As shown in Table 4.7, in an all-flash configuration, the capacity tier flash devices used in the design for persistent storage and to handle read requests, can use lower-cost, performance-class, solid-state flash drives, as opposed to the higher-cost, endurance-class devices. It is important to understand that a direct correlation exists between the SSD endurance-class drive and the level of Virtual SAN performance in an all-flash configuration. As is so often the case, the costliest devices are also the highest-performing hardware and will provide optimal performance for the design. As a result, cost can be the determining factor for a lot of IT organizations. Depending on the specific use cases and workloads, a lower class of endurance device maybe more appealing to the customer, even though overall performance may be affected.

Although the primary measurement mechanism used by SSD vendors to gauge device performance is write I/O metrics, there is no standard means across all manufacturers to provide a measure of endurance. Most producers of SSD hardware use either drive writes per day (DWPD) or petabytes written (PBW). A measurement of 1 DWPD is the equivalent of filling up the drive to its maximum capacity and then erasing the entire drive, with this process requiring a PEC for every NAND cell on the SSD.

VMware's Virtual SAN HCL requires that the following endurance metrics, over a five-year life expectancy of the device, can be met:

- For Endurance SAS and SATA SSD devices, the drive must support at least 10 full DWPD, and the drive must either support random write endurance of up to 3.5 PB on 8 KB transfer size per NAND module, or support random write endurance of up to 2.5 PB on 4 KB transfer size per NAND module.

- For Endurance PCIe SSDs, the device must support at least 10 full DWPD, or the drive must support random write endurance of up to 3.5 PB on 8 KB transfer size per NAND module, or 2.5 PB on 4 KB transfer size per NAND module.

In the absence of any formal guidance from the customer, a good starting point for an all-flash Virtual SAN design is to recommend the use of a Class C endurance device for write buffering, and performance Class B SSDs for persistent capacity and read I/O. This provides a good balance in order to gain a high level of performance from the Virtual SAN datastore, while not having to endure significant CapEx costs for the implementation of the environment.

PCIe-Based Flash Devices

PCIe-based flash devices are the most recent development of the solid-state media. For the most part, traditional SSD drives in the data center have used conventional SAS storage interfaces, designed to support mechanical drives. However, more recently drives have been also been developed for the high-speed PCIe bus interface.

When designing a Virtual SAN solution, as illustrated in Figure 4.15, make sure to verify that the flash devices satisfy the following criteria:

- Ensure that the model of the flash device (PCIe or SAS SSD) is listed on the Virtual SAN section of the VMware Compatibility Guide.

- PCIe devices often have a higher cost than SAS-based solid-state flash devices. However, a PCIe device generally has a greater capacity than an SAS SSD, and also typically has faster performance than SAS SSDs.

- Ensure that the write endurance of the flash device meets the requirements for the write buffer mechanism in an all-flash solution, or as a caching device in a hybrid configuration.

Mechanical Disk Drives

Mechanical disk drives are the traditional mechanism for storing information. They differ from solid-state flash media, as the data on the disk is recorded on tracks, which are concentric rings on the platter around a spindle (see Figure 4.16). The tracks are numbered, starting from zero, from the outer edge of the platter. The number of tracks per inch (TPI), or the track density, measures how tightly the tracks are packed together on the platter.

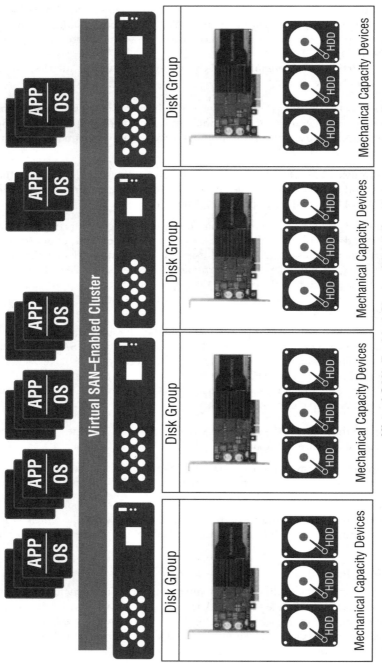

Figure 4.15
Virtual SAN configuration with PCIe-based flash devices

Each track is divided into smaller units called *sectors*. A sector is the smallest individually addressable unit of storage on the disk. The track and sector structure is written on the platter by the drive manufacturer, using a low-level formatting operation. The number of sectors per track varies according to the drive type as well as the physical dimensions and recording density of the platter.

FIGURE 4.16
Geometry of a mechanical disk

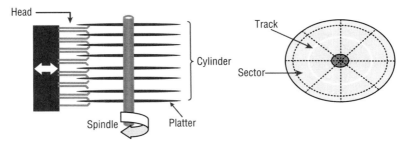

On mechanical disks, three components make up the total time it takes to perform a disk I/O operation:

- *Seek time*, or *positional latency*, is the time taken to position the read/write heads over the desired cylinder. This depends on how far the read/write heads have to move, but on average, this could be 4 to 9 milliseconds.

- *Rotational latency* is a function of the rotational speed of the drive. A 15,000 RPM drive takes 4 milliseconds to make one complete revolution, so the average is the time it takes to make half a revolution, or in the case of a 15,000 RPM drive, 2 milliseconds.

- *Transfer speed* is the time it takes to transfer data and is a function of both the rotational speed and the link protocol. When compared with the positional and rotational latency, the transfer speed is negligible.

In a Virtual SAN environment, mechanical disks are used in a hybrid configuration. They are used exclusively for persistent data storage and dirty reads, also known as *read misses*, where the block being read is not found in the read cache and therefore has to be read from a mechanical disk. Mechanical disks can also be a determining factor in the available stripe width for a virtual machine, if configured as part of a storage policy. For instance, if a specific stripe width is configured in the policy, you must confirm that the requested stripe width is available across all the disks in a host, to ensure that the cluster can comply with the policy. In addition, if the virtual machine is configured with a high FTT setting, additional mechanical disks are necessary as a result of each component being replicated to meet this policy requirement.

While the mechanical disk drive is often the slowest component in the Virtual SAN hybrid environment, the intelligent algorithms within the Virtual SAN mechanism are designed to service most read requests immediately from the read cache and through parallelism, where the workload is read over the multiple components that make up virtual machine objects concurrently. In addition, all virtual machine writes are handled by the endurance flash tier, meaning the mechanical disks should not, in a correctly designed hybrid solution, prove to be a performance bottleneck.

VMware supports the following three types of mechanical disk in a Virtual SAN solution:

- Serial Attached SCSI (SAS)
- Near Line Serial Attached SCSI (NL-SAS)
- Serial Advanced Technology Attachment (SATA)

NL-SAS can be thought of as an enterprise-class SATA disk with an SAS interface. For most workloads, the best results can be obtained with SAS and NL-SAS drives. SATA disks should typically not be used, unless the workload is capacity-centric and performance is not a factor in meeting service levels.

In addition, the rotational speed of the mechanical disks should be chosen to meet the specific requirements of the target workload. Mechanical disk speeds and use case characteristics are defined in Table 4.8.

TABLE 4.8: Virtual SAN mechanical disk characteristics and rotational speeds

DISK TYPE	CHARACTERISTIC	REVOLUTIONS PER MINUTE	AVG. IOPS/DRIVE
NL-SAS/SATA	Capacity	7,200	75
SAS	Performance	10,000	150
SAS	Additional performance	15,000	175

VMware recommends that in a hybrid Virtual SAN solution, the design uses SAS-based mechanical disks best suited to the workload characteristics of the environment being implemented. If high performance is not required, a larger number of lower-cost mechanical disks will allow for a higher FTT to be configured. If there are no specific customer requirements, consider selecting 10,000 RPM drives in order to achieve a balance between capacity, cost, and availability.

The significance of this design decision can easily be demonstrated through some simple calculations, using the metrics provided in the preceding table. For instance, if a customer requires 20 TB of capacity for a Virtual SAN hybrid configuration, this can be achieved in various ways, using different drive types. Two examples are shown here:

- Option 1: 20 TB / 2 TB NL-SAS = 10 NL-SAS disks (across disk groups) = 750 IOPS
- Option 2: 20 TB / 1 TB 10k SAS = 20 SAS disks (across disk groups) = 3,000 IOPS

As you can clearly see, there is a vast difference between the two options in terms of IOPS. Even though Virtual SAN predominately leverages flash devices to provide optimized read and write performance, these calculations remain significant. The primary reason is that in the hybrid model, capacity layer disk I/O performance impacts both destaging operations and latency when a read cache miss occurs, which results in data being retrieved from mechanical disks.

It is also noteworthy that VMware advises customers against mixing mechanical disk rotational speeds, in an attempt to achieve a blend of performance characteristics in the environment. This approach makes performance unpredictable and inconsistent, as there is only a single volume configured across a distributed Virtual SAN datastore. The recommendation is to select a single type of mechanical disk per hybrid cluster. If different performance characteristics are required to meet the needs of multiple workload types, the design should include several separate tiers of Virtual SAN clusters, as illustrated in Figure 4.17, to support the various service levels involved.

Virtual SAN Ready Nodes

Virtual SAN Ready Nodes are out-of-the-box, ready-to-go hardware solutions sold by server hardware vendors. Ready Nodes have been preconfigured and optimized to run the Virtual SAN hyper-converged platform in a prebuilt, hardware-certified form factor.

The VSAN Ready Nodes include a combination of compatible hardware components from the vendor, such as storage controller, solid-state flash devices, and mechanical drives, and may also in some cases include OEM-supplied vSphere and Virtual SAN software licensing. Opting to use VSAN Ready Nodes in a design provides a simple and preconfigured hardware solution, and avoids having to navigate VMware's HCL, as the vendor will have already certified compatibility. VSAN Ready Nodes are the ideal hyper-converged building blocks for medium and large Virtual SAN designs, which ensure hardware compatibility and support from the vendor and VMware.

Virtual SAN Licensing

Virtual SAN is not included in the cost of vSphere. Virtual SAN is a separate licensed product that sits on top of the VMware vSphere platform and therefore requires its own license. Because licensing is subject to regular changes based on market conditions, this topic is kept brief.

At the time of writing, as of the Virtual SAN 6.2 release, the product is distributed under three license types: Standard, Advanced, and Enterprise. Standard licensing is typically suitable for all implementations, except those requiring an all-flash configuration. Deduplication, compression, and erasure-coding features are available only through the Advanced license model. If the customer design is based around a stretched cluster or requires the Quality of Service (QoS) feature, Enterprise licensing is required.

Table 4.9 compares licensed features as of the release of Virtual SAN 6.2.

TABLE 4.9: Virtual SAN 6.2 feature licensing

FEATURE	VIRTUAL SAN STANDARD	VIRTUAL SAN ADVANCED	VIRTUAL SAN ENTERPRISE
Storage policy–based management (SPBM)	*	*	*
Read/write SSD caching	*	*	*
Distributed RAID (RAIN)	*	*	*
vSphere Distributed Switch (VDS)	*	*	*
VSAN snapshots & clones	*	*	*
Fault domains (rack awareness)	*	*	*
vSphere replication (with 5 min RPO)	*	*	*
Software checksum	*	*	*
All-flash		*	*
Deduplication & compression (all-flash only)		*	*
Erasure coding (RAID 5/6–all-flash only)		*	*
Stretched cluster			*
QoS–IOPS limits			*

NOTE Virtual SAN licensing bundles the vSphere Distributed Switch, and enables this feature, even when it is not included with the chosen vSphere edition. Therefore, in a Virtual SAN–enabled cluster, there is no requirement to obtain Enterprise Plus licensing just to enable the distributed virtual switch feature, which is optional but recommended for use in a Virtual SAN design.

However, there may be numerous other design factors that must be taken into account or that dictate which vSphere edition is required to be deployed within a specific design.

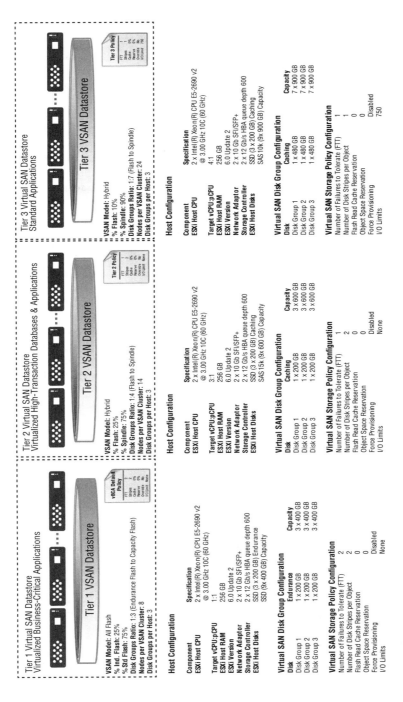

FIGURE 4.17
Tiered workload virtual SAN clusters

Virtual SAN Network Fabric Design

As with all other types of shared storage solutions, Virtual SAN is heavily dependent on the networking infrastructure to support I/O and data movement across the cluster. However, the vast majority of vSphere administrators do not provide operational responsibility for maintaining or designing networks, especially in organizations where there previously has been no IP SAN or NAS in place, with native Fibre Channel SAN typically being managed by storage engineers or a dedicated SAN team. Therefore, one of the challenges for an architect can be engaging with the right administrative and operational teams in order to develop a design.

Virtual SAN is a replication-centric technology that utilizes the network to transport all data, including internal virtual machine I/O operations, replication traffic, and management communication between the Virtual SAN cluster nodes. The vSphere transport is provided by a VMkernel port group, which is specifically created for Virtual SAN on the appropriate network VLAN. The VMkernel port must be configured on all hosts in the cluster, even nodes that are not contributing storage resources to the datastore.

To design the network transport for a Virtual SAN environment, you must consider key factors detailed in this section.

First, you must consider how much replication and communication traffic is traversing the network between hosts. In Virtual SAN, the amount of network traffic directly correlates to the total number of running virtual machines, the application workload I/O characteristics, configured storage policies, and whether an all-flash or hybrid model is being used.

As with other vSphere internal communications, such as vMotion or fault tolerance, it is highly recommended that Virtual SAN traffic be configured and isolated on its own layer 2 network. While this can be achieved with the use of dedicated switches and adapters, it is more commonly designed to use a dedicated VLAN in a shared 10 Gb/s architecture, as illustrated in Figure 4.18.

Because Virtual SAN network design covers numerous topics, this chapter divides them into two categories:

- vSphere network requirements
- Physical network requirements

However, it goes without saying that these two areas interact with one another in almost every way. To design a stable and technically robust environment, their integration is a critical architectural design factor.

vSphere Network Requirements

The vSphere network layer encompasses all network communications between the virtual machines and the physical network. Key qualities that are often associated with vSphere networking and are particularly pertinent in a Virtual SAN environment include performance, availability, and security. All of these key factors are covered in this section.

Virtual SAN VMkernel Network Configuration

To provide connectivity to the shared Virtual SAN storage, a dedicated VMkernel port must be configured on each host in the cluster for the replication and synchronization mechanisms to operate.

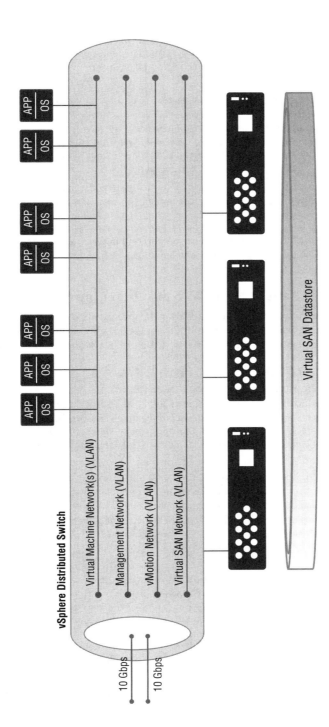

FIGURE 4.18
Virtual SAN logical network design

In all designs, the VMkernel port should be configured on a dedicated and isolated VLAN, and is used to pass traffic to the storage kernel. Virtual SAN traffic, if possible, should occur only over layer 2 and not get passed outside the cluster's dedicated storage VLAN. If 10 Gb/s network adapters are used, these interfaces can typically be shared with other traffic types. However, if 1 Gb/s networks are a design constraint, dedicated physical network interfaces should be assigned to the VMkernel port group.

An additional consideration, depending on the security or performance requirements of the design, is that you might need to recommend to the customer that they use dedicated network cards, even in a 10 Gb/s implementation. The design may even have to evaluate isolating Virtual SAN network traffic onto its own switching infrastructure, in order to meet compliance requirements in a security-centric design. This approach mitigates any risk of VLAN hopping and prevents other traffic from impacting the performance of the Virtual SAN environment.

NOTE Virtual SAN does not support multiple VMkernel adapters on the same subnet for load-balancing purposes. Multiple VMkernel ports on different network VLANs or on separate physical fabrics is, however, a supported configuration.

COMPARING THE vSPHERE STANDARD AND DISTRIBUTED VIRTUAL SWITCHES

When designing the virtual switch configuration, if not already dictated by other design factors, one required design decision may be whether to adopt the vSphere standard switch or to implement the vSphere Distributed Switch (VDS), with Virtual SAN deployment being supported on both options.

The major benefit of the vSphere standard switch is simplicity of implementation. However, as the Virtual SAN environment grows, the design might benefit from several features offered only by the VDS, including Network I/O Control (NIOC), Link Aggregation Control Protocol (LACP), and NetFlow. Another key consideration that factors into this design decision is whether VMware NSX is included in the overall environment architecture.

One of the key benefits of using the vSphere Distributed Switch in a Virtual SAN environment is that NIOC can be used, which allows for the prioritization of bandwidth when there is network contention. For instance, replication and synchronization activities that Virtual SAN will impose on the network can cause contention. Depending on the number of virtual machines, their level of network activity, and Virtual SAN network utilization, 1 Gb/s networks can easily be saturated and overwhelmed, particularly during rebuild and synchronization operations. Through the use of NIOC and QoS, vSphere has the ability to manage this contention, providing appropriate priority to traffic types, based on share value.

As vSphere Distributed Switch licensing is bundled with Virtual SAN, there is no requirement to license vSphere Enterprise Plus Edition in order to take advantage of these features in Virtual SAN–enabled clusters. Therefore, you should have no obstacles in deploying the vSphere Distributed Switch within a design.

VIRTUAL SAN NETWORK TEAMING DESIGN

Virtual SAN traffic is not designed to load-balance across multiple network interfaces, teamed together for the purpose of bandwidth aggregation. Therefore, teaming is implemented in a design purely to enhance availability and redundancy purposes, and not performance.

Even though Virtual SAN supports the use of IP-hash load balancing, this does not guarantee an improvement in performance for all designs. Virtual SAN benefits from the IP-hash algorithm only when used on shared interfaces, where IP-hash can perform load balancing. However, if Virtual SAN is used on dedicated interfaces, the design may not recognize any improvement. For instance, in a design that uses two dedicated 1 Gb/s physical adapters for the VMkernel port group, with IP-hash enabled for Virtual SAN, the network platform may not be able to use more than one 1 Gb/s link. This also applies to all other vSphere NIC teaming policies.

To provide stable and predictable performance, Table 4.10 outlines several options available to support the use of multiple network adapters implemented for availability and redundancy purposes.

TABLE 4.10: Virtual SAN network teaming

LOAD-BALANCING ALGORITHM	CONFIGURATION	DESCRIPTION
Route based on port ID	Active/passive with explicit failover	Basic and simple configuration
Link Aggregation Control Protocol (LACP) with route based on IP hash	Active/active with LACP port channel	Employed where the physical network is using a port channel configuration
Link Aggregation Control Protocol (LACP) route based on physical network adapter load	Active/active with LACP port channel	Employed where the physical network is using a port channel configuration

As a general design recommendation, assuming LACP can be used in the environment, configurations that use an active/active model with a *route based on physical adapter load* can ensure that idle network cards do not wait for a failure to occur and can aggregate bandwidth based on network interface load.

NETWORK I/O CONTROL

As outlined in Chapter 2, the NIOC feature of the vSphere Distributed Switch provides a QoS mechanism for network traffic within the ESXi hosts. This feature can help prevent traffic from flooding the network, causing contention and impacting Virtual SAN replication and synchronization activities. In addition, it is also recommended that tagging traffic types for 802.1P QoS, and configuring the physical upstream switches for support be used. If the QoS tagging is not implemented, the value of the NIOC mechanism is limited to within the hosts themselves.

In the sample design shown in Figure 4.19, two 10 Gb/s uplink interfaces carrying multiple traffic flows from each host in the Virtual SAN cluster have been configured from the vSphere Distributed Switch. NIOC monitors the network, and whenever it sees congestion, it automatically shifts resources to the highest-priority applications, as defined by the NIOC policy.

Figure 4.19 illustrates a Virtual SAN cluster with all nodes configured to use the same vSphere Distributed Switch, which is being used to carry all network traffic types. Each host is also configured with a single dvUplink group, which includes the two active 10 Gb/s Ethernet adapters.

In this example, all physical network switch ports connected to the host adapters are to be configured as trunk ports, as recommended by the switch vendor. The figure also shows the port groups that are used to segment traffic logically by VLAN, with traffic tagging occurring at the virtual switch level. Uplinks are to be configured as active/active with a load balancing algorithm of *route based on physical NIC load*, as the physical switches use EtherChannel capabilities. Both the virtual and physical switches are configured to pass traffic specifically for VLANs used by the design, as opposed to trunking all VLANs.

As shown, the two 10 Gb/s network interfaces carry all ingress and egress Ethernet traffic on all configured VLANs. The user-defined network resource pools have been configured port group by port group, as shown in Table 4.11.

TABLE 4.11: Sample Virtual SAN cluster Network I/O Control policy

PORT GROUP	VLAN ID	NO. SHARES	LIMIT (MB/s)	QOS TAG
VM networks	101	150	Unlimited	2
vMotion network	102	250	2,500	3
ESX management network	103	100	Unlimited	1
Virtual SAN network	104	500	Unlimited	5

In general, using limits is not recommended, as they put a hard cap on usage. As in the preceding table, if limits are used in a design, be sure to calculate the available MB/s, and set the limit based on the percentage of the available bandwidth.

The QoS tag (802.1P), shown in Table 4.11, is associated with all outgoing packets. This enables the compatible upstream switches to recognize and apply the QoS tags. The default setting is None, and a value between 1 and 7 is configurable.

By segmenting Virtual SAN, virtual machine, and vMotion traffic into separate VLAN-backed port groups, and using NIOC shares and the QoS mechanism, it is typically possible to sustain the level of performance expected for each traffic type in the design, even during possible periods of contention.

Physical Network Requirements

Until recently, typical enterprise and service-provider data-center networks were exclusively built in a three-tier—*access*, *aggregation*, and *core*—modular fashion. This traditional architecture was designed to serve predominantly north/south traffic, traversing in and out of the data center. Although this architecture offers high levels of availability, it can also, depending on the design, limit bandwidth. This limitation often comes about as a result of Spanning Tree Protocol (STP) being used to prevent network looping.

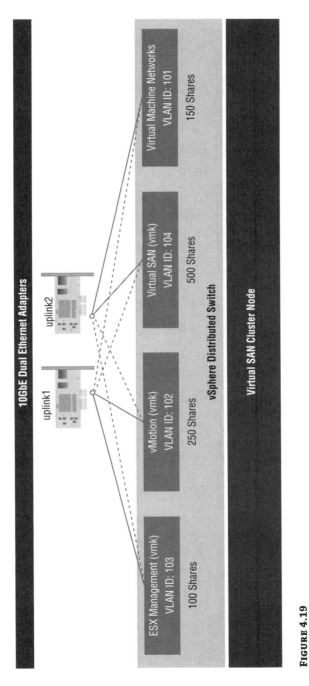

FIGURE 4.19
Network I/O Control

However, over recent years, as virtualization of the compute, network, and storage platform has evolved, more and more data centers have chosen to adopt the *leaf-spine* model. The leaf-spine topology simplifies scalability, bandwidth, availability, and quality of service.

Virtual SAN can operate without issue in either topology. Therefore, the key design factor is whether an existing network infrastructure is in place for the deployment of the new Virtual SAN clusters. In a greenfield environment, a physical network topology design recommendation can be made to the customer, and be implemented for the new Virtual SAN–enabled infrastructure.

THE CORE, AGGREGATION, AND ACCESS LAYER TOPOLOGY

This more traditional network model uses three layers, as illustrated in Figure 4.20. At the bottom of the tree is the access layer, where the vSphere hosts connect to the physical network. The aggregation layer, to which the access layer is connected in a redundant fashion, provides connectivity to adjacent access layers and rows of racks in the data center. The aggregation layer also provides connectivity to the core. The core layer provides high-speed routed connectivity to other parts of the data center, as well as to external destinations, via firewall services.

The core, aggregation, and access model can be subject to bottlenecks if uplinks between layers are oversubscribed, which can easily occur when scaling out top-of-rack access switches to expand the infrastructure. Problems can also come about as a result of latency being incurred as traffic flows through each layer, and from the blocking of redundant links through the use of the STP. For these reasons, among others, we are now starting to see more and more enterprises and service providers moving away from this three-tier model, in favor of simpler and more effective topologies.

THE LEAF-SPINE TOPOLOGY

In a *leaf-spine topology*, the leaf switches are fully meshed with the spine switches via routed or switched layer 3 links. The spine switches in the leaf-spine model are layer 3 switches, and the leaf layer typically represents a smaller layer 3 top-of-rack switch. Load balancing and failover is achieved through equal cost multipathing (ECMP), employing one of many protocol options, such as OSPF or ISIS. In this topology, illustrated in Figure 4.21, if packets need to be sent from one rack to another, it is always a single hop. This means the traffic can be load-balanced and distributed across all of the paths equally.

COMPARING THREE-TIER NETWORKS WITH A LEAF-SPINE TOPOLOGY

As already highlighted, Virtual SAN can operate across both network models without issue. As a result, other network architecture factors must come into play when recommending either a three-tier (core, aggregation, and access model) design or a leaf-spine topology.

Key design factors to consider when evaluating the physical network architecture include the following:

- Leaf-spine has no Spanning Tree Protocol; all links are functional and in use, with no unused ports.

- Leaf-spine is easy to scale. Additional leaf switches are added as rack numbers expand, and spine layer switches are added as required, in order to meet performance requirements.

- Leaf-spine requires no proprietary protocols. Network architects will use the ECMP protocol of choice, such as OSPF, BGP, or IS-IS, to provide uniform access and consistently low latency.

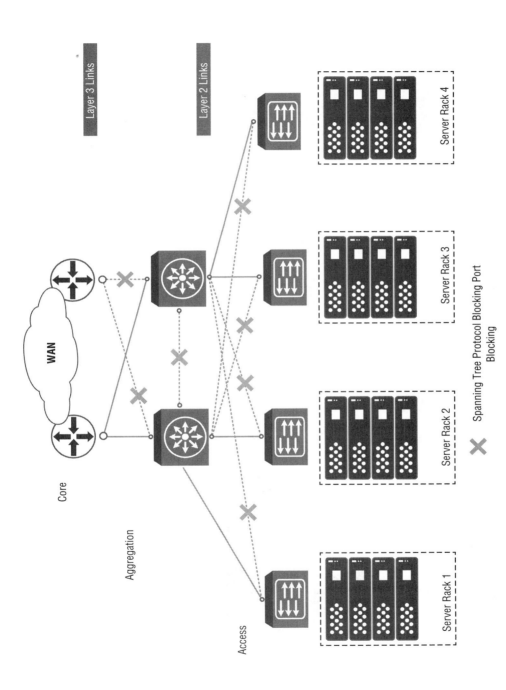

Figure 4.20
The core, aggregation, and access network model

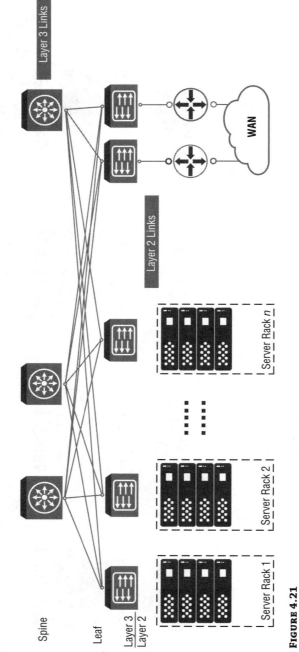

FIGURE 4.21
Leaf-spine network model

BUILDING-BLOCK RACK ARCHITECTURE

In Virtual SAN design, it is considered good practice to deploy multiple racks for the server hardware, and to span the vSphere hosts and clusters across racks, to minimize the impact of hardware failure. When wiring for redundant power, racks should have two power distribution units (PDUs), each connected to separate panels in the data center, with the assumption that the distribution panels are in turn connected independently to different uninterruptable power sources (see Figure 4.22).

This design is also particularly pertinent to the concept of Virtual SAN fault domains, which are presented later in this chapter.

BANDWIDTH REQUIREMENTS

VMware supports hybrid Virtual SAN configurations with both 1 Gb/s and 10 Gb/s Ethernet for network adapter uplinks. However, depending on the number of virtual machines and their workloads, Virtual SAN I/O can flood a 1 Gb/s network, causing contention, and therefore become a bottleneck in the design. Typically, network bandwidth performance has more impact on host evacuation and rebuild times than on workload performance. For instance, during intensive I/O scenarios, such as rebuild and synchronization operations, or other highly intensive disk operations, such as cloning or deploying a virtual machine, I/O may become a limiting factor in a 1 Gb/s environment.

It is highly recommended that enterprise or service provider production designs be based only within a 10 Gb/s infrastructure, in order to achieve the highest levels of performance for virtual machine workloads. Without 10 Gb/s support, it is highly likely that such a design will suffer significantly degraded performance.

In the all-flash Virtual SAN model, using 10 Gb/s Ethernet network uplinks is a mandatory requirement. The reason for this is the significantly increased performance provided through an all-flash configuration, which consumes much higher levels of network bandwidth as a result of the increased speed of the flash devices in the disk group. Also supported are 40 Gb/s Ethernet network adapters. However, even though Virtual SAN may not use this full bandwidth effectively in itself, this solution type may be an appropriate design option for extremely intensive I/O workloads, operating on an all-flash configuration across a shared network platform.

Another design consideration in a leaf-spine architecture is the oversubscription factor. As a result of the fully meshed topology and port density constraints, leaf switches are typically oversubscribed for bandwidth. For instance, a fully used 10 Gb/s Ethernet uplink, used by the Virtual SAN network, may achieve a throughput of only 2.5 Gb/s on each host, if the leaf switches are oversubscribed at a 4:1 ratio. The reason for this, as illustrated in Figure 4.23, is that the Virtual SAN traffic must traverse across the spine. The impact of this type of oversubscribed network topology, and the actual bandwidth available, should be considered a factor for Virtual SAN network design when using a leaf-spine architecture.

JUMBO FRAMES

As with the iSCSI and NFS protocols discussed in Chapter 3, "Fabric Connectivity and Storage I/O Architecture," the Virtual SAN port group (VMkernel port) should, if possible, be configured to use jumbo frames, with an MTU value of 9,000, and be carried end-to-end at that maximum transmission unit throughout the Virtual SAN network infrastructure. The design must ensure end-to-end configuration of jumbo frames. Otherwise, it may cause higher latencies, if any network component has to fragment the frame into 1,500-byte chunks.

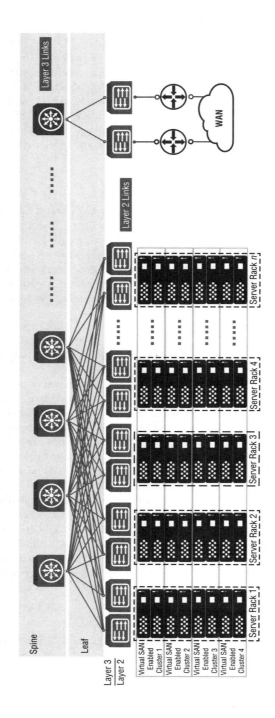

Figure 4.22
Virtual SAN optimum rack design

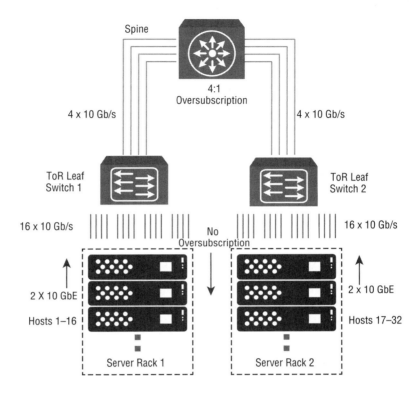

Figure 4.23
Leaf-spine network oversubscription

The principles involved in enabling jumbo frames in a Virtual SAN environment are identical to those set out previously for an IP SAN or NAS infrastructure. Jumbo frames enable you to squeeze more than 1,500 bytes into each frame in order to improve the efficiency of the transport mechanism. In testing, jumbo frames have shown to reduce CPU utilization and improve network throughput, although gains are minimal.

If jumbo frames are already being used or if you're working with a greenfield implementation, they can be recommended for use for the Virtual SAN environment. However, the additional operational overhead of potentially having to configure jumbo frames across a large number of network devices may outweigh any benefits gained from the minor performance enhancement added by including them in the design.

Virtual SAN Traffic Isolation

As with iSCSI and NFS storage protocols, discussed in Chapter 3, it is highly recommended that in any enterprise or service provider environment, isolating Virtual SAN traffic to its own dedicated VLAN should be stipulated as a design requirement. In addition, when multiple Virtual SAN–enabled clusters are used in a design, each cluster should utilize its own dedicated VLAN to isolate its own replication and workload traffic. Isolating Virtual SAN traffic in this way will prevent any interference between the traffic of different clusters and simplify troubleshooting scenarios with the cluster networking configuration.

Network Multicasting

The Virtual SAN 1 to 6.2 releases rely on multicasting to deliver metadata among hosts in the cluster, in order to provide efficient use of network bandwidth. Multicasting is somewhat like broadcasting, although hosts can be on a different subnet, but to only *members* or *subscribers* of a multicast group.

The implementation of multicasting in an environment can seem like a daunting task to some network administrators, as IP multicast must be configured on all the physical switches and routers that are in the data path of Virtual SAN traffic, which could potentially include a layer 3 path.

Multicasting is used by Virtual SAN for internal *intra-cluster* communication, and therefore can be limited to specific port groups and VLANs by using Internet Group Management Protocol (IGMP) version 3 snooping. In fact, it is considered good practice to not allow multicast traffic flooding across all networks and switch ports.

Typically, vendors do not enable multicasting by default, so there is no IGMP snooper configured to allow the passing of traffic. IGMP snooping provides an IGMP snooping querier that can be configured to limit the physical switch ports participating in the multicast group, to only the Virtual SAN VMkernel port uplinks. To ensure that multicast traffic can be passed, IGMP must be configured appropriately in line with the hardware vendor's configuration recommendations, as IGMP snooping varies by switch manufacturer. As noted previously, it is recommended that multicast traffic be enabled only on the network segments, or switch ports, that are being used by Virtual SAN.

Each Virtual SAN–enabled host is configured with a multicast address when it is assigned to the Virtual SAN cluster. In a design deploying multiple Virtual SAN clusters that reside on the same layer 2 network, the default multicast address must be changed on each host within the additional Virtual SAN clusters. Even though VMware supports the use of the same multicast addresses for more than one Virtual SAN cluster across the same network, configuring different addresses will prevent the multiple clusters from receiving all multicast data streams, and therefore reduce the overall networking overhead.

An additional related design consideration comes about if VMware NSX is being used in the design. If this is the case, multicast address ranges need to be considered as a wider part of the overall design, as other network services, including VXLAN can also use multicast traffic, depending on the NSX configuration.

Virtual SAN over Layer 3 Networks

Virtual SAN version 1, with vSphere 5.5 Update 1, did not support layer 3 routing. All designs were restricted to be configured within the same layer 2 subnet, as the Reliable Datagram Transport (RDT) could not be stretched into different network segments. With the release of vSphere 5.5 Update 4 and vSphere 6, Virtual SAN could be configured over layer 3 networks.

If the design requires Virtual SAN hosts to span across layer 3 boundaries, the design must ensure that the networks are configured correctly to enable multicast connectivity. The use of Virtual SAN over layer 3 networks is pertinent to both Virtual SAN failure domain design, and Virtual SAN stretched clusters. Failure domains are addressed later in this chapter, and stretched clusters are covered in Chapter 5, "Virtual SAN Stretched Cluster Design."

Physical Network Quality of Service

Quality of service (QoS) can be implemented alongside NIOC in a Virtual SAN environment, to allow an allocated amount of the network bandwidth, based on shares, to be assigned to

Virtual SAN traffic. This configuration is most useful in a design that uses shared 10 Gb/s network adapters, but does require that a vSphere Distributed Switch form part of the design. As addressed earlier in this chapter, when used in conjunction with NIOC, QoS can be used to prioritize network resources, as defined by the NIOC and QoS policies.

Internet Protocol Version 6 Support

To support Internet Protocol version 6 (IPv6), Virtual SAN 6.2 or later is required. Virtual SAN 6.2 can operate Virtual SAN through the vSphere Web Client in either native IPv6-only mode, with all network communications through IPv6, or a mixed IPv4 and IPv6 mode, for operations during the network migration process from one protocol to the other. using IPv6 as part of a design does not impact the Virtual SAN requirement for layer 2 or layer 3 multicast support across the network.

By providing IPv6 support, Virtual SAN aligns with vSphere support for IPv6, and addresses the requirements of customers, such as service providers or federal governments, who are moving toward this most recent version of the Internet Protocol.

Firewall Design Requirements

A layer 3 network design or a Virtual SAN stretched cluster configuration may require firewall rule sets to be configured to pass Virtual SAN traffic. If this is the case, appropriate firewall rules will be required to be configured on the security devices in order to ensure that Virtual SAN I/O and metadata traffic can traverse the edge boundaries freely. Table 4.12 defines Virtual SAN's network firewall requirements. However, also keep in mind that in such a design, other vSphere traffic types may also be required to traverse the same firewalls.

TABLE 4.12: Virtual SAN firewall port requirements

Virtual SAN Internal Component Service	Protocol	Port(s)	Description
Virtual SAN Vendor Provider (VSANVP)	TCP/Unicast	8080	vCenter to ESXi Inbound and outbound
Virtual SAN Clustering Service (CMMDS)	UDP/Multicast	12345, 23451	Inbound and outbound
RDT (Reliable Datagram Transport)	TCP/Unicast	2233	Inbound and outbound
VSAN Observer	TCP/Unicast	8010	Employed for capturing performance statistics in Virtual SAN

Virtual SAN Network Design Summary

Networking design is fundamental to the overall success of the architecture, implementation, performance, and stability of the Virtual SAN environment. It is critical that the Virtual SAN network be at the forefront of all key design decisions through the process of creating the proposed architecture. The following are the major design decisions required to be at the foundation of every Virtual SAN network architecture:

- The Virtual SAN network configuration.
  - Network speed requirements
  - Virtual switch version to be used
- Use of jumbo frames.
- Network multicasting requirements.
- Adapter teaming and failover considerations.
- The logical separation of network traffic and load considerations with 802.1Q VLAN tagging, to segment Virtual SAN traffic into dedicated VLANs.
- The use of QoS tagging and Network I/O Control.
- Redundancy, by using at least two active physical NIC adapters per host (preferably 10GbE). Where possible, provide adapter redundancy across physical devices, in order to protect against host PCIe slot failure.
- Physical adapters should be connected to redundant physical switches, which are designed and configured so as to tolerate a single physical switch failure.

Virtual SAN Storage Policy Design

We have already highlighted how VMware's next-generation storage products offer a new way to manage virtual machine performance and availability. When using Virtual SAN, in order to stipulate the characteristics and availability requirements of a virtual machine, we use *storage policies*. These storage policies employ rule sets that allow vSphere storage administrators to create instructions that can use the capabilities advertised by the storage. These rule sets allow for control over capacity, performance, and availability of individual virtual machines, or even virtual disks, as opposed to capabilities being defined at a datastore level, as when using the classic storage model, outlined in Chapter 2.

This section covers the SPBM framework capabilities offered by Virtual SAN. These capabilities are typically combined to create storage policies, which when assigned to a virtual machine, define the characteristics of its attached storage. Storage policies are assigned when deploying a new virtual machine, but can also be modified on the fly—for instance, when the workload's storage requirements change. These changes can be performed without impacting the virtual machine's running state.

Storage Policy–Based Management Framework

Virtual SAN leverages the storage policy framework that is employed in combination with vSphere APIs for Storage Awareness (VASA), to expose storage characteristics to vSphere.

This Storage Policy-Based Management Framework (SPBM) is a storage policy mechanism that enables policy-driven and virtual machine–centric provisioning. The SPBM framework (illustrated in Figure 4.24) is the foundation component of the software-defined storage management and control plane. SPBM allows vSphere storage administrators to simplify provisioning challenges, such as capacity planning, and allow differentiation between the service level and performance level of virtual machine workloads.

Figure 4.24
Storage policy–based management framework via the vSphere web client

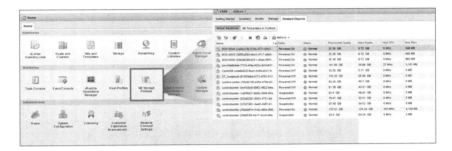

SPBM defines standard storage profiles for workloads, optimizes the virtual machine provisioning process, and eliminates the need to provision storage on a case-by-case basis. This mechanism also provides significant improvements in automation through tools such as PowerCLI, the vSphere API, vRealize Automation, vCloud Director for Service Providers, VMware Integrated OpenStack (VIO), or any other application that can leverage the vSphere SPBM API to automate storage management and operations in the software-defined storage-driven data center.

This framework functions by exposing the underlying available storage capabilities, which can then be assigned to specific virtual workloads at either a virtual machine or virtual disk level via the storage policy constructs, illustrated in Figure 4.25.

Virtual SAN Rules

A storage policy rule refers to a metadata tag for a vendor-specific, user-defined value that indicates storage capability being offered up for use by the storage system. This metadata tag and its associated value, when referenced together, ensure that the rule must be adhered to, in order for compliance of the object to be achieved.

For instance, Virtual SAN, as a storage provider, publishes several capabilities through VASA, one of which is the FTT. This capability is used by the SPBM mechanism to define the number of host, disk, and/or network failures that a virtual machine object, assigned with this policy, can tolerate. The minimum configuration value for the FTT capability is 0, the default value is 1, and the maximum value is 3.

If we created a rule that referenced this capability with a value of 2, this rule would require that for any virtual machine object assigned this capability, the resulting storage policy would look for a Virtual SAN datastore that can meet the requirements to support $n + 1$ copies of the virtual machine object, and $2n + 1$ hosts with the storage capacity required.

For vSphere storage administrators, after reviewing the storage capabilities published and identifying which capabilities and values are required in a storage policy, the next step is to group the rules into a rule set, which can be referenced in a storage policy.

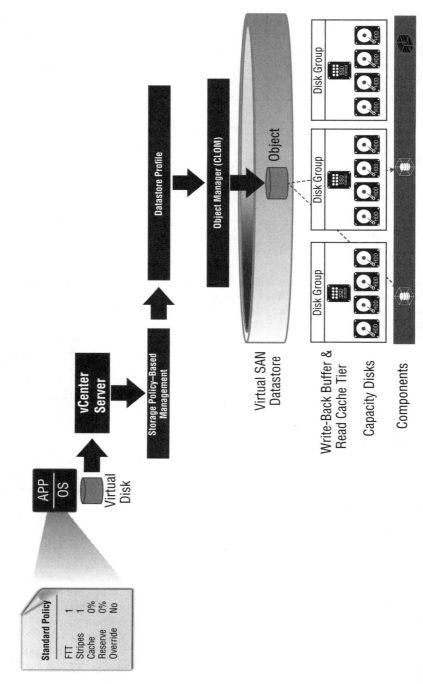

FIGURE 4.25
Virtual SAN storage policy object provisioning mechanism

Virtual SAN Rule Sets

A *rule set* is a group of rules that describes the storage requirements of virtual machine objects. A storage policy can include one or more rule sets that define the virtual machine's requirements for Virtual SAN storage resources (see Figure 4.26). Typically, the rule set is used to determine the configuration of the virtual machine on the Virtual SAN datastore. For instance, we can determine the redundancy level for the objects that make up the virtual machine, such as the VMDK objects. It is also possible to specify performance characteristics or requirements of the workload. These rules are then specified when creating a new virtual machine storage policy.

FIGURE 4.26
Storage profile rule sets

As illustrated in Table 4.13, each storage capability available through Virtual SAN contains a default value, a minimum value, and a maximum value. We can choose any combination of Virtual SAN capabilities to create a rule set for a storage policy.

In this example, the rule set contains two rules and targets a Virtual SAN datastore that has the capacity to create three replicas of the virtual machine object, with each replica existing on a minimum of two disks. The storage capabilities illustrated in the following table, which define the rule sets, are addressed in detail in the following section.

TABLE 4.13: Example Virtual SAN rule set

Storage Capability (Key)	Quantity or Quality (Value)	Options
Number of failures to tolerate	2	Default value is 1 Maximum value is 3
Number of disk stripes per object	2	Default value is 1 Maximum value is 12
Object space reservation	0	Default value is 0% Maximum value is 100%
Flash read cache reservation	0	Default value is 0% Maximum value is 100%
Force provisioning	No	Default value is No Optional Yes

Number of Failures to Tolerate

The first capability to be addressed is the *Number of Failures to Tolerate* (FTT). This policy is one of the configurations that is at the core of Virtual SAN's availability mechanism. It is this policy that controls the number of replicas or mirrored copies of virtual machine components existing across the distributed datastore. This policy can be applied either at a virtual machine level or to individual virtual disks. The FTT capability is also critical when planning and sizing Virtual SAN datastore capacity, as it directly relates to the consumption of physical disk space consumed by the virtual machines across the datastore.

As illustrated in Figure 4.27, for every n failures to tolerate, $n + 1$ copies of the data are required, and $2n + 1$ hosts (which are contributing storage to the cluster) are required. In this example the FTT is configured as 1 (FTT = 1), and therefore, the virtual machine disk has two replica copies of the components created, distributed across the Virtual SAN–enabled cluster nodes. If the number were to be modified to 2 (FTT = 2), three replica copies would be created. If a failure occurred on one of the host's mechanical disks that stores one of the mirrored copies, the replica would take over.

The key design factor associated with the FTT capability is that it can lead to a significant increase in storage capacity utilization when compared with the actual number of stored virtual machine disks, as replicas of the disks are created for each increment in the capabilities value. Another design consideration is that additional replica copies also increase the total number of components on a host, the maximum number of which can be 9,000 in a Virtual SAN 6 design, but only 3,000 in earlier releases.

The storage policies associated with this capability should be assigned based on the availability requirements of the virtual machine or disk in question. The default value is to have an FTT of 1, unless the default policy is modified. The maximum FTT value is 3, which requires a minimum of seven hosts or fault domains in the Virtual SAN cluster. The number of hosts required to meet an FTT policy requirement is set out in Table 4.14, and an example of the distribution of components across the Virtual SAN datastore is illustrated in Figure 4.27.

Table 4.14: The number of failures to tolerate capability host requirements

Number of Failures to Tolerate	Mirror Copies/Replicas	Minimum Number of ESXi Hosts/Fault Domains
0	1(single data copy)	1
1 (default)	2	3
2	3	5
3	4	7

In only a few use cases, a storage policy that includes an FTT of 0 would be used, particularly in a production environment. Typically, this policy is utilized only in a testing lab, where disk capacity is constringed and data loss is not of major concern, or in a virtual desktop infrastructure, where floating linked clones are being used, and therefore no data persistency is required.

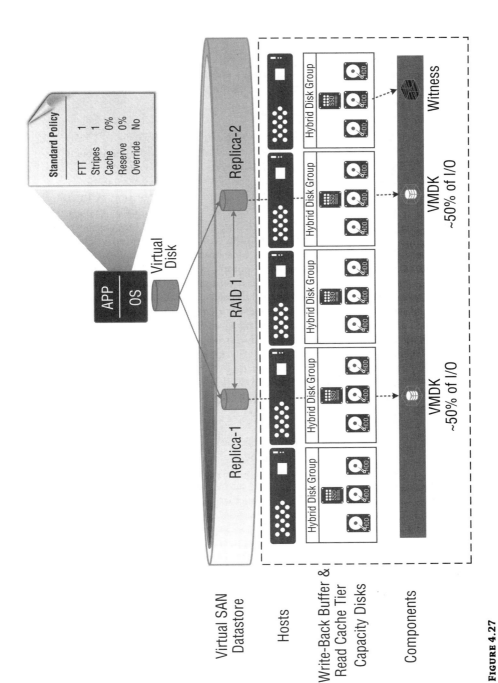

FIGURE 4.27
Number of failures to tolerate component distribution

From a design perspective, the FTT value of 1 should typically be sufficient, unless specific availability requirements exist for a particular workload or application. If this is the case, configure a separate policy that can provide the higher level of fault tolerance required to address that specific use case. This policy should be used sparingly with virtual machines, in order to limit the impact of disk capacity utilization across the Virtual SAN datastore.

All-Flash Erasure Coding (RAID 5/6)

Introduced in Virtual SAN 6.2, erasure coding adds to the existing FTT mirroring capability (RAID 1) by providing a RAID 5 or RAID 6 type of capacity reduction mechanism over the Virtual SAN network. By using erasure coding in an all-flash configured design, usable storage capacity can be increased, while retaining data resiliency and availability.

As addressed in Chapter 2, RAID 5 adds single parity protection, with double parity protection offered by the use of RAID 6. In a Virtual SAN–enabled cluster, erasure coding allows the striping of data and parity bits across nodes in the cluster. For instance, as shown in Figure 4.28, in a RAID 5 configuration, with the minimum of four hosts configured, erasure coding uses a 3 +1 logic, meaning one host can fail without data loss occurring. As a result, less capacity is required to host a virtual machine, when compared with the FTT mirroring mechanism (RAID 1). This can result in significant savings in usable capacity across hosts. For example, a virtual machine with a 20 GB virtual disk would normally, assuming FTT = 1, require 40 GB of raw disk capacity. However, by using RAID 5 erasure coding, this requirement is reduced to approximately 27 GB of raw disk capacity, realizing a savings of approximately 30 percent.

If an additional level of availability is required by the workload, the erasure coding mechanism also offers a RAID 6 configuration with dual parity, as shown in Figure 4.29. With the RAID 6 configuration, two host failures can be tolerated, in a similar way to FTT = 2, when using RAID 1 mirroring. In this same example scenario, where a mirrored FTT = 2 for a 20 GB virtual disk is being used, the disk space required would be 60 GB. However, with RAID 6 erasure coding, only 30 GB would be required.

From a design perspective, the parity data is distributed across all hosts in the cluster, and there is not a dedicated host for parity data. Erasure coding in a RAID 6 type of configuration requires that 4 + 2 hosts are used, meaning that 6 hosts is the minimum required number of nodes for a design using this type of availability architecture.

Unlike deduplication and compression, addressed previously, erasure coding provides a guaranteed capacity reduction. For designs that are not using thin-provisioning policies and that have data that is already compressed and deduplicated, this solution offers a predictable increase in storage capacity.

This Virtual SAN 6.2 feature uses SPBM in the same way as the RAID 1 mirroring functionality of FTT, and like the other capabilities, can be applied on a granular, per virtual disk basis, or alternatively can be assigned at the virtual machine level, as illustrated in Figure 4.30.

The key design factors associated with the choice of RAID technology, just as when using a classic approach to storage, comes down to ascertaining the appropriate balance between availability, disk capacity utilization, and performance. This ensures that the vSphere storage administrator can provide the most appropriate method of fault tolerance, which can be used workload by workload.

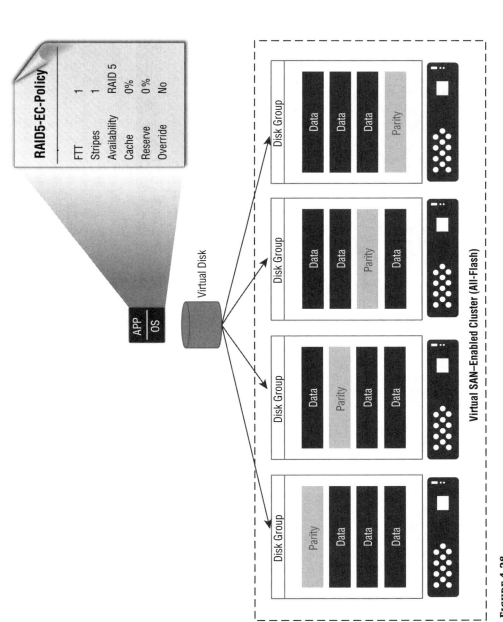

FIGURE 4.28
RAID 5 erasure coding

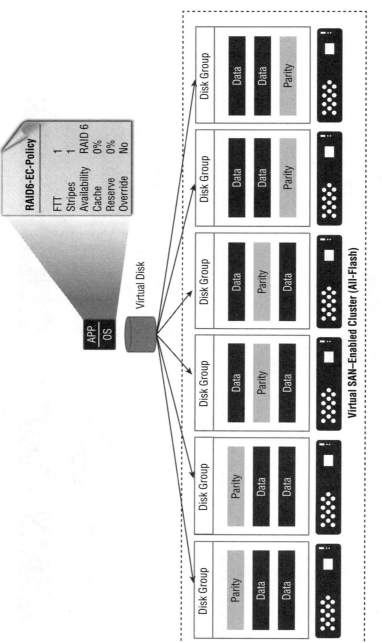

Figure 4.29
RAID 6 erasure coding

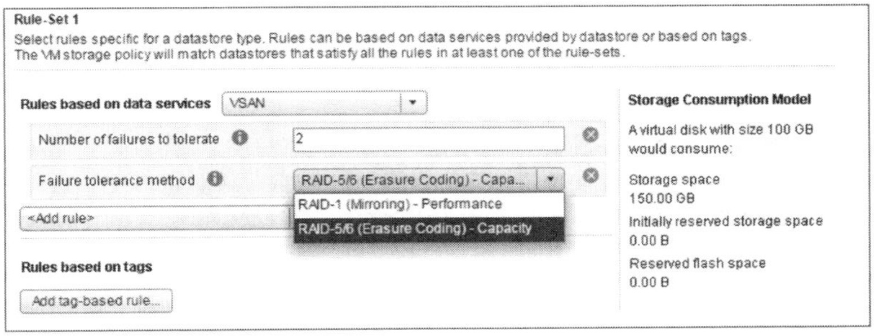

FIGURE 4.30
Erasure coding web client configuration

However, it is important to understand that RAID 1 mirroring provides optimum performance but is expensive in terms of used capacity. By using these RAID 5 and 6 erasure coding configurations, we can help ensure the same levels of availability for the workloads as mirroring, but by using these RAID types, we can achieve capacity reduction by as much as 50 percent over that of a similar RAID 1 design.

Having said that, as you would expect, and for the same reasons as found when using these RAID 5 and RAID 6 technologies within the classic storage model (addressed in Chapter 2), these methods of providing fault tolerance require an additional write overhead over mirroring, as a result of data placement and parity. For this reason, an erasure coding configuration is not supported (at the time of writing), when:

- Employing the hybrid disk group model
- In a Virtual SAN stretched cluster design
- In a remote office / branch office (ROBO) implementation, due to the lack of nodes

Table 4.15 and Table 4.16 present the host and capacity requirements for the number of FTT, in how it relates to failure scenarios and erasure coding.

TABLE 4.15: RAID 1 capacity and configuration requirements

SPBM FTT Value	Tolerated Failures	Minimum Required Hosts	Total Capacity Requirement *	100 GB Disk Capacity Utilized *
FTT = 0	0	3	1x	100 GB
FTT = 1	1	3	2x	200 GB
FTT = 2	2	5	3x	300 GB
FTT = 3	3	7	4x	400 GB

TABLE 4.16: Erasure coding capacity and configuration requirements

SPBM FTT Value	Minimum Required Hosts	Total Capacity Requirement	Tolerated Failures	100 GB Disk Capacity Utilized **	Parity	Space Savings over Mirroring *
FTT = 0	N/A	N/A	N/A	N/A	N/A	N/A
FTT = 1	RAID 5 3 + 1 = 4	1.33x	1	133 GB	One parity block	Minus 33%
FTT = 2	RAID 6 4 + 2 = 6	1.5x	2	150 GB	Two parity blocks	Minus 50%
FTT = 3	N/A	N/A	N/A	N/A	N/A	N/A

* Without deduplication and compression being enabled on the cluster.

** The space savings that come with RAID 5 and 6 are at the cost of performance overhead and I/O.

Finally, as erasure coding is supported in only an all-flash design, the effect on I/O performance is negligible and will therefore not impact workloads in all but the most latency-sensitive application use cases.

Number of Disk Stripes per Object

The second capability to be addressed is the *Number of Disk Stripes per Object*, also known as *stripe width*, which defines the number of mechanical or capacity flash disks across which each replica of a storage object is distributed. A higher configured value of this capability, in specific use cases, may result in better read performance in a Virtual SAN configuration, but may also result in higher utilization of system resources.

From a write-performance perspective, as all write requests are written to the endurance flash device in both hybrid and all-flash configurations, increasing the *stripe width* value across multiple capacity devices will not improve performance. This is the case because there is no way of ensuring that the additional stripe, or stripes, configured by the policy will result in using a different write-buffer flash device. The newly created stripe could equally be placed on a mechanical disk in the same disk group as the original component, and as such, will use the same write-buffer flash device.

The only way that write performance might be improved through an increased stripe width is via the destaging mechanism. For instance, if a large number of write operations are required to be destaged from the write buffer to capacity disk, then potentially, increasing the stripe width could improve the performance of the destaging process.

In the hybrid Virtual SAN model, an increase in stripe width could improve read performance in a scenario where a workload is experiencing a large number of dirty reads, or read cache misses, as those reads must be fetched from slower mechanical disks. For instance, if a virtual machine is performing 2,000 read operations per second and experiencing a flash cache hit rate of 90 percent, then 200 read operations per second are being serviced from mechanical disk at a much slower rate. In addition, assuming there is no other workload on the mechanical disk, and its performance specification allows it to provide only up to 150 read operations per second, then it is clearly unable to service all of these read request operations. As a result, increasing the stripe width in a use case such as this will help meet the virtual machine's read workload requirements.

As the all-flash configuration targets almost all read requests directly from the capacity flash layer, without the *hybrid model* read-buffer mechanism being used, spreading those read operations across multiple capacity flash devices could, with extremely read-intensive workloads, also improve read performance.

In general, for most designs, the default stripe width value of 1 should meet most, if not all, workload requirements. The Number of Disk Stripes per Object capability should be modified only if a specific use case, such as write destaging or read cache misses, has been identified as a bottleneck through performance monitoring.

Figure 4.31 depicts a scenario in which FTT has been configured with a value of 1, and the Number of Disk Stripes per Object value has been set as 2.

OBJECT SPACE RESERVATION

The *Object Space Reservation* (OSR) capability defines the percentage of the virtual machine's logical disk (VMDK) storage object capacity that should be reserved during initialization. All other objects associated with the virtual machine will remain thin provisioned, with the exception of the swap object, which is 100 percent reserved by default when the advanced host level option `SwapThickProvisionedDisabled` (referred to as *swap efficiency* and addressed earlier in this chapter) is not being used.

In Virtual SAN, with the exception of the swap object, all objects are by default deployed thin provisioned, similar to an NFS storage solution. This rule defines the percentage of the logical size of the storage object that is reserved during initialization, in order to prevent the overprovisioning of the Virtual SAN datastore. The OSR amount is specified with a value that is a percentage of the total object address space (see Figure 4.32).

In effect, this is the property that is used for specifying the thick provisioning of a storage object. In vSphere, there are two types of thick disk: lazy zeroed thick (LZT), where the initial zeroing of the disk blocks is delayed until the first write, and eager zeroed thick (EZT), where the disk blocks are preallocated with zeros at the time of disk provisioning.

If an OSR rule is set to 100 percent, the storage capacity requirements of the virtual machine are provided up front, but initial zeroing of the disk blocks is delayed until the first write, as in the LZT format. The purpose of this capability is to prevent vSphere storage administrators from overcommitting storage on the Virtual SAN datastore. The minimum OSR value is 0 percent, which equates to a thinly provisioned disk, and the maximum configurable value is 100 percent, which represents an LZT disk.

When considering the Virtual SAN design sizing, there is no reason not to overcommit capacity on a Virtual SAN datastore. After all, as addressed in Chapter 2, we've been doing it for years on thin disk pools on classic storage systems. However, a design must take this into account and ensure that sufficient monitoring mechanisms are in place to mitigate the risk of exceeding physical disk capacity as a result of the overcommitment of storage resources. In addition, you must take into account, when defining datastore capacity requirements, the impact of this capability if used on that sizing exercise.

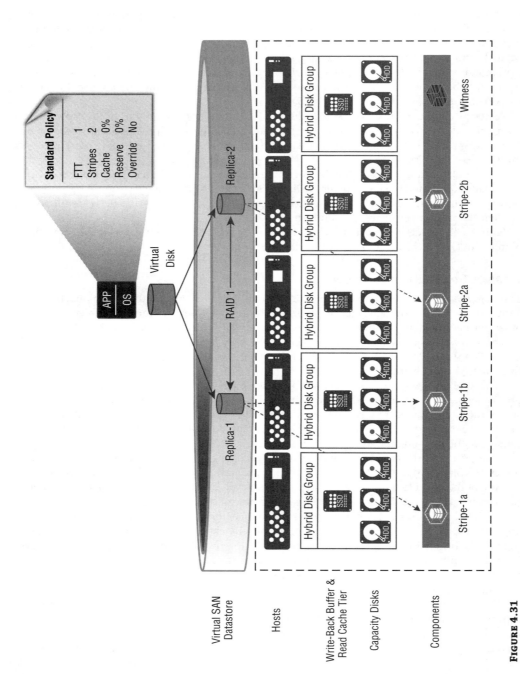

FIGURE 4.31
The Number of Disk Stripes per Object component distribution

FIGURE 4.32
Object space reservation capability

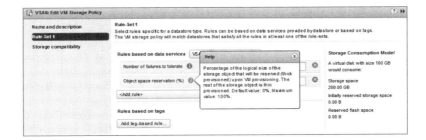

FLASH READ CACHE RESERVATION

The *Flash Read Cache Reservation* defines the proportion of logical address space to be reserved on the flash device as read cache. This capability can be used in a hybrid disk group design only.

In the Virtual SAN hybrid model, the amount of capacity reserved on the flash device as read cache for the storage object is specified as a percentage of the logical size of the storage object, or the virtual disk VMDK file. This value is specified as a percentage, with up to four decimal places, providing a granular unit size. By using such a deterministic approach to the Flash Read Cache Reservation configuration, vSphere storage administrators can express values of less than 1 percent as part of a storage policy assigned to a virtual machine or specific virtual disk (see Figure 4.33).

FIGURE 4.33
Flash read cache reservation capability

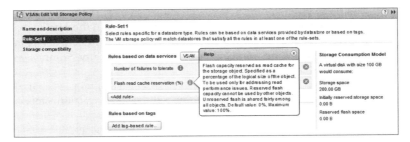

For instance, say a virtual machine has a 1.5 TB virtual disk and is assigned a storage policy that limits the read cache reservation to 1 percent increments. In the resulting configuration, the virtual machine read cache reservation would be 15 GB, which for most workloads is significantly higher than required. In addition, as you can see in this example, a user-defined Flash Read Cache Reservation policy can end up using significant amounts of the read cache, which is likely to result in an overall increase in read cache misses and therefore have a negative impact on workload performance.

It is important to understand that configuring this capability is not required in order for a virtual machine to be assigned a read cache reservation. By default, in a hybrid-configured Virtual SAN–enabled cluster, all virtual machines get allocated an equal share of the read cache capacity, which constitutes a maximum of 70 percent of the flash device. Therefore, this value should remain at the default configuration, unless performance monitoring has identified a workload problem that could be resolved by allocating a dedicated read cache reservation.

Even if this is the case, as you have seen, the vSphere storage team should use caution when configuring a read cache reservation requirement for a virtual machine. What appears to be a low value, in percentage terms, can easily allocate a significant amount of read cache resources. As a result of such a configuration, available read cache capacity will be quickly exhausted, and a significant impact will be felt on other workloads as the number of read cache misses increases, especially if the object space reservation value on all other virtual machines is defined as zero.

Finally, as already stated, Virtual SAN all-flash configurations do not use the read cache mechanism, but instead target all read requests directly to capacity flash devices to be serviced. Therefore, the Flash Read Cache Reservation capability is applicable in only a hybrid disk group design.

Force Provisioning

The *Force Provisioning* capability is used to force the application of the rules specified in the virtual machine's profile. If the value of this capability is not defined as zero, an object will be provisioned on the Virtual SAN datastore, even if the policies defined in the virtual machine's storage policy cannot be satisfied. Specifically, the Force Provisioning policy allows the Virtual SAN provisioning mechanism to violate capabilities defined in the *Number of Failures To Tolerate*, the *Number of Disk Stripes Per Object* and the *Flash Read Cache Reservation* policy settings, during the initial deployment of a virtual machine. However, if there is insufficient capacity on the Virtual SAN cluster to satisfy the reservation requirements of at least one replica object, the virtual machine provisioning process will fail, even if the Force Provisioning capability is enabled.

If these provisioning actions are allowed to occur, the virtual machine will be highlighted in the vSphere Web Client as being noncompliant in the virtual machine Summary tab, and also the storage policy views for that virtual machine. If the Virtual SAN datastore can satisfy the virtual machine storage policy, the virtual machine Summary tab will display the virtual machine as compliant.

However, if this is not the case, either because of failures or the Force Provisioning capability being used, then the virtual machine will be shown as noncompliant, as depicted in Figure 4.34.

FIGURE 4.34
Virtual machine compliance status

If the virtual machine is able to be deployed while being noncompliant, when additional resources become available within the cluster at a later stage, Virtual SAN's internal mechanisms will automatically deploy the required components, in order to ensure the object is brought into a compliant state. These operations would still occur, even if it meant that all of the newly provisioned storage resources were exhausted as a result.

Configuring the Force Provisioning capability also has an impact on the vSphere High Availability (HA) feature, as well as placing the Virtual SAN cluster into maintenance mode, and using the disk and disk group removal functions (see Figure 4.35).

FIGURE 4.35
Force provisioning capability

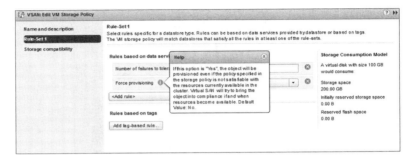

For instance, when a virtual machine is in a noncompliant state, as Forced Provisioning has been enabled, Full Data Evacuation when entering into maintenance mode will be as if you had chosen the Ensure Accessibility option, allowing the virtual machine objects to exist in a reduced state of availability, and therefore increasing risk. These are key design factors when considering enabling the Forced Provisioning capability as part of any solution, although it applies to only noncompliant virtual machine objects.

QUALITY OF SERVICE

The Quality of Service (QoS) feature was introduced into Virtual SAN in version 6.2, to help provide control, visibility, and reporting into the I/O per second consumed by each virtual machine disk object.

Like the use of Storage I/O Control, in a legacy storage environment, this feature assists in trying to eliminate noisy neighbor issues, and helps configure and control performance limits for business-critical workloads. Again, as in the case of Storage I/O Control, the term *noisy neighbor* is used to describe one or a small number of workloads that monopolizes the majority of available I/O resources and therefore negatively impacts other applications running on the same storage platform.

Of course, an ideal design has more than enough resources to meet the requirements of the running applications. However, over time environments grow, and applications change their behavior. Therefore, in some Virtual SAN designs, it may be necessary to use limits on I/O per second, which can be assigned to specific disk objects in order to maintain SLAs over the duration of the storage platform's life.

For instance, cloud service providers might wish to use this feature to create differentiated services on a single cluster of storage resources, by using the QoS capability as a means to help define service offerings, such as tier 1, tier 2, and tier 3 levels of service, as illustrated in Figure 4.36. In addition, customers who want to mix diverse workloads on a single Virtual SAN–enabled cluster may also be interested in being able to keep workloads from impacting one another.

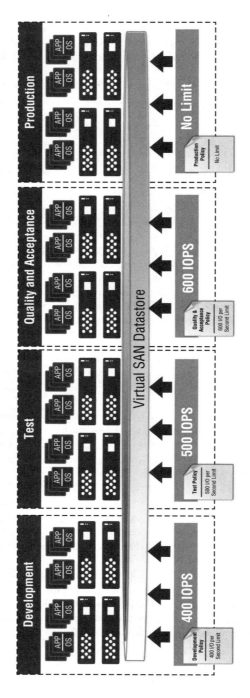

FIGURE 4.36
Quality of service (QoS) use case

Other use cases for this feature include environments that mix both high-I/O and low-I/O workloads, which could significantly impact the applications running with low disk utilization. Or alternatively, applications that have *bursts* in activity, such as in environments running quarterly business jobs that require high disk I/O, could result in starving other workloads using the same disk resources during this period of intense utilization, and therefore impacting SLAs only during these fixed periods.

The QoS feature is provided as an SPBM rule, linking this capability into the existing framework. This feature can be enabled on a per VMDK object basis, or be assigned to an entire virtual machine without impacting its running applications (see Figure 4.37).

Figure 4.37
Storage policy–based management quality of service rule

QoS, as implemented by Virtual SAN, applies the same mechanism to both reads and writes, and is normalized to a 32 KB block size. Therefore, in the preceding example, where a rule assigning an IOPS limit of 500 is being applied, this is by default normalized down at a 32 KB block size, so this 500 IOPS limit will result in 250 64 KB blocks being used, in order to limit the targeted object.

Default Storage Policy

Virtual SAN provides a default storage policy that is automatically created in SPBM when the cluster is first created. The default storage policy is assigned on any virtual machine that is created without an explicit SPBM storage policy being configured.

It is also possible for vSphere storage administrators to designate a custom virtual machine storage policy as a preferred default policy. This provides the ability in a large platform of multiple Virtual SAN clusters for each Virtual SAN datastore to have a different default policy assigned.

The Virtual SAN default storage policy is shown in Table 4.17. These values can be modified by using the `esxcli vsan policy setdefault` command, if required to be changed for a tiered storage design.

Table 4.17: Default storage policy values

Capability	Value
Number of Failures to Tolerate	1
Number of Disk Stripes per Object	1

TABLE 4.17: Default storage policy values *(CONTINUED)*

CAPABILITY	VALUE
Flash Read Cache Reservation	0
Object Space Reservation	0
Force Provisioning	Disabled

Application Assessment and Storage-Policy Design

Typically, a Virtual SAN storage policy is made up of one or more rule sets, used in combination to create the policies required for the design. Understanding the customer's workload requirements is key to designing appropriate policies for their virtual machines.

The storage policies, which define the characteristics and capabilities of virtual machine storage, are created to allow different levels of service to be provided to workloads. If no specific policy is applied to a virtual machine, the default policy, tolerating a single failure and having a single disk stripe, is associated during provisioning. Creating multiple custom policies for Virtual SAN workloads allows for virtual machine–centric policies to be created and assigned, and allows for changes to be made to running virtual machine capabilities nondisruptively as required. Although using the default policy exclusively for all virtual machines can be perfectly acceptable in a design, this does not demonstrate the granular policy-driven storage concept being offered up by Virtual SAN, and as addressed in Chapter 8.

Unless the Force Provisioning capability has been enabled, Virtual SAN guarantees the storage policies that have been configured and assigned to virtual machines. If a storage policy cannot be met during deployment, virtual machine provisioning will fail. Although almost any combination of the available capabilities can be configured in a storage policy, some settings in an all-flash model, such as Flash Read Cache Reservation, may not apply. Explaining the available policies to vSphere storage administrators is critical in order to ensure that designed and configured policies are optimized.

The storage policies are enforced by Virtual SAN to ensure that the appropriate virtual machine receives exactly the capabilities that the vSphere storage administrator has defined. Virtual SAN also verifies that if the virtual machine falls out of compliance, corrective action can be automatically taken by its internal mechanisms to bring it back in line with its assigned storage policy.

An additional benefit of Virtual SAN storage policies is provided by the ability to modify the associated virtual machine policy at any time, by switching it *on the fly*. Changes such as these might result from the workload or availability requirements of the virtual machine changing, after the provisioning process has been completed, or as part of day-two operations, such as where application owners have identified a need for higher levels of redundancy.

Storage policies should be configured based on the application's requirements for availability and performance. It is this storage policy feature that gives Virtual SAN its SDS credentials, as it allows storage administrators the ability to adjust how a disk performs based on its assigned policies. Therefore, it should be seen as a critical component of any Virtual SAN design to define customer application requirements and create storage policies based on business needs that address performance, capacity, and the availability requirements of the targeted application workloads.

As you have seen, SPBM allows for any configuration to become as customized as needed, in order to meet the workload requirements. A storage policy design should start with an assessment of business and application requirements, which is the advised strategy for the architect to take. This means taking a granular approach that addresses requirements such as read, write, disposability, redundancy, and performance to create a storage policy strategy. If you are unable to identify specific use case requirements with the customer, using a general storage policy strategy is advisable. These policies may include the following:

- A performance-centric storage policy
- A capacity-centric storage policy
- A balanced storage policy

This approach may also provide a good storage policy strategy for use cases such as test, development, and proof-of-concept environments, where the application requirements are not yet defined or fully understood by their owners.

For an application assessment, you should start by assessing and defining the application availability requirements, and understand the impact of downtime on stakeholders, application owners, and most important, end users.

To identify these availability requirements, you can use some or all of the following questions. The answers to these questions can help the vSphere storage architect, to gather, define, and clarify the deployment goals of the applications and services that will be running on the Virtual SAN–enabled clusters:

- Are the applications considered business critical to the organization's central purpose? What applications and services do end users require when working?

- Are there any SLAs or similar contracts that define service levels for the applications in question?

- For the application end users, what defines a satisfactory level of service for the applications in question?

- What increments of downtime are considered significant and unacceptable to the business (for example, 5 seconds, 5 minutes, or 1 hour) during peak and nonpeak hours? If availability is measured by the customer, how is it measured?

In addition, is there a business requirement for 24-hour, 7-days-a-week availability, or is there a working schedule, for instance, 9:00 a.m. to 5:00 p.m. on weekdays? Do the services or applications that you are focusing on have the same availability requirements, or are some of them more important than others?

Business days, hours of use, and availability requirements can typically be obtained from end-user leadership, application owners, and business managers. For instance, Table 4.18 provides a simple business application list along with the end-user requirements for availability and common hours of use. These requirements are important to establish, so that downtime when an application is not being used (for example, overnight) may not negatively impact the application service-level agreement.

TABLE 4.18: Example application uptime requirements

Business Application	Business Days	Hours of Use	Availability Requirements
Customer-tracking system (CRM)	7 days	0700–1900	99.999%
Document management system	7 days	0600–1800	99.999%
Microsoft SharePoint (collaboration)	7 days	0700–1900	99.99%
Microsoft Exchange (email and collaboration)	7 days	24 hours	99.999%
Microsoft Lync (collaboration)	7 days	24 Hours	99.99%
Digital-imaging system	5 days	0800–1800	99.9%
Document-archiving system	5 days	0800–1800	99.9%
Public-facing web infrastructure	7 days	24 hours	99.999%

It is also important to establish and understand application dependencies. Many of the applications shown in the previous table consist of components such as databases, application layer software, web servers, load balancers, and firewalls. To achieve the levels of availability required by the customer's business, storage policies and a range of Virtual SAN clusters, with different performance and availability capabilities, may have to be used in the design to meet the requirements.

Other key design factors that may be pertinent to developing a new Virtual SAN architecture may also include the following:

- Do the applications in question have variations in load over time or in the business cycle (for instance, 9:00 a.m. to 5:00 p.m. on weekdays, monthly, or quarterly)?
- How many vSphere host servers are available, and what type of storage is being used for the applications?
- Is having a disaster-recovery option for the services or applications important to the organization?
    - What type of infrastructure will be available to support the workload at the recovery site?
    - Is the recovery site cold/hot or a regional data center used by other parts of the business?
    - Is any storage replication technology in place?
    - What steps must be taken to ensure that the application is accessible to users or customers if failed over to the recovery site?
- Is it possible for some of the Virtual SAN cluster nodes to be placed in a separate site or adjacent data center or data center zone, in order to provide an option for disaster avoidance or recovery, if a serious problem occurs at the primary site?

- What is the current I/O performance and characteristics of the targeted application workload, on a per virtual disk basis?
- Are there any specific application best practices associated with the workload, such as block size?

It is important to examine all options, and carefully consider and understand the impact that the Virtual SAN design will have on the business application or service. Figure 4.38 combines what you have learned about vSphere SPBM and provides key design factors for vSphere storage policies that reflect real-world application-provisioning scenarios.

Even though these policies are assigned either at a virtual machine level, or more granularly at an individual virtual disk level, when assigned, these capabilities are applied differently depending on the object type. Table 4.19 outlines the default policy options and how they are applied to types of objects in Virtual SAN.

TABLE 4.19: Object policy defaults

OBJECT	POLICY	COMMENTS
Virtual machine namespace / VM home	1 Failures to Tolerate	Configurable. However, changes are not recommended.
Swap	1 Failures to Tolerate	Configurable. However, changes are not recommended.
Virtual disks (VMDKs)	User-Configured Storage Policy	Can be any storage policy configured on the system.
Virtual disk snapshots/deltas	Uses Virtual Disk Policy	By default, the same as the virtual disk policy. Changing is not recommended.

Policy defaults for the Virtual Machine Namespace and Swap object types are set by the Virtual SAN system and are not configurable. This ensures that the appropriate protection for these critical virtual machine components is consistently maintained throughout the platform.

If no user-defined policy is configured against the object, the default storage policy of Number of Failures to Tolerate equals 1, and Number of Disk Stripes per Object equals 1, for virtual disks, and virtual disk snapshots will be assigned.

Virtual SAN Datastore Design and Sizing

vSphere architects have over the years become adept at designing and sizing compute clusters based on aggregate CPU and memory utilization, alongside physical-to-virtual CPU ratios. The aim has always been to locate the *sweet spot*, where you can achieve the maximum utilization and consolidation ratio of the hardware, with the correct balance of CPU, memory, and cost. However, when designing a cluster that also includes Virtual SAN, you have a whole range of new sizing considerations to calculate and take into account, for both host and cluster sizing.

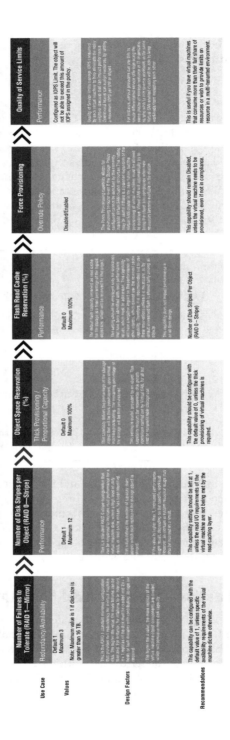

FIGURE 4.38
Storage capabilities and recommended practices

One of the key principles associated with the designing and sizing of Virtual SAN datastores is to optimize storage for consistent performance across the aggregate volume. One of the ways in which Virtual SAN hybrid disk groups minimize the risk of inconsistent performance is through its hybrid tiered architecture, which uses a flash acceleration layer that acts as a read cache and write buffer, combined with mechanical disks for capacity. With this internal architecture in a properly sized Virtual SAN solution, the majority of reads and all writes are serviced by flash, allowing for excellent performance in vSphere environments, which typically perform with highly randomized I/O. As a result, the Virtual SAN architecture also helps minimize what is known as the *I/O blender effect*, commonly seen in virtualized environments.

The I/O blender effect, illustrated in Figure 4.39, occurs when multiple virtual machines all simultaneously start sending I/O to the storage. This typically causes I/O to occur in a highly randomized pattern, which can increase latency in some classic shared storage systems. This usually occurs because data is stored wherever the capacity exists on the disks, in a random fashion. As the number of virtual machines trying to access and write data simultaneously increases, it creates this I/O blender effect. As a result, data gets stored in an increasingly random fashion, forcing the read and write arm of a mechanical disk to move constantly, causing latency. This problem increases as additional virtual machines are deployed, and the amount of data being searched grows, increasing the number of slow responses to I/O requests, and therefore creating a bottleneck in the storage layer, resulting in a slower responsiveness of applications.

However, in Virtual SAN, this is mitigated by all virtual machine write requests being targeted to the caching flash device. Later, when data requires destaging to mechanical disks, from the flash acceleration layer, assuming the hybrid model is in place, the destaging operation occurs predominantly made up of sequential I/O. This allows Virtual SAN to efficiently take full advantage of the full sequential I/O capability of the underlying mechanical disks.

The second design principle used for optimizing Virtual SAN, for consistent aggregate performance across the distributed datastore, is its architecture not depending on data locality to guarantee performance. This was addressed in more detail earlier in this chapter when focusing on data locality and caching algorithms.

Hosts per Cluster

Virtual SAN currently supports a minimum of 3 hosts in a cluster and a maximum of 64 hosts in both hybrid and all-flash configurations, although two node clusters with an external witness are also now supported in a ROBO design. In Virtual SAN 1, a maximum of 32 nodes per Virtual SAN cluster is supported, which aligns with vSphere's maximum configuration limitations.

However, despite supporting three nodes in a cluster, this would not typically be recommended for a production environment or running business-critical workloads. The reason for this recommendation is that three-node clusters have redundancy and availability limitations that are mitigated when the fourth node is added to the Virtual SAN cluster.

For instance, in a three-node cluster configuration, with an FTT equal to 1, the default, there are two replicas of the data components and a witness, which must all exist on different nodes in the cluster. A three-node configuration can tolerate only a single failure, and if this occurs, Virtual SAN cannot rebuild data components; nor is it able to provision new virtual machines. In addition, in this configuration, Virtual SAN cannot reprotect virtual machine objects after the failure, until such time that the failed components are restored. Also note that in three-node cluster configurations, maintenance mode is unable to perform full data migrations to free a host for operational maintenance. For these reasons, the strong design recommendation from the architect must be to consider at least four or more nodes for a business-critical environment, in order to maximize availability and flexibility of the Virtual SAN–enabled cluster.

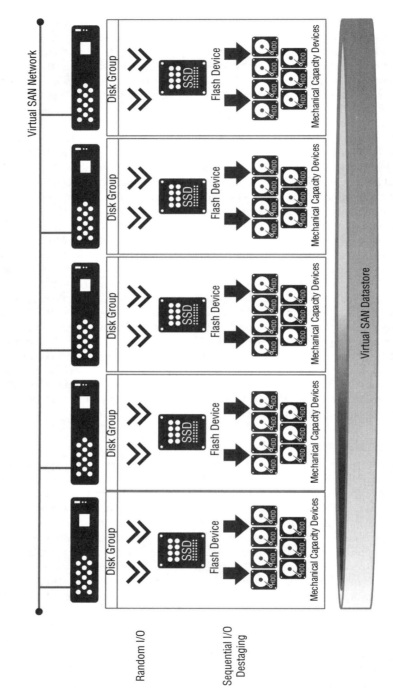

FIGURE 4.39
I/O blender effect

To determine the number of hosts required to meet a customer's storage requirements, you must address the following key design factors:

- The amount of usable space on the Virtual SAN datastore
- The number of hosts or hardware component failures that the Virtual SAN cluster is to tolerate in the design

You also have to address finding the *sweet spot* between the number of nodes in the cluster and their specification in either a scale-up or scale-out strategy. For instance, increasing the number of vSphere hosts in the Virtual SAN cluster, a scale-out strategy, is likely to mean higher hardware costs. However, decreasing the number of vSphere hosts and increasing the number of disk groups and resources available per host in the same cluster, a scale-up strategy, could result in resource availability being reduced.

Storage Capabilities

Several of the storage policy capabilities previously discussed can have a significant impact on Virtual SAN datastore sizing, with the FTT capability having the biggest impact. It is critical, that in order to accurately size an environment, you must understand how these storage capabilities impact the consumption of storage capacity in the Virtual SAN cluster.

In addition, several of the SPBM capabilities also have an impact on the Virtual SAN datastore design. The Number of Failures to Tolerate, the Number of Disk Stripes per Object and the Object Space Reservation, and the way they are used in the design must all be considered as part of a Virtual SAN datastore sizing exercise.

Sizing Impact of the Number of Failures to Tolerate Capability

The Number of Failures to Tolerate capability provides availability and redundancy, which is assigned either at the virtual machine level or to individual virtual disks. The impact of this policy on Virtual SAN datastore sizing needs to be considered when designing and calculating required storage capacity. This rule may increase the storage requirements for virtual machines by a factor of four times the original required disk capacity, if the FTT setting is configured at its maximum of 3.

If the FTT value is configured as 1, two replica mirror copies of the virtual machine or virtual disk are created across the cluster distributed datastore. If the value is increased to 2, three replica copies are distributed across the cluster; then, likewise, a value of 3 leads to four copies being created and distributed across the Virtual SAN datastore.

However, as addressed earlier in this chapter, if the all-flash erasure coding mechanism is used to facilitate either a RAID 5 or RAID 6 type of capacity reduction within the FTT capability, it is possible to achieve a capacity utilization savings of as much as 50 percent over that of a similar RAID 1 mirroring design.

Despite the potential capacity savings offered by an all-flash erasure coding configuration, from this you can clearly understand why the FTT value defined in the Virtual SAN storage policies has the largest impact on design sizing. There are no specific guidelines related to implementing this feature; it is purely based on workload availability requirements, as set out by the customer or application owner, the number of nodes that make up the Virtual SAN cluster, and having an understanding of the impact on hardware costs when using this capability beyond its default value.

Sizing Impact of the Number of Disk Stripes per Object Capability

When considering the Number of Disk Stripes per Object capability, two Virtual SAN datastore sizing considerations should be taken into account.

First, ensure that there are sufficient physical disk devices in the hosts and across the Virtual SAN cluster to accommodate the additional stripe width requested during the provisioning process. This can be particularly pertinent to design sizing if a higher-than-default value for the FTT capability is also being configured in the same storage policy.

The second design consideration is to verify that the chosen stripe width value is not going to significantly increase the host component count, for which a maximum of 9,000 exists in Virtual SAN 6, 6.1, and 6.2.

These are both design factors that need to be addressed during the Virtual SAN datastore sizing process. However, because the component count in Virtual SAN 6 increased significantly with the introduction of the VirstoFS on-disk format, this is now far less likely to be a design constraint.

Sizing Impact of the Object Space Reservation Capability

The *Object Space Reservation* capability is used for specifying the thick provisioning of a storage object in the Virtual SAN datastore. If an OSR rule is set to 100 percent, the storage requirements of the virtual machine are allocated up front, with the purpose of preventing the vSphere storage administrator from overcommitting storage on the Virtual SAN datastore.

The minimum OSR value is 0 percent, which represents a thinly provisioned disk, and the maximum configurable value is 100 percent, which allocates all disk capacity in advance. If this capability is to be used as part of a design, Virtual SAN datastore sizing must reflect this in terms of the increased capacity required.

Configuring Multiple Disk Groups

Disk groups pool disk resources into management constructs, which are created in the Virtual SAN hybrid model by combining a caching flash device and mechanical disks. In the all-flash configuration, an endurance flash device and performance-based capacity flash devices are used to create a similar two-tiered approach. The top-tier flash device, in both instances, provides disk groups with a distributed flash tier, and the capacity drives in both configurations provide the storage capacity of the Virtual SAN datastore.

vSphere hosts that form part of the Virtual SAN cluster can comprise up to five independent disk groups, each made up of the same or different configurations. However, it is strongly recommended that Virtual SAN designs support a consistent disk group configuration across the cluster, composed of the same hardware types, in a uniform and repeatable building block approach, as illustrated in Figure 4.40.

When considering the design of Virtual SAN disk groups, the vSphere storage architect should calculate the ratio between the flash-based caching device and mechanical or capacity flash drives, based on workload performance requirements. In a hybrid configuration, the higher the ratio of flash to persistent capacity storage, the larger the size of the cache reservation per object, which will provide overall enhanced performance.

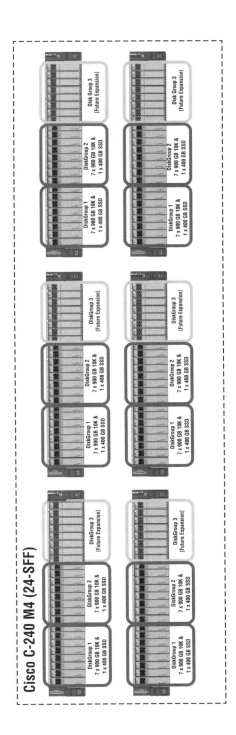

FIGURE 4.40
Multiple disk group building-block configuration

An additional design factor in considering multiple Virtual SAN disk groups is the ability to define, and potentially reduce, the size of failure domains. In both hybrid and all-flash configurations, if the flash-based caching device fails, all capacity drives within that disk group will become unusable, and in turn, the storage capacity provided by the affected disk group will become inaccessible to the Virtual SAN cluster. If a design stipulates multiple smaller disk groups over a single larger disk group, not only is performance improved, but the failure domain is limited to a smaller number of capacity disks within the affected disk group.

It is a critical design factor to take into account that as a consequence of the distributed datastore architecture, a single disk group failure can impact a number of virtual machines, if the Virtual SAN cluster does not have sufficient storage capacity to provide for the affected workloads. Restoring large numbers of virtual machines affected by a failure, after hardware has been replaced, can take a significant period of time, particularly in a design that uses a 1 Gb/s network infrastructure.

Endurance Flash Sizing

In the Virtual SAN hybrid model, 30 percent of the flash device's capacity in each disk group acts as a write buffer. The purpose of this, as addressed earlier in this chapter, is that each write first gets targeted onto the flash device; it is also preserved on as many flash tier devices as there are replicas for the corresponding object across the datastore. Even though this 30 percent value is not hard-coded, and the parameter can be changed, in order to safeguard the required availability of virtual machine data, it is highly recommended that this configuration is not modified unless specifically directed to do so by VMware support services as part of performance troubleshooting.

In the hybrid model, the remaining 70 percent is allocated as a read cache. However, a data block will never be placed in the read cache of more than one endurance flash device within the Virtual SAN datastore. Unlike the write cache mechanism, Virtual SAN maintains the same percentage of the endurance flash device as read cache for an object, regardless of the number of replica objects that exist. As a result of this architecture, increasing the availability of an object, through the FTT capability, does not increase the object's usage of the read cache capacity on the flash device.

In the all-flash configuration, 100 percent of the endurance flash device is allocated as a write buffer, as all read requests are targeted directly to the capacity flash devices.

As a starting point, for flash sizing, VMware recommends that in both the hybrid and all-flash models, a design should use a minimum of 10 percent of the total anticipated capacity of the Virtual SAN datastore, before the FTT calculation is taken into account.

However, this should be considered as a starting point. The optimal percentage of the endurance flash should be based on actual virtual machine storage workload characteristics, in particular, the size of the working data set of the applications, and any specific read or write requirements of the workload, as defined by the application owners.

Endurance Flash Sizing Example

In this example, the vSphere administrator has been asked to deploy 250 virtual machines for a new project, each with a single 100 GB, thin-provisioned virtual disk. However, the business anticipates that over the 36-month life cycle of the project, the actual consumed storage capacity per virtual machine will be 40 GB.

Table 4.20 illustrates this example sizing scenario, based on the VMware 10 percent recommendation of the total anticipated capacity of the Virtual SAN datastore.

TABLE 4.20: Flash capacity sizing example

CUSTOMER REQUIREMENT	VALUES
Anticipated virtual machine space	40 GB
Number of virtual machines	250
Total anticipated storage capacity	40 GB × 250 = 10,000 GB = 10 TB
Target endurance flash capacity (%)	10 percent
Total endurance flash requirement	10 TB × 0.10 = 1 TB

In this example, the total expected storage consumption, before object replicas, equates to 250 × 40 GB = 10 TB. If the storage policy assigned to the new virtual machines to increase availability defines the FTT as equal to 1, the resulting configuration creates two replicas for each virtual machine instance, generating just over 20 TB of total consumed storage capacity for the project, including replica data.

As shown in Table 4.20, the flash sizing requirement for the project will be equal to 10 percent of the consumed capacity, before replica data is taken into account, which equates to 10 percent of 10 TB = 1 TB. Therefore, for this project, the design would require a minimum of 1 TB of endurance flash, based on the aggregate capacity of the cluster, where the new virtual machines are to be deployed.

NOTE Currently, the maximum supported size of flash in a hybrid configuration is 600 GB per disk group, which would not result in being a constraint in this sizing example.

Objects, Components, and Witness Sizing

As already addressed earlier in this chapter, a Virtual SAN datastore is not based on the classic VMFS construct, but instead on an Object Storage File System (OSFS). A single Virtual SAN datastore is created, using the aggregate storage components from across the multiple hosts in the cluster. The storage is then mounted as a single distributed datastore using OSFS.

Virtual SAN stores and manages data in the distributed datastore in the form of flexible data containers called *objects*. An object can be thought of as a logical volume that has its data and metadata distributed across a number of components and accessed across the entire Virtual SAN cluster. In the vSphere storage stack, these objects appear as devices.

An object can also be thought of as a *volume*, the same term used by Amazon for EC2 and OpenStack technologies. Virtual SAN objects are *mutable*, in that their fields can change after they are created, and are strongly consistent objects, unlike *Blob* storage objects in Amazon S3 or Microsoft Azure. In a Virtual SAN–enabled cluster, the only supported object types are virtual machine–based files, such as VMDKs, with Virtual SAN being capable of storing and managing tens of thousands of these object types in a single cluster.

For each virtual machine deployed on the Virtual SAN datastore, an object is created for each virtual disk, plus a container object, the virtual machine namespace, which holds a VMFS volume and stores all of the metadata files associated with that virtual machine. Virtual SAN provisions and manages each object individually—for instance, before creating the object for a virtual disk, Virtual SAN considers the following:

- The SPBM policies specified by the vSphere storage administrator for the specific virtual disk
- The cluster resources and their utilization at the time of provisioning

Then, based on these considerations, Virtual SAN decides how to distribute the object's components across the cluster.

Objects

Each Virtual SAN object that exists on the datastore is made up of multiple components distributed across the disk groups configured on the cluster's host servers. Each object is provided with performance and availability parameters through the SPBM mechanism. As highlighted previously, four object types exist in Virtual SAN, as shown in Table 4.21.

TABLE 4.21: Virtual SAN object types

Object Type	Description
Virtual machine namespace / VM home	The location, or container, where all virtual machine configuration files reside. These include the .vmx file, log files, and others.
Swap	A unique storage object type that is created when the virtual machine is powered on and then deleted when it is powered off.
Virtual disks (VMDKs)	The virtual machine disk files.
Virtual disk snapshots/ deltas	Unique storage objects, created when virtual machine snapshots are performed by administrators or backup solutions. A memory object is also created when the snapshot memory option is selected, when creating or suspending a virtual machine.

Components

Virtual SAN objects are composed of multiple components that are distributed across hosts in the cluster. Components are stored intelligently across disk groups within the Virtual SAN distributed datastore, in order to maximize availability. Each component is assigned caching and buffering capacity on the disk group's flash-based devices automatically, in line with the mechanism being used by that disk group type, hybrid or all-flash, with their data residing persistently on the mechanical or flash-based capacity disks. Virtual SAN 6.0, 6.1, and 6.2 all support a maximum of 9,000 components per host.

Also, objects that are greater in size than 255 GB automatically are broken up into multiple components. This is a configurable value via the `VSAN.ClomMaxComponentSizeGB` parameter, and may be required to be reduced so that it does not exceed 80 percent of the smallest capacity disk size that the design is using. For instance, if the smallest capacity disk in a Virtual SAN disk group is 200 GB, and you expect virtual machine objects such as VMDKs to grow to 400 GB, adjust the `VSAN.ClomMaxComponentSizeGB` option to 160 GB (which is 80 percent of 200 GB). The `VSAN.ClomMaxComponentSizeGB` default value may be particularly pertinent in an all-flash configuration, where disk capacities remain lower than those available on mechanical disks.

Also, if the Number of Disk Stripes per Object capability is increased beyond its default value of 1, each stripe created is a separate component.

WITNESS METADATA

Even though it is a minor factor, here's a final consideration for component sizing: for every component created in the Virtual SAN datastore, an additional 2 MB of disk capacity is consumed for witness metadata. *Witnesses* are components that consist only of object metadata, but they are an important part of each object. The witness component acts as a tiebreaker, in order to avoid split-brain scenarios when availability decisions must be made by the Virtual SAN cluster. Each Virtual SAN witness component consumes 2 MB of storage capacity. However, this metadata consumption of capacity is so small, it typically is not included in datastore-sizing exercises.

Datastore Capacity Disk Sizing

Virtual SAN sizing calculations can be achieved in various ways, depending on the requirements and other factors, such as the SPBM capabilities included in the design, which can significantly impact the capacity requirements, and result in additional host hardware or disk groups being configured.

VMware has kindly provided sizing and total cost of ownership (TCO) tools on its public website, to help reduce the complexities of estimating the required size of the Virtual SAN datastore, based on design parameters such as number of virtual machines, number and size of virtual disks, anticipated snapshot count, and the read and write I/O ratio.

However, it is critically important that you fully understand and appreciate how these sizing calculations are made. After all, it is typically the architect who has to justify the storage costs to business leaders. Therefore, you must be able to answer the question, "Why do we need so many disks again?" in order to get projects approved. Simply answering, "That's what the VMware sizing calculator said" may not help.

In addition, even though VMware's online tools will calculate an optimal Virtual SAN configuration, on some occasions vendor hardware constraints may require the calculations to be modified, in order to satisfy specific design factors. This might result in a configuration that is not optimal, yet meets the customer's specific requirements.

Figure 4.41 illustrates the calculations provided by a typical sizing exercise from the VMware online tools.

Figure 4.41
Virtual SAN total cost of ownership (TCO) and sizing calculator

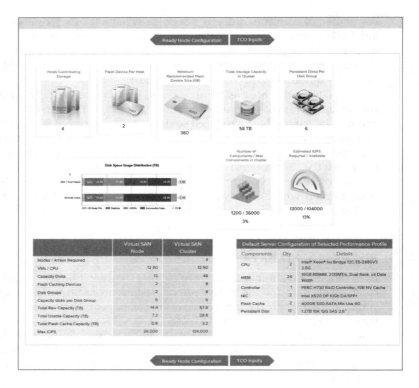

Before we address the manual sizing formulas for calculating Virtual SAN datastores, it's important to recognize that the end datastore size and number of hosts must always be based on the customer's workload requirements, and it should be sized accordingly, founded on empirical data or estimated sizing requirements.

Capacity Disk Size

This section presents the formulas that you can use to perform manual calculations for producing the sizing design for Virtual SAN datastores. These calculations can be approached in various ways.

The first example is based on the hardware configuration provided in the Table 4.22, effectively calculating against predefined hardware. This first formula takes into account the FTT capability setting, allowing us to ascertain how much actual capacity is available for virtual machine workloads. This formula is particularly useful when determining how much usable capacity is available in the Virtual SAN datastore when preconfigured hardware such as VSAN Ready Nodes are used in the design.

Table 4.22 lists the values required in order to calculate the actual capacity available on the preconfigured hardware. Some information is based on design assumptions as opposed to empirical data from the customer. The values shown in the table are used as the basis for all of the following sizing calculations.

VIRTUAL SAN DATASTORE DESIGN AND SIZING

TABLE 4.22: Sizing factor values

DESIGN FACTOR	ABBREVIATION	VALUE
Number of hosts per cluster	Hst	16
Number of disk groups	DskGrp	3
Number of disks per disk group	DskPerDskGrp	4
Size of capacity disks	SzHDD	1,200 GB (1.2 TB)
Estimated % of VM disk consumption	%VMSzCAPDsk	~50%
Number of failures to tolerate	ftt	1
Number of virtual machines	No.VMs	250
Number of disks per virtual machine	NumOfVMDK	1
Memory per virtual machine	vmSwp	8 GB

In the Virtual SAN calculations shown in the following formulas, you will notice that the flash devices used exclusively for read caching and write buffering do not participate, as they do not contribute to the available datastore capacity.

CLUSTER CAPACITY

The Virtual SAN raw storage capacity of preconfigured cluster nodes can be calculated based on the following formula:

Formula 4.1: Raw Cluster Capacity
Hst × NumDskGrpPerHst × NumDskPerDskGrp × SzHDD = RawClusterCapacity

Example:
16 × 3 × 4 × 1,200 GB = 230,400 GB = 225 TB

OBJECT COUNT

The formula to calculate the number of objects takes into account the virtual machine files, which include the virtual machine home namespace, virtual machine swap file, and virtual disks. The number of objects can be estimated based on the number of virtual machines and assigned capabilities by using the following formula.

Note that even though snapshots must be included in the Virtual SAN object count, in this example scenario, no requirement for snapshots was identified by the architect, and they're therefore not included in the calculation.

Formula 4.2: Calculating Number of Objects
No.VMs × (VMnamespace + vmSwap + NumOfVMDK)
= NumberOfObjects

Example:
250 × (1 + 1 + 1) = 750 objects

COMPONENT COUNT

The virtual machine's performance and availability requirements effectively dictate the number of components associated with that entity. Therefore, calculating the effective number of components as part of a design is prudent, in order to ensure that the number does not exceed the host's hard-coded limit of 9,000. The number of components can be estimated based on the number of objects per virtual machine, with use of the following formula.

This estimation can then be used to calculate the total number of components per virtual machine, which can then be aggregated across the environment to ensure the maximum component count per host is not exceeded. This formula takes into account the replicas and witnesses created based on an FTT value of 1. The components that result from this capability will be distributed across all hosts in the Virtual SAN cluster.

If the Number of Disk Stripes per Object capability value is increased above the default of 1, each stripe is also counted as a separate component. However, in this example scenario no requirement for this additional capability was identified by the architect; therefore, the number of disk stripes remains at its default value of 1, and as a result does not affect the following calculation.

Formula 4.3: Calculating Component Count
Object × (FTT × 2 + 1) = ComponentCount

Example:
750 × (1 × 2 + 1) = 1,500 components = average 94 components per host

SWAP FILE OBJECTS AND SIZING

All vSphere storage designs require swap file capacity to be taken into account, as a certain amount of disk capacity is typically used by virtual machine swap-file space in all vSphere environments.

A `.vswp` file is created every time a virtual machine is powered on, and is equal to the size of the unreserved memory configured on the virtual machine. For instance, if a virtual machine is configured with 8 GBs and the memory reservation is set to 0 MB (the default), the virtual machine swap file will be 8 GB. However, if the memory reservation was 4 GB, the `.vswp` file will be 4 GB (8 GB minus 4 GB).

In this example scenario, the architect did not identify any customer requirement for memory reservations, so this will not be taken into account in the following formula. In addition, while swap efficiency / sparse swap, introduced in Virtual SAN 6.2 will not impact object count, the following calculations presume that the advanced host-level option, `SwapThickProvisionedDisabled`, is not being used in the design.

Virtual SAN will by default always store swap-file space with two replica components, regardless of the FTT setting. However, as previously noted, swap storage will consume raw capacity, which must be taken into account as part of the storage design calculations, shown here:

Formula 4.4: Calculating Swap File Object Capacity
No.VMs × vmSwp × 2= SwapFileObjectCapacity

Example:
250 × 8 GB × 2 = 4,000 GB

Usable Capacity Calculation

The Virtual SAN datastore usable capacity is the actual amount of space available that can be used to store the virtual machine objects. This usable capacity is calculated by subtracting the Virtual SAN overhead from the raw disk space, and then dividing the remaining amount by the FTT value, plus 1:

Formula 4.5: Calculating Usable Capacity
(DiskCapacity − DskGrp × DskPerDskGrp × Hst × VSANoverhead) / (FTT+1) = UsableCapacity

Example:
(230,400GB − 3 × 4 × 16 × 1) / FTT+1 = 230,208 GB / 2) = 115,104 GB
(approximately 112 TB)

NOTE   A general guideline is to allow 1 GB of storage capacity per virtual disk in the calculations for a combination of Virtual SAN component and VMFS metadata overhead. This is referred to as VSANoverhead.

Based on the preceding calculations, approximately 225 TB of raw capacity exists across the 16-node Virtual SAN– enabled cluster. Users can create virtual disks that in total consume as much as 112 TB. The difference in capacity is consumed predominantly by replicas created for availability purposes and virtual machine swap space.

A general good practice in vSphere storage design is to ensure that operational teams allow no more than 80 percent of usable capacity to be allocated to virtual machines. This common good practice acts as a safeguard to allow for vSphere storage overhead factors to be accounted for, such as snapshots and working space.

In addition, the total number of components should always be kept in mind, although with its increase from 3,000 to 9,000 per host in vSphere 6, this is unlikely to continue to constitute a design constraint, and is therefore unlikely to prove to be a limiting factor moving forward. In the example component calculation provided earlier, there were approximately 94 components per host. However, an increase in the number of virtual disks per virtual machine, stripes per object, snapshots, or FTT value, could potentially increase component count significantly.

Calculating Virtual Machine Capacity Requirements

As addressed previously, despite the availability of an official Virtual SAN sizing calculator, it is critically important to understand how sizing calculations can be performed. Here we outline how these calculations can be carried out based on the customer's virtual machine requirements, which of course first requires the architect to fully understand the configuration of the targeted workloads.

As with any design and sizing exercise, it should start by gathering requirements, such as the following:

- Number of vCPUs per virtual machine
- Virtual machine memory size

- Virtual machine disk size
- Number of virtual disks per virtual machine
- Estimated percentage of disk consumption

The following calculations are based on the requirements outlined in Table 4.23, as gathered from the customer during a series of design workshops. In this scenario, the customer has stated that because of ongoing growth, a total of 500 workloads should be accounted for in the design.

TABLE 4.23: Design scenario customer requirements

DESIGN FACTOR	PER VM	500 WORKLOADS
Avg. no. of vCPUs per VM	3 vCPUs	1,500 vCPUs
Avg. VM assigned memory	8 GB	4,000 GB
Avg. VM disk (single disk)	100 GB	50,000 GB
Estimated % of VM disk consumption	~50%	25,000 GB
Avg. VM memory reservation	None	None

Based on a 5:1 vCPU to pCPU ratio, the compute design will require 300 cores and a little under 4 TB of memory, assuming no overcommitment of resources is being used. For the storage calculations, we will take into account the additional design factors listed in Table 4.24.

TABLE 4.24: Design scenario additional storage factors

DESIGN FACTOR	PER VM
Number of failures to tolerate	1
% of capacity for snapshots (SnapSpace)	10%
Virtual SAN component and VMFS metadata overhead (VSANoverhead)	1 GB
Swap capacity (default thick)	100% of assigned memory

The Virtual SAN storage capacity required to meet the customer's requirements can be calculated based on the following formula:

Formula 4.6: Calculating Required Capacity
(((No.VMs × AvgVM SzHDD) + (No.VMs × AvgVM SzMEM)) × FTT+1) + 10% SnapSpace + (No.VMs × VSANoverhead) = RequiredCapacity

Example:
(500 × 100) + (500 × 8) = (50,000 + 4000) × 2 = 108,000 GB + 10% + (500 × 1 GB) = 119,300 GB
119,300 / 1024 = 116.5 TB

TABLE 4.25: Customer compute and storage requirements summary

DESIGN FACTOR	PER VM
Raw storage capacity	117 TB (rounded up)
Memory	4 TB
Physical CPU cores	300

Based on this raw information, derived from Table 4.25, you can now define the options available to the customer, in determining the final vSphere host configuration, in terms of number of hosts required, physical CPUs and memory required, disk group configuration, number of disk groups per host, per disk capacity, and the percentage of flash per disk group.

Designing for Availability

Virtual SAN is designed from the ground up with the ability to handle failure events and to ensure that data is not lost as a result of a hardware outage. Virtual SAN leverages replication as its primary mechanism to ensure ongoing availability of virtual machine objects. In the event of a hardware failure, or an engineer accidently pulling out the wrong power cord or disk, a properly designed solution will suffer minimal impact.

This is made possible as the replication process makes multiple copies of the data and stores them on various disk groups or hosts in the cluster. Therefore, if a host, disk group, or disk fails, one or more replicas of that data are still stored on other nodes in the Virtual SAN cluster (see Figure 4.42).

This mechanism is referred to as a redundant array of interdependent nodes, or RAIN, and uses RDT for communication between hosts. RDT is optimized to send very large files and delivers datagrams (potentially very large datagrams, if necessary) between logical endpoints that which act conceptually like a fault-tolerant client and server mechanism, typically over multiple paths. Based on link health-status changes published by the Cluster Monitoring, Membership, and Directory Services (CMMDS), the RDT sets up and tears down transport connections very quickly, to minimize delay in datagram transport due to link failures.

The Virtual SAN RAIN mechanism operates by ensuring that when an object is created, the CLOM checks to see whether there are enough disk groups to satisfy the assigned storage policy. The CLOM does this by communicating with the other nodes through CMMDS. When a suitable disk group has been identified, the object components are created by the DOM, both locally, using the Local Log-Structured Object Manager (LSOM), and remotely via the RDT to the DOM clients on other hosts in the cluster, who in turn use their local LSOM to write the data to the disk group. All subsequent reads/writes for a virtual machine go directly through the DOM. Each of these internal Virtual SAN services and mechanisms are addressed in more detail later in this chapter.

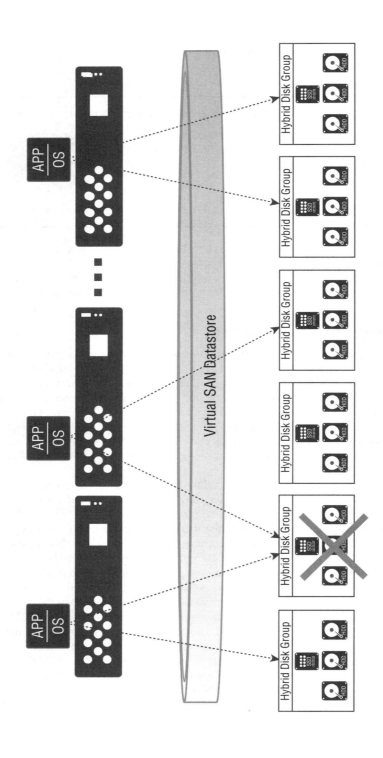

FIGURE 4.42
Virtual SAN availability by design

Designing for Hardware Component Failure

This sophisticated and resilient replication process, provided by the RAIN mechanism, allows Virtual SAN to be able to handle both transient and permanent hardware outages, such as a network misconfiguration or a full device failure, including a complete vSphere host, without impacting virtual machine workload operations.

Virtual SAN classifies a failure into two distinct scenario types: absent failures and degraded failures, which have different default behaviors. The way that Virtual SAN handles the outage depends on the type of failure event.

ABSENT FAILURES

An *absent failure* event occurs in Virtual SAN when I/O failures are detected that affect physical switches, network adapter interfaces, vSphere hosts, or a fault domain. If an absent failure event occurs, Virtual SAN does not immediately begin resynchronizing all data onto different hosts in the cluster. In fact, the resynchronization operation is not initiated until 60 minutes after the detection of the absent failure event, although this value is configurable.

The rationale behind this 60-minute delay is that if a failure occurs on physical switches, network adapter interfaces, vSphere hosts, or a fault domain, these are typically brief or transitory issues. For instance, a network switch or host could be rebooting as part of planned operational maintenance. Therefore, when the failure is detected by Virtual SAN, the system waits to see whether the failed component comes back online before taking remedial action and initiating a resynchronization operation of all the data on the failed or disconnected host.

As noted previously, the default delay of 60 minutes is a configurable value that can be increased or decreased by an administrator, based on business or design requirements. For instance, an operational design may stipulate that this value be increased before prolonged operational maintenance takes place overnight, or a window longer than 60 minutes is required during a soft maintenance window to replace hardware. The modification of this parameter can be carried out without the hosts requiring a reboot.

DEGRADED FAILURES

The other type of failure event scenario identified by Virtual SAN is referred to as a *degraded failure*. A degraded failure occurs when Virtual SAN I/O failures are detected with flash-based devices, mechanical disks, or storage controllers.

In contrast to absent failures, a degraded failure event is actioned immediately by a resynchronization of the data across the surviving hosts in the Virtual SAN cluster. Again, unlike absent failures, this function does not offer any configurable parameter, as failures of hardware such as those that initiate a degraded failure event are typically permanent. There is no operational benefit to postponing the resynchronization, as the device is unlikely to come back online.

When a degraded failure occurs, the resynchronization operation is initiated for objects affected by the failure, and as a result, are no longer in compliance with their associated storage policies.

RESYNCHRONIZATION OPERATIONS

A resynchronization operation has a high resource overhead on the Virtual SAN cluster. New replica copies of the virtual machine data that previously existed on the failed device must be re-created using the surviving data as the source. Thanks to the RAIN mechanism, this data exists, distributed across the other hosts or capacity disks in the cluster. The extent of this distribution depends largely on the configured FTT capability, which creates object replicas distributed across the entire Virtual SAN cluster by the CLOM userspace daemon.

During a resynchronization operation, virtual machine I/O activity can be impacted, as it has to contend with the I/O generated by the rebuild process. This process has the potential to starve virtual machines of IOPS, and may limit performance during the operation to recover lost components. As you would expect, this could have a detrimental effect on the overall capabilities of the workload running on the Virtual SAN– enabled cluster, and therefore, sizing and designing the environment should allow for these resynchronization operations to occur.

In a worst-case scenario for Virtual SAN, from a data-availability perspective, the resynchronization operation would be unable to rebuild objects because of an insufficient amount of available capacity on the cluster. If this scenario occurs during a rebuild operation, accessibility to the data by virtual machines may be at risk. For instance, in a three-node Virtual SAN cluster configured with the default FTT value of 1, a single host failure would result in the two remaining hosts not being included in the resynchronization operation, as they would already be hosting components for the impacted workloads. In a design such as this, the resynchronization operation would have to wait until the failed host is brought back online, or a new host added to the existing cluster. This is yet another reason that a three-node cluster should not be recommended for production environments or for workloads that require business-critical availability to meet operational service levels.

Rebalance Operations

In a Virtual SAN–enabled cluster, a rebalance operation occurs to proactively redistribute component data throughout the nodes, in order to maintain a balanced consumption of storage resource use across the cluster, as shown in Figure 4.43.

This rebalancing operation occurs by default when the utilization of the capacity disks reaches 80 percent, but can also be triggered manually by initiating maintenance mode on a host, by selecting either the Ensure Accessibility or Full Data Migration option. By placing a host into maintenance mode, you essentially force the rebalancing of available storage capacity. However, adding resources to a cluster does not in itself automatically activate a rebalancing operation. Storage resources that are added to an existing cluster are identified by Virtual SAN as *additional capacity*, and this new capacity is used only when new virtual machines are provisioned. Otherwise, initiated resynchronization operations occur.

Unplanned rebalancing operations can also take place when hardware failure events occur that force capacity utilization over the 80 percent threshold, after the CLOM-driven resynchronization operations are completed. Once again, the purpose of a rebalance operation is to redistribute the virtual machine component data evenly across the Virtual SAN cluster's storage devices, and as highlighted previously in this chapter, 10 Gb/s networking will provide optimum performance for these operations.

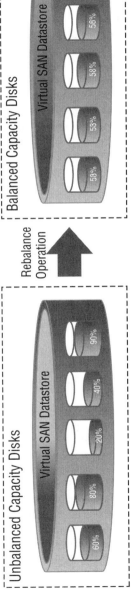

FIGURE 4.43
Rebalance operations

Because of the I/O overhead incurred by a rebalance operation, when you add new storage devices or additional hosts with new capacity, Virtual SAN does not automatically distribute data to the newly added devices. In order for Virtual SAN to distribute data to the new storage devices, an administrator must perform a manual rebalance operation on the cluster, by using the Ruby vSphere Console (RVC) `vsan.proactive_rebalance` command. When a proactive rebalance is initiated, the administrator is also able to specify thresholds, such as how long the rebalance operation is able to run for, and the maximum amount of data that may be moved per hour for each node in the Virtual SAN cluster. If values are not specified, default parameters are chosen. It is typical that a manually initiated rebalance operation would be performed during a soft maintenance window, due to the I/O overhead incurred on the cluster during its execution.

Host Cluster Design and Planning for Host Failure

Grouping ESXi hosts into clusters for Virtual SAN also facilitates the use of traditional vSphere availability and load-balancing technologies such as vMotion, DRS, HA, and SMP-FT. These technologies and their integration in a Virtual SAN–enabled environment are critically important design considerations when planning for vSphere host failure or planned operational maintenance tasks.

VIRTUAL SAN AND VSPHERE HIGH-AVAILABILITY INTEROPERABILITY

An important consideration for Virtual SAN cluster design is planning for host failure or planned maintenance. vSphere clusters are typically used to provide increased availability and resource load balancing, and one of the key design factors for cluster configuration is the proportion of the cluster's total capacity, which is to be reserved for a failure event, or to undertake operational maintenance tasks. Determining the number of compute nodes in each cluster and the total compute and storage capacity you want to reserve, which makes it unavailable for day-to-day workloads, will provide the percentage of capacity that must be allocated for failure scenarios.

Virtual SAN does not interact with vSphere HA to ensure that there is enough free storage capacity if a host failure occurs. It is the FTT capability and its associated storage policies that must ensure that sufficient objects and components are distributed across the Virtual SAN cluster in order to tolerate host failures.

For instance, Figure 4.44 illustrates an eight-node cluster, in which a design would typically recommend reserving the equivalent of a single host's CPU and memory resources as failover capacity, thus allowing for a single server failure within the cluster without impacting the performance of the virtual machines, once restarted on remaining hosts in the cluster.

As illustrated in the figure, the Virtual SAN datastore, and its object and component distribution across the hosts, ensures that failures can be tolerated. This mechanism is operated separately from that of vSphere HA's admission control policy, which takes into account only host compute CPU and memory resources into its availability calculations.

vSphere HA operates at the cluster level, and when enabled, provides the capability to monitor vSphere hosts for failures and automatically restart virtual machines if necessary, typically as a result of a host failure or network isolation event.

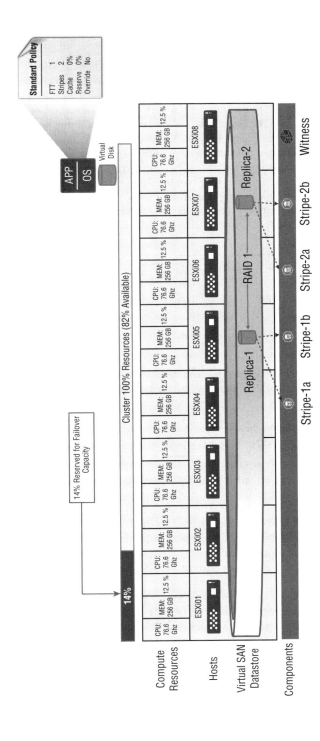

FIGURE 4.44
Calculating vSphere HA admission control policy and the number of failures to tolerate capability

In a vSphere compute cluster, with classic VMFS storage volumes attached, the HA mechanism monitors all hosts in the cluster through both network and datastore heartbeats. In the event that the network heartbeat between a slave node and master node in a cluster is lost, the master node attempts to use the datastore heartbeat system to verify that the slave node is still operational. If the datastore heartbeat has stopped as well, the slave node is determined to have failed, and the master node begins restarting the appropriate virtual machines on other nodes in the cluster. This is the purpose of the vSphere HA admission control policy, and the incentive to reserve unused capacity within the vSphere cluster for failure events.

Virtual SAN is fully interoperable with the vSphere HA mechanism. However, in a Virtual SAN–enabled cluster, this mechanism works slightly differently, as outlined in Table 4.26. On a Virtual SAN platform, datastore heartbeats become irrelevant, and the vSphere HA agent uses the Virtual SAN network to communicate instead of the host management network. However, the management gateway is still used by the host to detect whether it has become isolated.

TABLE 4.26: vSphere HA operational comparison

Availability Mechanism	Virtual SAN Cluster	Nonvirtual SAN Cluster
Network used by HA	Virtual SAN storage network	Management network
Heartbeat datastores by HA	Any external datastore mounted to more than 1 host, but not the Virtual SAN datastore	Any datastore mounted to more than 1 host
Host declared isolated	Isolation addresses not accessible by ping, and storage network inaccessible	Isolation addresses not accessible by ping and management network inaccessible

In a Virtual SAN–enabled cluster, the HA agents communicate over the same network as Virtual SAN uses for internal datastore communication, and therefore the host management network is not used by the Fault Domain Manager (FDM), which is the case for an HA agent in a cluster without Virtual SAN enabled. As shown in Figure 4.45, when performing its inter-agent communication, when Virtual SAN is enabled on the cluster, HA will use the Virtual SAN network for keep-alive communications.

This change in architecture is a result of Virtual SAN needing to ensure that it has the same view of what is occurring on the network as the vSphere HA mechanism, in the case of a network partition, and is able to have a single view of what the partition looks like.

Specifically, during a partition event, when the partitioned subclusters form, the same membership needs to be seen by both Virtual SAN and HA. If this is not the case, we risk failure scenarios in which HA is trying to control a virtual machine on one side of a network partition, but the other side of the partition contains the master replica datastore components (see Figure 4.46).

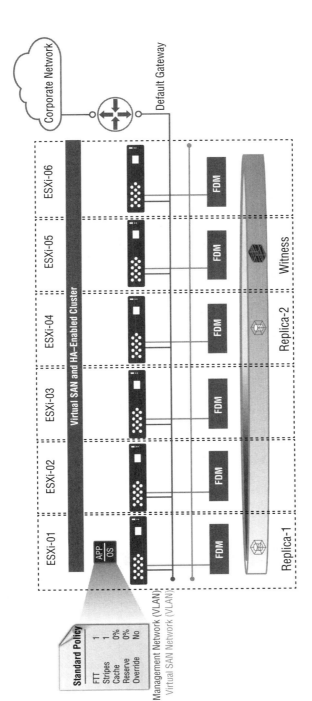

Figure 4.45
vSphere high availability network communication

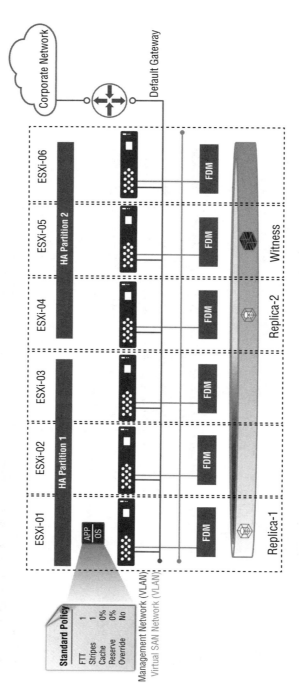

FIGURE 4.46
Virtual SAN network partition scenario

vSphere High-Availability Heartbeat Datastores

Another important difference with a Virtual SAN implementation of HA is that storage heartbeats cannot be used on a Virtual SAN datastore. Therefore, you will find that when configuring a Virtual SAN cluster via the Web Client and you select to choose heartbeat datastores, the Virtual SAN datastore cannot be selected. Also, if you attempt to bypass the user interface to use the vSphere API to try to configure a Virtual SAN datastore for storage heartbeating, then you will receive a configuration fault.

Heartbeat datastores are not required for HA to operate with virtual SAN. However, if you have external storage that is accessible by an alternate network path independent of the Virtual SAN network, that storage can be configured to enable heartbeat datastores. This configuration might provide additional benefits, but at a financial cost.

If the design already includes storage presented via a different protocol, across a different physical network, the cost of configuring small dedicated heartbeat datastores would be relatively small. The design must use only storage that is accessible to all the hosts in the Virtual SAN cluster during a network partition or host isolation event. If the design does already include non–Virtual SAN datastores accessible to the entire Virtual SAN cluster, such as NFS mount points for vSphere catalogs and/or maintenance logs, then there is no requirement to add datastores simply for heartbeating, as long as the existing accessible datastores are isolated from the Virtual SAN network.

Enabling the use of heartbeat datastores provides the following benefits to the operational management of the environment:

- Heartbeat datastores allow vCenter Server to report the specific state of a partitioned or isolated host, rather than simply reporting that it appears to have failed.

- Heartbeat datastores increase the likelihood that for virtual machines not being stored on a Virtual SAN datastore, that the FDM master will respond to a virtual machine that fails, after the host becomes partitioned or isolated.

- Heartbeat datastores can also help prevent vSphere HA from causing virtual machine MAC address conflicts on the virtual machine network, after a host isolation or partition event.

Host Isolation Addresses Recommendations

The vSphere High Availability agent, configured on each host, monitors network traffic for isolation events. If it observes no HA network agent-to-agent traffic, it attempts to ping the configured isolation address, which by default is the gateway address on the host management network. This is true even though, as noted previously, the heartbeat mechanism in a Virtual SAN environment uses the Virtual SAN network to communicate, as opposed to the host management network in a non–Virtual SAN design. However, despite this, the vSphere HA mechanism in a Virtual SAN–enabled cluster continues to use the management network default gateway as its default isolation address.

The purpose of the isolation address is to prevent the HA agent from falsely declaring that the host is isolated, if for some reason agents deployed across the cluster are unable to communicate with one another. However, if attempts to reach the isolation address, or addresses, also fail, the host is declared as isolated.

The vSphere HA configuration allows the administrator to configure up to 10 isolation addresses. In a Virtual SAN environment, consider configuring an isolation address, or addresses, that can enable hosts to determine whether they have lost access to the Virtual SAN network, such as the Virtual SAN network's default gateway.

This is configured in the vSphere HA advanced settings by configuring HA not to use the host management network's default gateway, with the vSphere HA advanced option `das.useDefaultIsolationAddress=false`, and also determining a new isolation address, with the advanced option `das.isolationAddress0 = VSAN Network IP Address`. A second, or multiple, isolation addresses can be configured by amending the last number of the advanced setting incrementally, for instance, `das.isolationAddress1` and `das.isolationAddress2`.

If isolation and partition events are a possibility in the design, ensure that one set of isolation addresses is accessible by the hosts in each of the potential network segments. In addition, if the Virtual SAN network is nonroutable and the design is broken into segments creating potential for partitions, then configure pingable isolation IP addresses on the Virtual SAN network from a subset of addresses from any potential partitions in the environment.

Finally, ensure that each Virtual SAN–enabled cluster network is on a unique and isolated network segment. Utilizing the same network subnet for more than one Virtual SAN cluster can cause unforeseen consequences, which may impact on the ability of operational teams to troubleshoot problems.

In summary, vSphere High Availability should always be configured on all Virtual SAN–enabled clusters as part of the design, to provide local data-center recovery of virtual machines in the event of a vSphere host failure or network isolation event. If a vSphere host fails, the virtual machines running on that server will go down but will be restarted on another host, typically within a couple of minutes. Although a service interruption perceivable to users would occur, the impact is minimized by the automatic restarting of these virtual machines on other hosts. Storage resource availability within the Virtual SAN datastore is handled separately by the defined storage policy assigned to workloads, specifically the FTT capability.

In addition to HA, in order to continuously balance workloads evenly across available vSphere hosts, and to maximize performance and scalability, DRS should be configured as part of the design on all Virtual SAN–enabled clusters. The DRS mechanism will work with vMotion to provide automated resource optimization and virtual machine placement. However, the VMware Distributed Power Management (DPM) feature, which enables a DRS cluster to reduce its power consumption by powering hosts on and off based on cluster resource utilization, is not supported for use in a Virtual SAN–enabled cluster.

In a multicluster Virtual SAN platform design, it is good practice to configure all HA and DRS parameters consistently, across all clusters, in all data centers, to help limit variability and to simplify operational management. Table 4.27 provides design guidance for a Virtual SAN implementation, by addressing the standard settings and options configured for vSphere HA and DRS attributes.

TABLE 4.27: Example Virtual SAN HA and DRS parameters

PROPERTY	SETTING	CONFIGURATION
Cluster features	High Availability (HA)	Enabled
	Distributed Resource Scheduler (DRS)	Enabled
vSphere High Availability (HA)	Host Monitoring Status	Enabled
	Admission Control Policy	Host failures in the cluster to tolerate = 1 (dependent on number of cluster nodes)
	Virtual Machine Options ➤ Virtual Machine Restart Policy	Medium
	Virtual Machine Options ➤ Host Isolation Response	Leave powered on
	VM Monitoring	Disabled
	Datastore Heartbeating	N/A
vSphere DRS	Automation Level	Fully automated (apply 1, 2, 3 priority recommendations)
	DRS Groups Manager	Applicable in stretched cluster configuration only
	Rules	Applicable in stretched cluster configuration only
	Virtual Machine Options	Default
	Power Management (DPM)	Off (not supported)
	Host Options	Default (disabled)
Enhanced vMotion capability		Enabled
Swap-file location		Store in the same directory as the virtual machine
Advanced settings	das.usedefaultisolationaddress	False
	das.isolationaddress0	VSAN Network IP Address/Gateway
	das.isolationaddress1	VSAN Network IP Address/Gateway

Resource Balancing and Transparent Maintenance

An additional benefit of DRS is realized when there is an operational requirement to place a host in maintenance mode. For instance, when servicing is required in order to install additional memory into a host, all virtual machines running on the targeted host will be automatically migrated off to other hosts within the cluster, assuming that DRS policies are configured appropriately to allow automatic migration. This mechanism provides a significant additional operational benefit gained from the vSphere HA admission control policy reserving spare capacity for both planned and unplanned outages, which can prove invaluable when performing rolling hardware maintenance or orchestrated patching. These types of actions can be carried out without any need to disrupt application workloads, and all maintenance remains completely transparent to the users, allowing for nonstop IT services, which in turn, provides transparent maintenance without downtime and greatly improved availability.

Virtual SAN Maintenance Mode Operations

When configuring a vSphere host to enter maintenance mode on a Virtual SAN–enabled cluster, three options are presented to the administrator that directly relate to the Virtual SAN's data-evacuation mechanism: Ensure Accessibility, Full Data Migration, and No Data Migration (see Figure 4.47). The most appropriate option to select depends on several operational factors associated with that cluster, and the type of and length of maintenance that is to take place. The option selected is operationally significant in order to ensure data availability, the minimum I/O impact, and the minimum amount of time is taken to complete the task.

Figure 4.47
Virtual SAN maintenance mode evacuation options

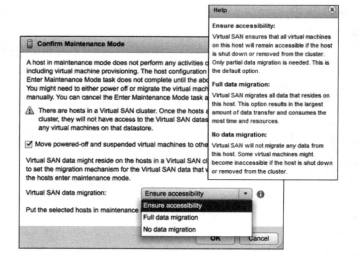

Before making the selection, the operator must consider that the Virtual SAN datastore has objects and components associated with virtual machines distributed throughout. Therefore, even though a specific virtual machine does not reside on the host being entered into maintenance mode, it is still likely to be impacted. For this reason, if Ensure Accessibility or Full Data Migration is selected, components associated with the virtual machine may have to be migrated, which will significantly slow the task of the host entering maintenance mode, and increase I/O operations on the Virtual SAN datastore during this period.

Ensure Accessibility Option

Ensure Accessibility is the default option. When you place a host into maintenance mode via Ensure Accessibility, Virtual SAN ensures that all virtual machines on this host remain accessible on the remaining nodes in the cluster. This option is typically most appropriate if the host is being temporarily placed into maintenance (in order to install new hardware, for example). This option should not be used if the intention is to permanently remove the host from the Virtual SAN cluster.

With this option selected, typically only partial data evacuation occurs. However, the virtual machines might no longer be compliant with its associated storage policy during the maintenance period. As a result, Virtual SAN may not have access to all replica components. Therefore, if an outage occurs during the maintenance period, and the FTT capability has been defined as 1, virtual machines in the environment are at risk of experiencing data loss and therefore very likely to fail. Ensure Accessibility is the only available option for evacuating a host in a three-node cluster, or in a Virtual SAN design that uses three fault domains.

A small operational change occurred between Virtual SAN 1 and Virtual SAN 6 that relates to the behavior of the Ensure Accessibility maintenance mode setting. In Virtual SAN 1, when the hosts entered maintenance mode with this option selected, the host continued to contribute its disk group storage toward the Virtual SAN datastore. As a result, virtual machines whose components were stored and were accessible on the host in maintenance mode remained in compliance, with all components available. In contrast, in Virtual SAN 6, when the host enters maintenance mode, it no longer continues to contribute its disk groups to the Virtual SAN datastore. Any components that are hosted on the disk groups in that node are marked as absent, and as a result, the vSphere Web Client will show these virtual machines as being out of compliance.

Full Data Migration Option

The Full Data Migration option evacuates all data onto the remaining hosts in the Virtual SAN cluster. With this option selected, the Virtual SAN datastore maintains, or fixes, availability compliance issues for the affected components. As a result, data is protected, assuming of course that sufficient storage resources exist across the Virtual SAN cluster.

This option is most typically selected when the administrator plans to migrate workloads off the host, in order to permanently remove the host from the cluster. This evacuation mode may result in large quantities of data traversing the Virtual SAN network, and may also see a significant increase in Virtual SAN datastore I/O, due to all of the components stored in the disk groups of the selected host being migrated elsewhere in the cluster. Virtual machines will continue to have access to their storage objects and components throughout the operation and remain in compliance. If virtual machine data is not accessible for some reason, and the host cannot be fully evacuated, the host will be unable to enter maintenance mode and the operation will timeout with an error.

No Data Migration Option

If selected, the No Data Migration parameter does not evacuate any data from the host before it enters maintenance mode. As a consequence, there is a risk that some virtual machine object components may become inaccessible, and the application or service could fail. This option is not generally recommended for production or business-critical environments, as it could result in data loss.

Quorum Logic Design and vSphere High Availability

A Virtual SAN–enabled cluster uses a quorum-based system with witness components to ensure consistent operations across the distributed datastore. The *quorum* is the minimum number of votes that a distributed system must be able to obtain, in order to be allowed to perform an operation. In Virtual SAN, 50 percent of the votes that make up a virtual machine's storage object must be accessible at all times for that replica to be active. If less than 50 percent of the votes are accessible to the host, the object is not available and is marked as inaccessible in the Virtual SAN datastore.

This can become a problem for Virtual SAN that can affect the availability of virtual machines, if after a host failure, the loss of quorum for a virtual machine object results in vSphere High Availability not being able to restart the virtual machine until the cluster quorum is restored. vSphere HA can guarantee that a virtual machine will restart only when it has a cluster quorum and can access the most recent copy of the virtual machine object.

For instance, Figure 4.48 shows a three-node cluster that has a single virtual machine running on host 1, which has been assigned a storage policy with an FTT of 1. If all three hosts fail in sequence (with host 3 the last host to fail), when host 1 and host 2 come back online, vSphere HA will be unable to restart the virtual machine because the last host that failed, host 3, retains the most recent copy of the virtual machine object components and is currently inaccessible.

In this scenario, either all three hosts must recover at the same time or the two-host quorum must include host 3. If neither of these conditions is satisfied, vSphere High Availability will attempt to restart the virtual machine again when host 3 comes back online.

Fault Domains

Virtual SAN 6 introduced the concept of the *fault domain* into its architecture, which allows the grouping of hosts making up a Virtual SAN–enabled cluster into different logical failure zones. This mechanism ensures that all replicas of a virtual machine object are not all provisioned onto the same logical failure zone. Virtual SAN architects can use this feature to design a fault domain–aware platform, which can provide tolerance of environmental failures (such as a data-center server cabinet failure or switch failure), or any group of hosts, that are likely to fail together, such as an availability zone, has redundancy, rather than just allowing for single host outages.

In Virtual SAN 1, the CLOM daemon assumes that every host exists in a separate fault domain, which results in node-independent failure behavior, as illustrated in Figure 4.49.

This example depicts how Virtual SAN 1 object placement may occur when a storage policy that includes an FTT of 1 is used in a six-node cluster across three server cabinets. In this example, there are two nodes per rack, and therefore, as illustrated, an FTT of 1 may not be able to protect the virtual machine's objects if a full rack failure occurs.

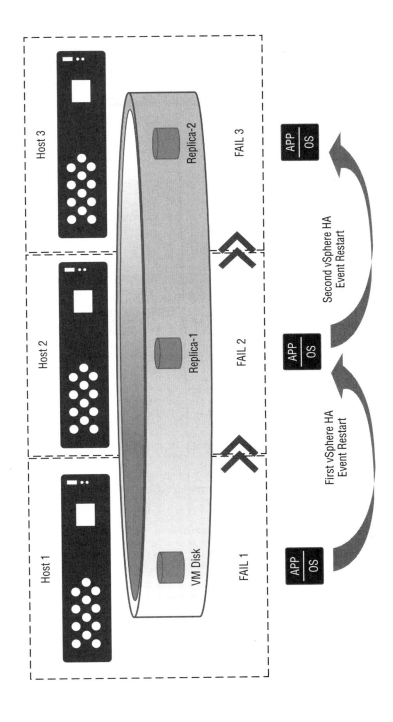

FIGURE 4.48
Quorum logic failure scenario

FIGURE 4.49
Virtual SAN 1 object placement

Figure 4.50 shows another example, employing an architecture that enables a vSphere 6 fault domain with the same policy of FTT of 1 in a six-node cluster; again, the design has employed two nodes per server cabinet. In this second example, each rack is configured as a fault domain, so the FTT policy of 1 would effectively protect the virtual machine, even if a full rack outage were to occur.

FIGURE 4.50
Virtual SAN 6 object placement (fault domain–enabled environment)

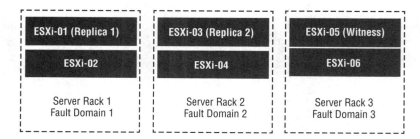

Figure 4.51 illustrates a typical design using fault domain architecture across four data-center server cabinets.

As shown, the fault domains provide the ability to group hosts within a cluster and define failure zones. This is achieved through Virtual SAN's fault domain architecture, ensuring that replicas of virtual machine data are spread across the defined fault domains, and can therefore tolerate a failure of a single fault domain. The use of fault domains in a design is primarily targeted at the ability to tolerate the following:

- Servers placed within the same server cabinet
- Servers sharing the same power supply
- Data-center availability-zone failures
- Network device failures, such as servers that share the same top-of-rack switches
- Flash device and mechanical disk failures

In a fault domain design, the FTT capability must be applied based on the fault domain architecture, and no longer based on hosts in the Virtual SAN cluster. If FTT = n, then $2n + 1$ fault domains are required. If a fault domain solution is not correctly configured to allow the distribution of objects and components across failure zones, then provisioning failures can occur because of this misconfiguration or because there are not enough fault domains to satisfy the virtual machine's assigned storage policy.

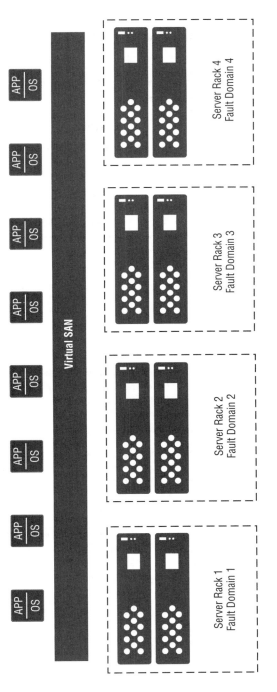

FIGURE 4.51
Fault domain design

An example of Virtual SAN 6 utilizing the fault domain feature with four server cabinets, with two hosts residing in each, is illustrated in Table 4.28 and Figure 4.52. In this example, we have four fault domains configured.

TABLE 4.28: Fault domain sample architecture

FAULT DOMAIN	HOSTS
FD1	ESX-1 & ESX-2
FD2	ESX-3 & ESX-4
FD3	ESX-5 & ESX-6
FD4	ESX-7 & ESX-8

The customer's design requirement is to mitigate against single rack failure, so the FTT capability is configured to 1.

A fault domain architecture (Figure 4.52) can be configured by administrators via the vSphere Web Client or ESXCLI, and are also configurable through host profiles, in order to simplify ongoing operational management of the environment.

Key design factors associated with including a fault domain architecture as part of an enterprise design include the following:

- In a design utilizing fault domains, the FTT capability must be configured to support n number of fault domain failures. Therefore, rather than needing $2n + 1$ hosts, the design will require $2n + 1$ fault domains. This ensures that replica data is distributed across the fault domain architecture, as opposed to other hosts, and therefore increases the Virtual SAN cluster's ability to handle a failure of greater than one host.

- Virtual SAN requires a minimum of three fault domains to be configured, with at least six hosts being used for a fault domain architecture to be a viable design option. However, it is recommended to employ four or more fault domains where possible, in order to increase redundancy. Therefore, as a design best practice, a fault domain architecture should employ a minimum of eight hosts across four fault domains.

- A fault domain architecture is used to further protect an infrastructure from single points of failure, assuming there are sufficient hosts configured in the design to properly support the environment.

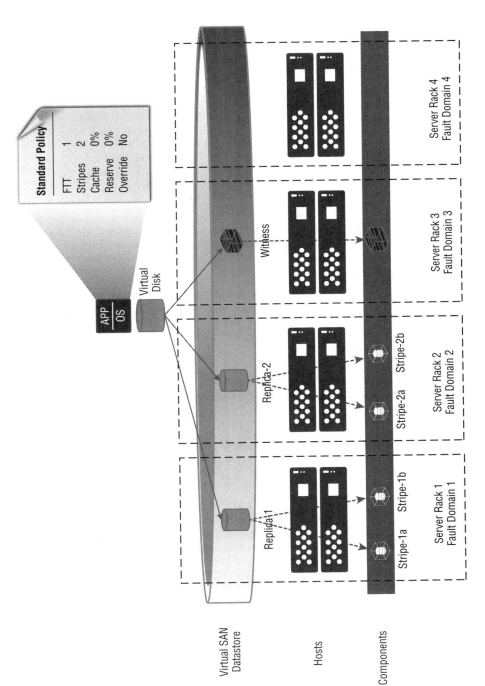

FIGURE 4.52
Fault domain sample architecture

Virtual SAN Internal Component Technologies

Throughout this chapter, several Virtual SAN internal services and mechanisms have been highlighted in conjunction with the availability architecture. Even though these internal component technologies don't directly relate to Virtual SAN design, it is important for you to maintain a complete understanding of the technologies used within the customer's solution. It is commonplace for architects to have to explain technologies to customers and provide a justification for design decisions through their knowledge and understanding of a solution. This section presents these key component technologies that operate in the background and how they interact with one another to make Virtual SAN operate seamlessly across the environment.

As you would expect, Virtual SAN operates in conjunction with various component services, mechanisms, and technologies, all of which must interact seamlessly to provide the distributed datastore storage platform, which offers the continuous availability and low latency for virtual machine workloads running across the entire cluster (see Figure 4.53).

The following is a brief overview of each of Virtual SAN's services and mechanisms, and how they operate as part of the overall VMware hyper-converged storage solution.

Reliable Datagram Transport

Reliable Datagram Transport (RDT) is used for cluster network communication between nodes. RDT is optimized to send very large files to move object data between hosts, by delivering datagrams between logical endpoints, typically over multiple paths. RDT is able to set up and tear down transport connections very quickly, depending on link health-status changes that are published by the CMMDS, in order to minimize any delay to the datagram transport as a result of link failures.

Cluster Monitoring, Membership, and Directory Services

Cluster Monitoring, Membership, and Directory Services (CMMDS) is responsible for discovering, establishing, and maintaining a cluster of networked node members, and detecting failures in nodes and network paths. CMMDS manages the inventory of items such as nodes, devices, and networks, and stores metadata information such as policies and RAIN configuration. CMMDS is also responsible for providing a consistent database as a service to other Virtual SAN component services.

Many of the other Virtual SAN component technologies (described later in this section) browse this directory and subscribe to updates to learn of changes in cluster topology and object configuration. For instance, the DOM and LSOM can use the content of the directory to determine the nodes storing the components of an object and the paths by which those nodes are reachable.

The cluster will also appoint a CMMDS master, with the master node being the main node in the cluster. These CMMDS roles are applied during the cluster discovery; the nodes elect a master by using a distributed consensus protocol. The master is responsible for discovering, establishing, and maintaining the cluster directory, as well as for managing the physical cluster resources. These roles are always system defined; the role configured for a node cannot be set or modified by a vSphere storage administrator.

If the master node is lost and there is no backup, all nodes in the Virtual SAN cluster must reconcile their entire view of the directory with the new master node to assure consistency of the directory. This results in all cluster nodes sending all their metadata, in their respective views, to the new master node. Having an already elected CMMDS backup node helps speed up this process.

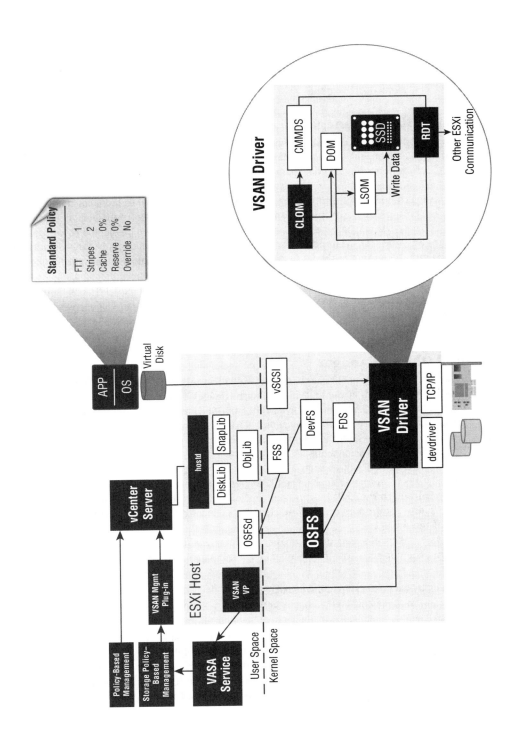

FIGURE 4.53
Virtual SAN internal component technologies and driver architecture

If the CMMDS cannot communicate or update information, problems result in the Virtual SAN cluster. When something is published into the CMMDS directory, information about it can be queried from any node in the Virtual SAN cluster by using `cmmds-tool`. High-level CMMDS log messages can be viewed in `vmkernel.log`.

Cluster-Level Object Manager

Cluster Level Object Manager (CLOM) is responsible for orchestrating objects and placing objects during create operations, and is the component service responsible for ensuring that objects and components have a configuration that meets their assigned storage policy requirements.

CLOM schedules rebuild operations when a component needs to be replaced, and also makes sure that there are enough disk groups to store the virtual machine files and that policies are being satisfied before performing placement operations. In addition, if an object is reconfigured with a new storage policy, CLOM handles the placement of any new components that need to be created. This daemon also talks to the CLOM on other hosts to see what space is available. The DOM then applies the configuration, as dictated by the CLOM daemon.

CLOM exists as a user-space daemon on each host in the Virtual SAN cluster. The status of the daemon can be checked or the service restarted by using `/etc/init.d/clomd <status /restart>`. Log messages for the CLOM daemon can be located and viewed in `/var/log/clomd.log`.

Distributed Object Manager

The Distributed Object Manager (DOM) is responsible for handling object availability and initial I/O requests. Objects exist at the DOM layer, with the DOM implementing and providing distributed access to Virtual SAN objects or VMDK virtual disks built from LSOM object components. The simplest cluster-level object type is a RAIN-1 mirror, of two or more components, illustrated in Figure 4.54. The DOM is responsible for ensuring that all sides of these mirrors are consistent by handling the synchronous I/O.

There is only one DOM client per Virtual SAN–enabled host; with all I/O being directed to the DOM client, I/O is then forwarded to the DOM owner. Likewise, there is only one DOM owner per object, although the DOM has no concept of locality. The DOM owner may be a host that does not own any component of that object, and the DOM owner of an object can change during the life of that object.

The DOM exists in kernel space, and there are no daemons that can be monitored or restarted without rebooting the vSphere host. High-level DOM messages can be viewed in `vmkernel log`.

Local Log-Structured Object Manager

The Local Log-Structured Object Manager (LSOM) is responsible for handling I/O and for ensuring the consistency of the components that are resident on local disks. Components exist at the LSOM layer. The LSOM receives I/O from the DOM, and subsequently returns acknowledgments to the DOM when write operations have been completed. The LSOM also returns payloads for read operations and handles write buffering, read caching, and the destaging of data to capacity tier disks. The LSOM is unaware of distribution, quorum, and I/O synchronization, with the DOM being used to handle those tasks. The LSOM is responsible only for handling the I/O.

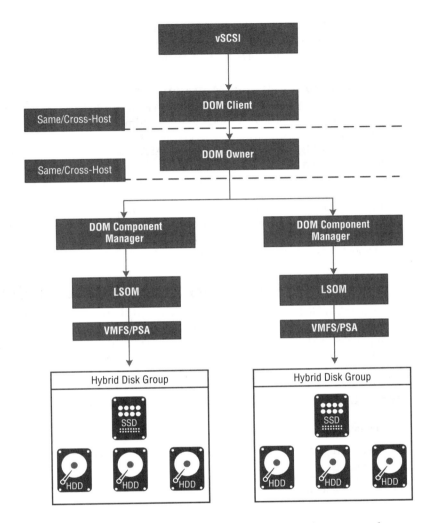

FIGURE 4.54
Distributed Object Manager object mirror I/O path

The LSOM exists in kernel space, and no daemons can be monitored or restarted without rebooting the vSphere host. High-level LSOM messages can be located and viewed in `vmkernel log`.

Object Storage File System

The Object Storage File System (OSFS) is responsible for creating the Virtual SAN datastore. OSFS provides a filesystem-like construct used for backward compatibility with vSphere. The Virtual SAN datastore has no directories as such. Directories in a Virtual SAN architecture are objects formatted with the VMFS, specifically, the Virtual Machine Home Namespace object type. The OSFS handles the initial formatting of these namespace objects and maps friendly names. In addition, the OSFS is responsible for mounting the namespace objects and making them available in the Virtual SAN datastore cluster.

The OSFS is a user-space daemon on each host in the Virtual SAN cluster. The status of the daemon can be checked, or the service restarted using /etc/init.d/osfsd <status/restart>. Log messages for the OSFS daemon can be viewed and queried in /var/log/osfsd.log.

Storage Policy–Based Management

Storage Policy–Based Management (SPBM) has already been addressed in this chapter. The SPBM framework provides vSphere storage administrators with control through the use of storage policies assigned on a per object basis in the Virtual SAN datastore. SPBM operates by using VMware's APIs for Storage Awareness (VASA) through vCenter and each vSphere host, utilizing a policy-driven storage service.

The SPBM mechanism sends the defined storage policy to the CLOM during the provisioning process. When a new policy is assigned to a virtual machine or virtual disk, the policy is modified and those changes are then reflected in SPBM. Object status, including distribution, reconfiguration, and resynchronization, are all reflected in the vSphere Web Client through SPBM

SPBM log messages can be viewed in the sps.log file on vCenter Server and in /var.log.vsanvpd.log on the ESXi provider.

Virtual SAN Integration and Interoperability

Virtual SAN is fully integrated and fully interoperable with most vSphere advanced storage features. However, several storage technologies have become irrelevant or unviable, or are not supported by VMware in a Virtual SAN–enabled environment. Table 4.29 and Table 4.30 reflect the design factors associated with these vSphere storage features.

TABLE 4.29: Integrated and interoperable vSphere storage features

FEATURE	STATUS
vSphere High Availability (HA)	Fully interoperable with Virtual SAN.
vSphere vMotion	Fully interoperable with Virtual SAN.
Enhanced vMotion	Fully interoperable with Virtual SAN.
Distributed Resource Scheduler (DRS)	Fully interoperable with Virtual SAN.
vSphere svMotion	Virtual SAN operates within a single cluster-wide distributed datastore, so a storage vMotion operation within the cluster is not an action that can occur. However, svMotion migration to externally attached datastores, such as those presented via a traditional block or NFS mechanism, is supported and fully interoperable with Virtual SAN.
Symmetric Multiprocessing Fault Tolerance (SMP-FT)	SMP-FT is supported on Virtual SAN clusters including ROBO two-node deployments. However, SMP-FT is not supported on a Virtual SAN stretched cluster because of the additional latency incurred across the data-center interconnect.

TABLE 4.29: Integrated and interoperable vSphere storage features *(CONTINUED)*

FEATURE	STATUS
vSphere Replication	Fully interoperable with Virtual SAN, with enhanced features provided over other storage solutions, such as a 5-minute RPO.
Host Profiles	Fully interoperable with Virtual SAN.
Data Protection	Fully interoperable with Virtual SAN through VADP.
Windows Server Failover Clusters (WSFC)	Fully supported in Virtual SAN from 6.1 release, including ROBO and stretched cluster deployment types. Microsoft Failover Cluster Instances with single-copy storage are not supported, because of the lack of support for raw device mappings (RDMs) within a Virtual SAN–enabled cluster.
Oracle RAC (Real Application Clusters)	Fully supported with Virtual SAN from 6.1 release onward.
SAP	Fully supported with Virtual SAN from 6.2 release onward. All SAP applications, including Business One and NetWeaver-based products, are supported on Virtual SAN in production environments. At the time of writing, SAP does not support HANA on any HCI-based solutions, including Virtual SAN.
Snapshots	Fully interoperable with Virtual SAN.
Virtual Machine Thin Provisioning	Fully interoperable with Virtual SAN. Thin provisioning is facilitated through Virtual SAN's SPBM mechanism via the Object Space Reservation capability.

TABLE 4.30: Irrelevant, unviable, or unsupported vSphere storage features

FEATURE	STATUS
Storage Distributed Resource Scheduler (SDRS)	Not employed for Virtual SAN datastore storage. SDRS operates within the constructs of the vSphere cluster. Therefore, as a consequence of Virtual SAN operating within a single distributed cluster-wide datastore, this feature cannot be employed.
Distributed Power Management (DPM)	Unsupported vSphere feature on a Virtual SAN–enabled cluster. DPM is an optional feature that cannot coexist with Virtual SAN because the distributed datastore requires continuous access to storage components on the host servers.
Storage I/O Control (SIOC)	Not employed for Virtual SAN datastore storage. As Virtual SAN does not utilize shared storage resources, this advanced storage feature is not employed in a Virtual SAN–enabled cluster.
Raw device mappings (RDMs)	Cannot be configured through a Virtual SAN datastore, but can be presented to virtual machines via an external block storage device.

Chapter 5

Virtual SAN Stretched Cluster Design

Introduced with the 6.1 release of Virtual SAN was the capability to configure a stretched, metro-style cluster across two physical locations, as illustrated in Figure 5.1. As highlighted in Chapter 2, "Classic Storage Media and Constructs," the concept of a stretched cluster isn't new, and storage vendors have offered hardware-based solutions for years. However, although these have worked well and met the needs of many customers, the cost, proprietary nature of the hardware, and their complexity, which has made them operationally difficult to maintain, has up until now rendered them out of reach for many enterprise IT customers and service provider environments.

Stretched clusters are based on a two-site active/active architecture, with the primary use cases based on increasing enterprise application availability and data protection. For stretched clusters to operate, mirrored storage must be available to both data centers, and the storage must be completely synchronized at all times, allowing virtual machines to be live-migrated from one location to the other. In order for data to be synchronous, there must be sufficient bandwidth, and also minimal latency between the two data centers.

In order for vSphere to perform live migrations between sites, the hypervisors must have access to the Virtual SAN distributed datastore from both locations. Bandwidth and latency are significant factors in this type of design. In a Virtual SAN stretched cluster architecture, all read requests are served from the copy of the data that resides on the site where the virtual machine is running. However, the hypervisor writes to storage across the WAN interconnect, and therefore requires acceptable levels of performance to function properly. As a consequence, the location to which the data is being mirrored and the available connectivity between the two locations are key design considerations.

As the data is being written to two locations from one host, this can significantly complicate the network configuration. In addition, another key design consideration is the way that the solution handles communication failures between sites in order to avoid a split-brain scenario if the two sites become disconnected for a period of time.

You also have to consider what happens if a host fails and virtual machines need to be restarted. Do you restart them on the local site or the remote site? Also, the use of affinity rules is key to avoiding unnecessary vMotion events between data centers, which can affect read cache performance and flood cross-site interconnects needlessly. There may also be a customer requirement to keep interrelated multitiered applications together, to optimize workload performance, reducing the potential additional latency that might result from distributing the applications across the two physical locations.

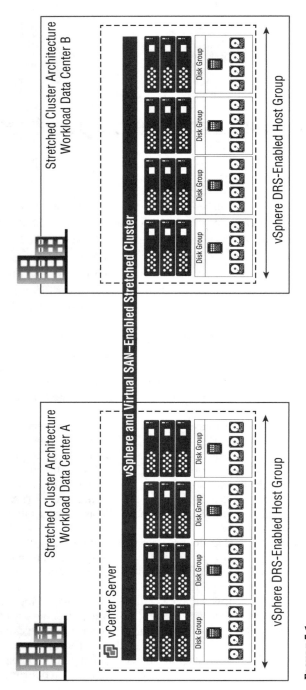

Figure 5.1
Virtual SAN stretched cluster

Use cases for stretched cluster environments must be defined clearly in the customer's requirements. Just because you can, doesn't mean you always should, particularly if it increases costs exponentially or overcomplicates the environment in ways that the operational team does not fully understand.

The following are key business requirements that must be considered when evaluating the use of stretched clusters in a design:

- The vSphere hosts that form the highly available Virtual SAN stretched cluster should be evenly distributed across the two sites, with sufficient capacity to host migrated or failed-over virtual machines.
- Mirrored Virtual SAN storage configurations must be provided across both data centers.
- Layer 2 adjacency, with sufficient bandwidth, and the low latency required to maintain a synchronous storage platform must be available as part of the cross-site architecture.
- Stretched clusters are supported on both hybrid and all-flash disk group architectures.
- Stretched clusters require Virtual SAN 6.1 or newer, and therefore, vSphere 6 as its core platform.
- Virtual SAN stretched clusters require that the VirstoFS on-disk format (v2) be used.
- The maximum round-trip latency on the IP network, between the two sites, must not exceed 5 milliseconds (ms).
- Any IP subnet used by virtual machines, which resides on the stretched cluster, must be accessible from vSphere hosts in both locations. This requirement ensures that virtual machines are able to function smoothly after being migrated, as a result of either a vSphere High Availability (HA) triggered event or a vMotion operation.
- The vCenter Server must be able to connect to and manage vSphere host servers in both data centers.
- The maximum number of hosts in a highly available Virtual SAN stretched cluster cannot exceed 30 hosts, with 15 hosts on each site.
- Erasure coding RAID 5 and 6 configurations are not supported with Virtual SAN stretched clusters.
- Stretched clusters require Virtual SAN enterprise edition licensing.

Stretched Cluster Use Cases

Typical customer use cases for Virtual SAN stretched clusters include a transparent maintenance requirement for an entire data-center site, disaster avoidance, and the automated recovery of workloads.

The ability to perform maintenance at an entire physical data center, which is transparent to the end users and application owners, provides significant operational and business advantages. For instance, being able to perform planned maintenance of one data-center location, without any service interruption, helps avoid lengthy and complex approvals from the application owners and removes the need for planned outages.

Another typical use case for stretched cluster environments is disaster avoidance. Using a disaster-avoidance strategy for applications can prevent outages before forecasted disasters occur, such as hurricanes or floods. Therefore, using a stretched cluster strategy can help IT organizations to provide zero application downtime and data loss to the business. This provides IT operational advantages, as well as a competitive business advantage, through the provisioning of a solution that delivers continuous availability to revenue-generating applications.

The final traditionally adopted use case for a stretched cluster strategy is the delivery of a cross-site automated recovery mechanism for virtual machines, through the use of vSphere's HA restart process. This solution type provides a close-to-zero recovery time objective (RTO) and recovery point objective (RPO) for the majority of unplanned failure scenarios. Providing this type of automated approach to virtual machine recovery allows the IT operations team to focus on the health of applications after the loss of a data-center infrastructure, as opposed to platform or operating system availability.

Fault Domain Architecture

Stretched clusters are built on Virtual SAN's preexisting *fault domain architecture*. However, a stretched cluster can be configured with a maximum of only three fault domains. As illustrated in Figure 5.2, the first fault domain is the *preferred* site for workloads, the second fault domain must host only the stretched cluster *witness virtual appliance*, and the third fault domain is the *nonpreferred* site. This architecture provides support for a storage policy based on a Number of Failures to Tolerate (FTT) capability of 1, allowing virtual machine objects to remain accessible as a result of preferred or nonpreferred fault domain failure.

The *witness virtual appliance* fault domain is used in the stretched cluster architecture to host only the witness virtual machine, and cannot be used to host other stretched cluster protected workloads.

Witness Appliance

Virtual SAN stretched clusters require the use of a single *witness appliance*, which must be hosted in the third fault domain. This location can either be provided by a third data center, be managed by the IT organization, or alternatively be deployed onto a public cloud platform.

The witness is a nested ESXi appliance, specifically modified to provide quorum functionality to a stretched cluster architecture. The witness is deployed from an OVA onto a vSphere host with two vCPUs, and between 8 GB and 32 GB of assigned memory, depending on the environment size, which is configured during the deployment.

As a nested Virtual SAN host, the witness appliance has a requirement for both flash and mechanical disks. These are fully configured within the appliance during its deployment, with one of the appliance's VMDKs being tagged as a flash device. The Virtual SAN administrator does not need to manually configure the appliance, and there is no requirement for a physical flash device in the vSphere host on which the witness appliance is deployed. In addition, all of the witness's virtual disks can also be thin provisioned if required.

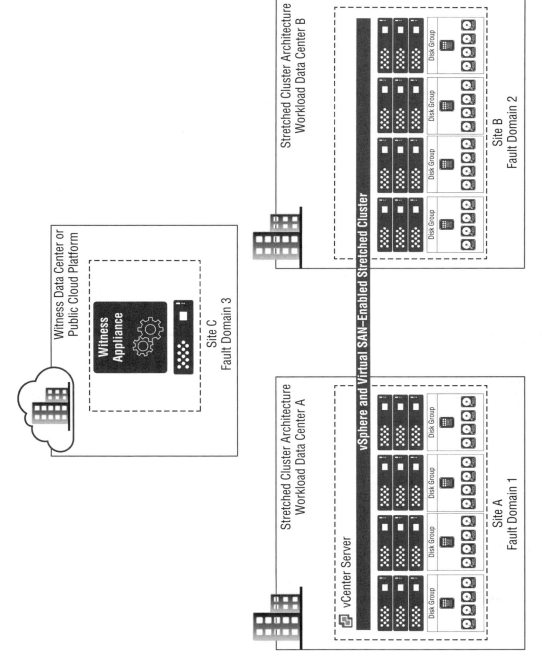

FIGURE 5.2
Stretched cluster fault domain architecture

Even though the virtual appliance witness typically provides the most flexible and operationally efficient configuration, a Virtual SAN stretched cluster design also has the option of using a physical host as a witness. However, this typically reduces the flexibility of the deployment of the witness and the design in general, and increases costs, as the physical host will require an appropriate vSphere license to be assigned as well as server hardware and its associated operational expense. The witness virtual appliance has an embedded license preconfigured with no additional cost associated.

The witness appliance stores only metadata and is not designed to host any virtual machines. As a result, the witness appliance requires only 16 MB of capacity storage for each witness component stored. One witness component is required per object hosted on the Virtual SAN stretched cluster.

The number of components stored on the witness directly reflects the number of objects associated with the virtual machines running on the stretched cluster. Each virtual machine requires at least one virtual disk (VMDK), one namespace, and one swap file. Therefore, there are a minimum of three objects per virtual machine, with each snapshot also adding one object per VMDK. However, adding additional stripe width, through the Number of Disk Stripes per Object capability, does not add to the number of objects associated with a virtual machine.

During the deployment of the witness appliance, three sizing options are presented to the vSphere administrator. These relate directly to environment sizing, as shown in Table 5.1. Selecting the most appropriate option ensures that sufficient storage and memory capacity are provisioned on the witness appliance for each of the virtual machines that will be running across the stretched cluster platform.

TABLE 5.1: Witness appliance sizing configuration options

Environment Configuration	vCPUs	Memory	Flash Device	Mechanical Disks	Supported Components
Tiny	2	8 GB	10 GB	1 × 15 GB	750
Medium (Default)	2	16 GB	10 GB	1 × 350 GB	21,000
Large	2	32 GB	10 GB	3 × 350 GB	45,000

Network Design Requirements

A Virtual SAN stretched cluster requires the interconnection of all three sites in the topology. Several key design requirements must be met for a stretched architecture to be implemented. The following key network specifications are required between the primary and secondary workload data-center sites. In addition, these design requirements must be met between the two workload data-center sites, and the location supporting the witness appliance:

- There must be vSphere management network connectivity between all three sites.
- There must be Virtual SAN network connectivity between all three sites.
- vMotion network connectivity between the primary and secondary workload sites is required to facilitate the live migration of virtual machines.
- Layer 2 stretched network connectivity is required for all virtual machine networks between the primary and secondary workload sites.
- Layer 3 network connectivity is required between the workload sites and the location hosting the witness appliance.

One of the key requirements for any stretched cluster configuration is the layer 2 connectivity between data centers. Some flexibility around this configuration is available, with layer 3 supported for some traffic types. But in a typical stretched cluster design, workload sites should always operate under the same stretched layer 2 segment for all virtual machine networks. This simplifies the design and reduces the operational overhead, as outlined in Table 5.2.

TABLE 5.2: Virtual SAN stretched cluster layer 2 and layer 3 network requirements

Traffic Type	Supported Configurations	Design Recommendation
Management network traffic	Layer 2 stretched or layer 3 routed	For vSphere management traffic, either option is viable. Where possible, to simplify design and operations, a layer 2 configuration should be deployed.
Virtual SAN network traffic	Layer 2 stretched between workload sites and layer 3 between workload sites and witness	It is recommended that Virtual SAN network traffic be stretched across layer 2, between the primary and secondary workload sites, and layer 3 between workload sites and the witness appliance.
vMotion network traffic	Layer 2 stretched or layer 3 routed	For vSphere vMotion traffic, either option is viable. To provide the ability to migrate running virtual machines without disruption, a layer 2 configuration should be deployed.
Virtual machine network traffic	Layer 2	To provide the capability to move virtual machine workloads between primary and secondary data centers, without the modification of IP addresses, a layer 2 configuration is required.

It is an absolute requirement that the virtual machine networks must sit on a stretched layer 2 extension (see Figure 5.3), as the IP addresses of the guest operating systems will not change during a vMotion operation or a vSphere HA recovery event. If the destination port group is not on the same layer 2 address space as the source, network connectivity to the guest operating system will be lost. Stretched layer 2 technology is therefore a key design requirement for a Virtual SAN stretched cluster platform.

VMware does not recommend any specific stretched layer 2 technology. Any mechanism that can present the same layer 2 networks to the vSphere hosts at each physical location will work, as the actual network configuration is irrelevant to the hosts in the stretched cluster design. Some examples of valid technologies include Virtual Extensible LAN (VXLAN), VMware NSX layer 2 gateway services, Cisco Overlay Transport Virtualization (OTV), and GIF/GRE tunnels. Also, there is no defined maximum distance between the source and destination networks that will be supported by VMware, as long as the stretched network meets the requirements described in Table 5.3.

TABLE 5.3: Network bandwidth and latency requirements

	BANDWIDTH REQUIREMENT	LATENCY	CONNECTIVITY TYPE
Network requirements between workload data sites	10 Gb/s connectivity or greater is strongly recommended	< 5 millisecond latency round-trip time (RTT)	Layer 2 and/or layer 3 network connectivity with multicast support
Network requirements to witness fault domain	100 MB/s connectivity	< 200 milliseconds latency RTT	Layer 3 network connectivity without multicast

As illustrated, Virtual SAN stretched clusters maintain strict requirements around interconnect bandwidth and latency. Figure 5.4 depicts the bandwidth and latency requirements set out in Table 5.3.

Distance and Latency Considerations

The interconnect used between the two workload facilities is a significant factor in the stretched cluster solution design. Therefore, when designing a continuously available and disaster-recovery solution, as provided by a Virtual SAN stretched cluster, it is important to understand factors such as link capacity, latency, and use.

One of the stretched cluster's architectural requirements states that the vSphere source and target hosts must be within 5 milliseconds round-trip time (RTT) from one another. Therefore, we must consider the practical implications in defining stretched cluster use cases with respect to how far the physical data centers can be located from one another. Typically, network latency and the RTT correspond to distance, although various factors can vary these calculations significantly. The < 5 milliseconds RTT response requirement must be translated into a physical distance between locations, and the many variables accounted for.

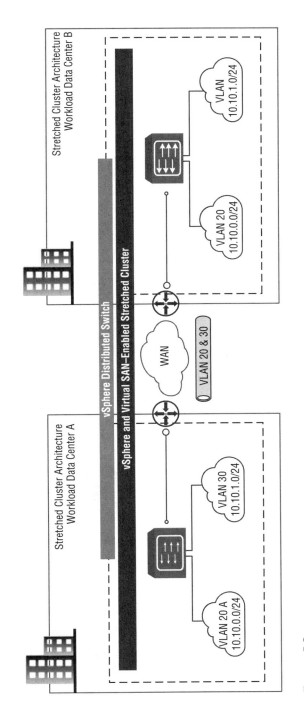

FIGURE 5.3
Layer 2 extension

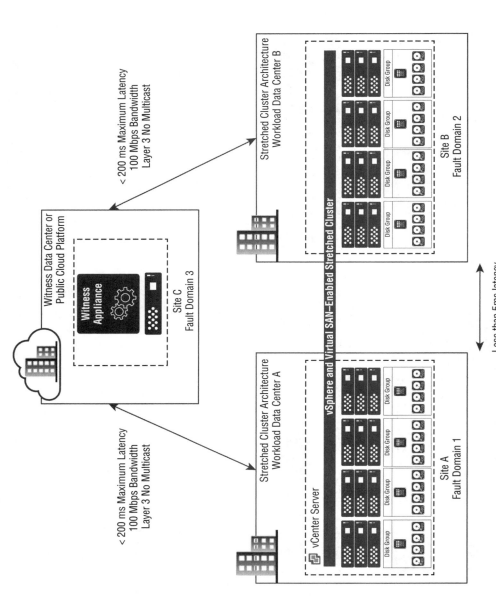

FIGURE 5.4
Virtual SAN stretched cluster overview

Table 5.4 provides sample distances and their corresponding link latencies. Note that the values shown here are estimates and can vary significantly depending on interconnect quality, network hops, and physical infrastructure. Based on this table, which does not take into account other environmental factors, the absolute distance limit between sites for a Virtual SAN stretched cluster design is likely to be less than 500 km.

TABLE 5.4: Distance and estimated link latency

APPROXIMATE DISTANCE IN KILOMETERS	ONE-WAY LATENCY IN MILLISECONDS	ROUND-TRIP LATENCY IN MILLISECONDS
50 km	0.25 ms	0.5 ms
100 km	0.50 ms	1 ms
200 km	1 ms	2 ms
400 km	2 ms	4 ms
500 km	2.5 ms	5 ms

Bandwidth Requirements Calculations

As with calculating capacity storage or endurance flash-based requirements, calculating bandwidth prerequisites across a stretched cluster interconnect requires using various formulas, collecting a significant amount of data through a current state analysis of existing workloads, and examining performance metrics for the endurance flash device hardware. Even then, because of variabilities in workloads and interconnect performance, these figures should be treated with caution. Design estimates should always be considered conservative.

As noted earlier, it is strongly recommended that the interconnect between the two workload data centers provides a layer 2 extension. This recommendation is based on the fact that reconfiguring operating system IP address values as the virtual machines fail over adds significant operational overhead, cannot be easily automated, and is not practical from an application perspective in most environments. In addition, configuring multicasting over a layer 3 network is complex and provides a difficult configuration to maintain operationally. Therefore, providing the technology to stretch the required layer 2 networks in a stretched cluster is considered essential to the design.

As already mentioned, bandwidth requirements are dictated by the workload, specifically the number of write operations performed by each host in the Virtual SAN cluster. Although other factors may also need to be considered as part of the calculation, such as rebuild operations, this exercise focuses specifically on providing metrics for write actions, which of course are targeted at only the endurance flash devices.

Although 1 Gb/s networks are supported as a minimum requirement, 10 Gb/s networks are strongly advised. For most enterprise workloads, 10 Gb/s networks are the only viable option available.

To deliver a single 4 KB write I/O, it is calculated that 125 Kbps of bandwidth is required across the layer 2 interconnect, between the two workload data centers. We can take this derived figure, and based on the following formula, calculate the bandwidth required. In the following example formula, Hst is equal to the total number of nodes in the stretched cluster, and No. of 4 KB IOPS is equal to the number of 4 KB IOPS per node.

Interconnect bandwidth calculations can be developed from the following example formula.

Formula:
Hst × No. of 4 KB IOPS × 125 Kbps = Required Bandwidth in Kbps
Example:
12 nodes × 1,000 IOPS × 125 Kbps in bandwidth = 24,000,000 Kbps (or 2.4 Gb/s)

For our example calculation, we will be assuming that the maximum utilization of a high-end endurance flash device is able to achieve approximately 1,500 × 256 KB IOPS of sequential writes. Each flash device is to be configured in a single hybrid disk group on each of the six hosts in the cluster (6 + 6 +1), where 1 represents the witness.

This suggests that the maximum number of 4 KB write IOPS per node is approximately 16,000, with this flash device being used. As illustrated in the preceding example formula, with 12 nodes in the cluster (6 + 6 + 1), this equals 12 nodes × 16,000 IOPS × 125 Kbps. As a result, the design would require 24,000,000 Kbps, or 2.4 Gb/s, of consistently low-latency network bandwidth between the two workload data centers.

However, this calculation does not take into account other environmental variables or other factors, such as whether read locality has been modified or the interconnect was used for other network traffic types, such as *bursty* vMotion network traffic, which is likely to be the case.

In addition, read requests always occur locally in a stretched cluster by default. But if a host or disk failure takes place, and read requests need to be served over the data center interconnect, this temporary increase in utilization could cause contention across the WAN link if it is not sized appropriately to take these factors into account.

Furthermore, stretched cluster read locality can be disabled as part of a design if required, for specific use cases such as intra-data-center availability zones or a campus design that doesn't have available bandwidth or latency as significant design factors. If a stretched cluster design does require that read locality functionality be disabled, the same formula, based on 125Kbps for every 4 KB read IOPS, will provide a good level of guidance in calculating additional read request bandwidth requirements.

An additional bandwidth design factor is the traffic associated with resynchronization and rebuild operations that can occur after a physical component fails or is removed from the cluster. These I/O-intensive operations take place as Virtual SAN rebuilds or updates components that have been lost or have fallen out of synchronization. The design should aim for allowing these operations to occur without impacting running workloads. The following formula can be used as part of the design calculations to allow for resynchronization and rebuild traffic across the data center interconnect.

For example, if the design requires a single 1.2 TB SAS disk failure to be catered for, and requires that the rebuild operation should take no longer than 1 hour, the network bandwidth required during the rebuild operation should be at least 2.4 Gb/s, based on a maximum disk rebuild of 300 MB/s, as illustrated in the following calculation.

1.08 TB per hour = 18 GB per minute
18 GB / 60 = 300 MB/s
300 MB/s = 2.4 Gb/s

Stretched Cluster Deployment Scenarios

To avoid network loops within the design, it is not recommended to implement layer 2 network extensions between all three sites. This configuration would add unnecessary complexity to the network design. The optimal layer 2 and layer 3 network topology is illustrated in Figure 5.5.

The optimal design provides a layer 2 configuration between the two workload sites, and a layer 3 configuration between each of the workload sites and the witness location. In the event of an outage on either of the workload data centers, this design prevents any network traffic from workload data center A being routed to workload data center C, via the witness site (illustrated as route C in Figure 5.5). If this was allowed to occur, it would result in significant performance degradation due to the higher-latency and lower-bandwidth link typically used to provide connectivity to the witness location.

Default Gateway and Static Routes

Stretched cluster hosts that are connected over a layer 2 network, and communicate with one another over the Virtual SAN network, must also be configured to communicate with the witness appliance over layer 3. The recommended solution is to apply *static routes* on all hosts in the Virtual SAN cluster, enabling hosts that are located in the workload sites to be able to route network traffic to the witness appliance in the third site, and as such, also allow the witness appliance to communicate with hosts in the workload data centers.

For vSphere ESXi hosts located in the workload data centers, a *static route* must be configured on the Virtual SAN network, in order to route traffic to the witness appliance via the specified gateway. In the case of the witness appliance, the Virtual SAN network must have *static routes* added that direct network traffic to the destination workload data centers. This configuration can be completed on each host included in the stretched cluster design and on the witness appliance, via the `esxcli network ip route` command.

In addition, depending on the design, it may also be required to add static routes to facilitate ESXi management network connectivity. The single vCenter Server must be able to manage all vSphere hosts via the management network, at both workload data centers and the witness virtual appliance located at the third site. The design should not require any additional configuration to enable vMotion network traffic, as vMotion operations will occur only between workload data centers and will not interact with the witness site.

Stretched Cluster Storage Policy Design

In a Virtual SAN stretched cluster configuration, the Number of Failures to Tolerate (FTT) capability value must be configured as 1. This is the maximum and only value that a stretched cluster design can comply with, because only three fault domains are supported. The FTT capability will be *Force Provisioned* during an outage, when only two of three sites are available. This non-compliance will be addressed by the Cluster Level Object Manager (CLOM) daemon after the third site becomes available again, bringing affected objects back into compliance.

> **NOTE** An all-flash configuration using erasure coding is not supported in a stretched cluster design, as an exact mirrored copy of data is required at both data centers.

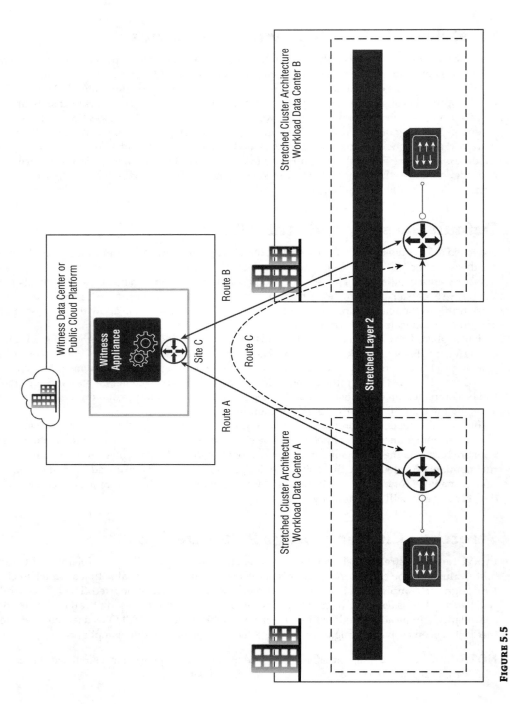

FIGURE 5.5
Stretched cluster optimal layer 2 and layer 3 configurations

Preferred and Nonpreferred Site Concepts

The concept of a preferred site is required by a stretched cluster, in order to align with Virtual SAN's fault domain mechanism. One of the workload data-center sites is designated as the *preferred* fault domain, to handle a *split-brain* scenario if a link failure between the active sites occurs that results in a network partition. The preferred fault domain determines which active site the witness joins, which remains running. Typically, the preferred site is the data center that is expected to have the highest overall levels of availability.

For instance, a virtual machine in a stretched cluster can potentially run on either of the two sites. If network connectivity between the two workload data centers is lost, but both sites retain connectivity to the witness site, the preferred data center remains active, while the nonpreferred data center is marked as down, allowing the vSphere HA process to be initiated.

Stretched Cluster Read/Write Locality

With virtual machine components distributed across two physically separated data centers, you'd typically expect an increase in latency for read and write requests that must traverse that interconnect. However, as illustrated in Figure 5.6, in a virtual SAN stretched cluster, a new read algorithm ensures that all read requests are serviced on the local site, where both the compute and storage components reside. This is unlike the typical Virtual SAN read algorithm, which services read requests by using a round-robin mechanism across all available component replicas.

This new algorithm, specifically designed for Virtual SAN stretched cluster implementations, mitigates the risk of increased read latency resulting from the cross-site configuration. However, in a design that incurs less than 1 ms of latency and has sufficient bandwidth available between workload data centers, this new read locality algorithm could be disabled, reverting to the non-stretched-cluster, round-robin mechanism. This parameter change would result in 50 percent of read requests being handled by the remote data center, which as a consequence, would see a significant increase in cross-site interconnect bandwidth utilization. Therefore, modifying this parameter value is not recommended, except in specific use cases, such as a stretched cluster configuration across two distinct availability zones within the same physical data-center building, resulting in ultra-low latency and no design concerns over bandwidth utilization.

The read locality algorithm is enabled by default when a Virtual SAN stretched cluster is configured, but can be disabled with the advanced parameter `VSAN.DOMOwnerForceWarmCache`. This setting is hidden from the vSphere Web Client, and is available only in the CLI.

One consequence that must be considered, as a result of this change in the read locality mechanism used by a Virtual SAN stretched cluster, is what happens when a virtual machine is migrated to the alternate data center. In this scenario, after the vMotion operation had completed, all reads would now be served from the copy of the data at the new site. A performance penalty might be incurred, as a result of the time it takes for the read cache to be rewarmed. However, as you will see next in this chapter, the design should also use mechanisms to prevent cross-site vMotion events from occurring under normal operating conditions, ensuring that live migrations between data centers do not become a common daily occurrence.

Write requests have no such locality mechanism in place (see Figure 5.7). To maintain data consistency across both workload data centers, writes are acknowledged from both sites before the write operation is acknowledged back to the virtual machine's operating system or application. This ensures data availability, even if a minimal amount of impact on performance is incurred as a result of cross-site latency.

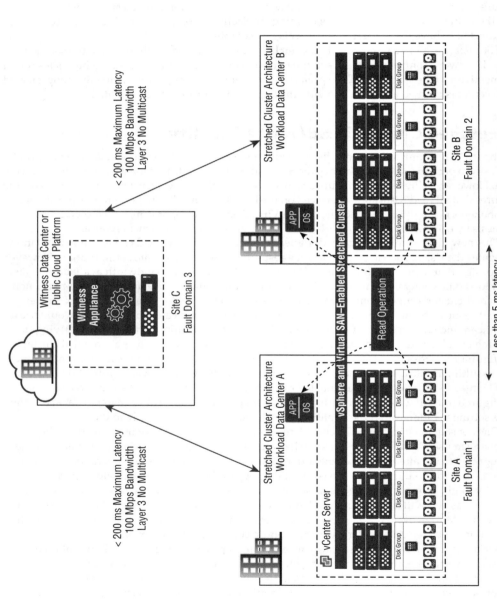

FIGURE 5.6
Anatomy of stretched cluster local read operation

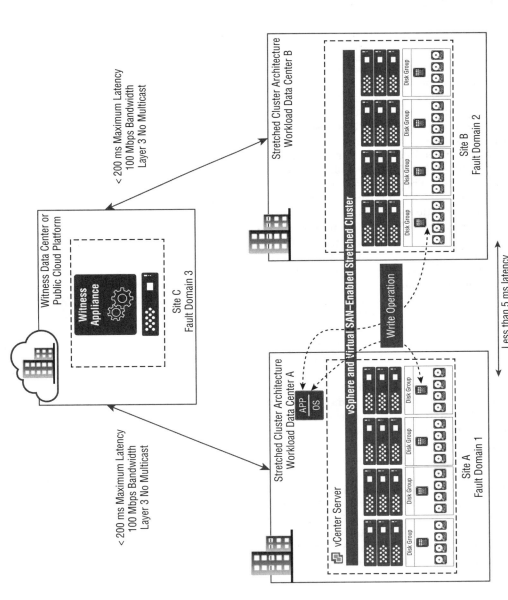

FIGURE 5.7
Anatomy of stretched cluster write operation

Finally, a write operation does not typically include any communication with the witness appliance. Therefore, any increase in latency between the virtual machine workload data centers and the witness site will not affect write performance.

Distributed Resource Scheduler Configurations

In a Virtual SAN stretched cluster, vSphere Distributed Resource Scheduler (DRS) should be used to define affinity rules, help prevent unnecessary downtime, and reduce intersite network traffic overhead by enforcing desired site affinity rules for workloads and ensuring that unnecessary cross-site vMotion operations do not occur.

It is recommended to set virtual machine host affinity rules to ensure that a virtual machine, under normal operating conditions, always runs on the same host, configured as the primary read node for a given component. In a stretched cluster design, this approach results in aligning the virtual machine host affinity rules with the Virtual SAN storage configuration across both sites.

As illustrated in Figure 5.8, in a Virtual SAN stretched cluster design, a virtual machine's host affinity rules should be configured to ensure that the workload remains local to its components, and not be migrated unnecessarily across sites during normal operating conditions, resulting in all read I/O staying local. In addition, this also ensures that in the case of a temporary or intermittent network connection failure between sites, virtual machines will not lose their connection to that particular host's storage.

It is strongly recommended that the stretched cluster design also implement *should rules* and not *must rules*. This ensures that if these rules are violated by vSphere HA, during a failure the availability of services would always prevail over performance. If *must rules* were used, vSphere HA could not violate the rule set, so virtual machines would not be restarted in the alternate data center, should a vSphere HA event occur; this could lead to service outages. This behavior results from vSphere DRS communicating these rules to vSphere HA, which are then stored in a compatibility list, governing how virtual machines are allowed to power on. In an HA scenario where a full data center fails, *must rules* make it impossible for vSphere HA to restart the virtual machines, as they do not have the required affinity to be allowed to be powered on across hosts in the alternate data center.

It is possible for vSphere DRS, under certain circumstances such as massive host saturation coupled with aggressive recommendation settings, to also violate *should* rules, although this is rare. It is important for operational teams to monitor for the violation of these rules, as a violation could affect the availability and performance of the Virtual SAN stretched cluster during a disaster-recovery scenario.

To configure the required affinity rules, you must first define *sites* by creating groups of hosts that align with each workload data center, as illustrated in Figure 5.8. The next task is to nest the virtual machine groups inside the host groups, based on the affinity required. It is possible to automate this process of defining site affinity for workloads by using tools such as vRealize Orchestrator or vSphere PowerCLI. If automating the process is not an option, using a generic naming convention is recommended, to enable simplifying the creation of these groups, as illustrated in Figure 5.9. It is advisable that these groups be validated on a regular basis by operational teams, to ensure that all virtual machines belong to a group with the correct site affinity.

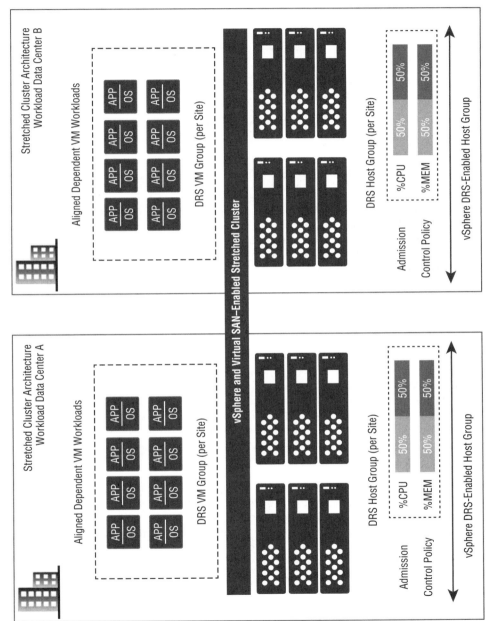

Figure 5.8
Stretched cluster vSphere DRS affinity rule configuration

FIGURE 5.9
Configuring a DRS affinity rule set for a Virtual SAN stretched cluster

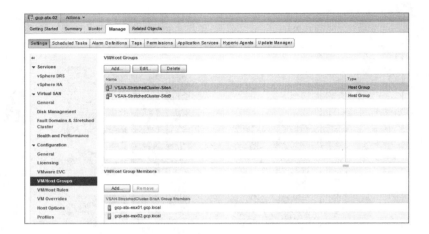

One additional operational consideration is that you must have hosts and virtual machines configured before it is possible to create the groups. Also, as noted previously, ensure that the rules are configured as *should run on hosts in group*.

The primary aim of this affinity group configuration is to ensure that a group of virtual machines remains within its preferred site. The only reason they should be migrated, under normal operating conditions, is as part of a disaster-avoidance strategy or a disaster-recovery event at one of the data center locations.

A secondary benefit of this affinity group configuration is that it also prevents virtual machines from migrating, through the normal DRS load-balancing placement mechanism, to the nonpreferred data center. Of course, nothing is stopping a vSphere administrator from manually initiating the migration of a virtual machine to the nonpreferred data center. However, the machine will of course migrate back via vMotion shortly afterward, in order to comply with its affinity rule constraints. Ensuring that these processes are well documented and communicated in the operational and change management procedures is worthwhile, in order to prevent vSphere administrators from making this mistake. In addition, operational policies should be put into place to ensure that newly provisioned virtual machines are added to the appropriate data centers' virtual machine group. Failing to do so would result in the new workload being unprotected by this mechanism.

By default, DRS is invoked every 5 minutes to perform corrective actions over rule violations and load-balancing operations (although in this design type, load-balancing occurs only within the user-defined host group). However, DRS is also triggered if changes in the cluster configuration occur, such as a host reconnecting to the cluster or new hosts being added. In this scenario, DRS is invoked immediately to generate recommendations to correct rule violations, typically within 30 seconds. Also, when it comes to correcting affinity rule violations, DRS assigns a higher priority than it does to load-balancing operations, allowing these to be actioned before load-balancing migrations or recommendations can occur, in order to correct rule infringements first.

To migrate a virtual machine from one data center to the other, you must modify its virtual machine affinity group membership, as the group-associated *should* rule will migrate it back. In a disaster-avoidance scenario, implementing maintenance mode on all hosts at the data center facing imminent failure will ensure that all workloads are migrated to the alternate location.

The next DRS-based design decision for a Virtual SAN stretched cluster platform is whether to use a *fully automated* or *partially automated* virtual machine placement mechanism for workload load balancing.

VMware provides a clear design recommendation for this configuration in a stretched cluster environment: partially automated. The reason for this relates to virtual machine behavior during and after a site failure. If the site has been down for a significant period of time, a major resynchronization of component data will likely be required when the hosts become available again. If DRS is configured as fully automated, virtual machines would begin rebalancing across the two sides of the stretched cluster in order to comply with their affinity configuration immediately.

However, as there is no synced local copy of the data, all I/O from these virtual machines would have to traverse the site interconnect in order to reach the active replicas of the components required. This will cause latency as well as bandwidth and performance issues in the environment, and is therefore not an optimal configuration choice for a stretched cluster design. In addition, because vSphere DRS is limited by the overall capacity and number of concurrent operations over the vMotion network, it could take multiple DRS processing cycles before all rule violations are corrected. Partially automated mode is the recommended design configuration in a Virtual SAN–enabled stretched cluster.

In partially automated mode, vSphere storage administrators can wait until full resynchronization of the environment is complete before temporarily enabling fully automated mode, allowing the environment to rebalance in line with its compute affinity configuration.

From an operational perspective, in partially automated mode, vSphere administrators will be informed about recommended changes to workload placement in order to balance the host affinity groups within each data center, but changes will not be executed automatically. The DRS recommendations will need to be actioned manually by vSphere administrators as part of daily operational change management of the stretched cluster environment.

A final design factor associated with affinity group configuration relates to the colocation of multivirtual machine, or tiered, applications. Virtual machines that have dependencies on one another and traverse a lot of network traffic between each other should be configured in the same virtual machine group. This ensures that they remain local to one another, at the same physical data center, avoiding any additional unnecessary use of the cross-site interconnect from traffic that would otherwise be generated. However, at the same time, other application types might best be placed across the two locations in order to provide an additional layer of availability within this active/active design.

High Availability Configuration

vSphere High Availability (HA) is not site aware. Therefore, a Virtual SAN stretched cluster requires several considerations and additional configuration in order to simulate site awareness.

To ensure that vSphere HA can restart all virtual machines during an outage at one of the workload data centers, its admission control policy must be configured to reserve 50 percent of both CPU and memory resources for failure events, as illustrated in Figure 5.10. Because vSphere HA is not site aware, this configuration is essential in order to ensure that the design reserves enough resources in each data center to handle the failover of all of the virtual machine workloads on the crashed site. Many customers are surprised and concerned by this requirement, because they are effectively unable to use half of the compute resources they are procuring. However, if their business intends on using the configured stretched cluster to provide the application continuity benefits it is designed for, this is the only way. After this conversation has taken place, you often must readdress a customer's sizing requirements, so it is recommended to have this discussion as early as possible during the design phase.

FIGURE 5.10
Admission control policy configuration

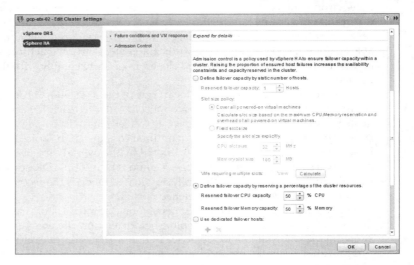

It is also important to remember that Virtual SAN itself is not aware of vSphere HA or the admission control policy, and does not have any mechanism for reserving capacity for 100 percent of the workload being migrated to a single site. Therefore, in a scenario where one of the three fault domains is unavailable, virtual machines will remain in a noncompliance state until the environment has been fully restored. If maintaining a fully compliant state at all times is a design requirement for the stretched cluster, this will require manual calculation as part of the design sizing stage for the environment, and additional operational steps for vSphere administrators in the event of a failure.

The next design recommendation for a stretched cluster is to specify two isolation addresses—one in each data center—for response to isolation events. This is required when network connectivity is lost between the two data centers. In this scenario, each site will still be able to reach its own vSphere HA isolation address, and will not invoke an HA isolation response that could cause the virtual machines in a particular site to shut down or power off unnecessarily, depending on the cluster's HA isolation response configuration.

In a Virtual SAN environment, even though the heartbeat mechanism uses the Virtual SAN network to communicate, as opposed to the management network in a non–Virtual SAN design, vSphere HA continues to use the management network default gateway for isolation response purposes. In a stretched Virtual SAN design, this should be modified to an address on the Virtual SAN network, using the `das.usedefaultisolationaddress`, with the value configured as `false`.

In addition, it is recommended to configure two additional isolation response addresses, one residing in each physical location, and both configured on the Virtual SAN network. This can be achieved by using the advanced setting `das.isolationaddress0` on the preferred site, configured with an appropriate IP address value on the Virtual SAN network. Then configure the second isolation address on the nonpreferred site, using the advanced setting `das.isolationaddress1`, with an IP address value associated with that data center (see Figure 5.11).

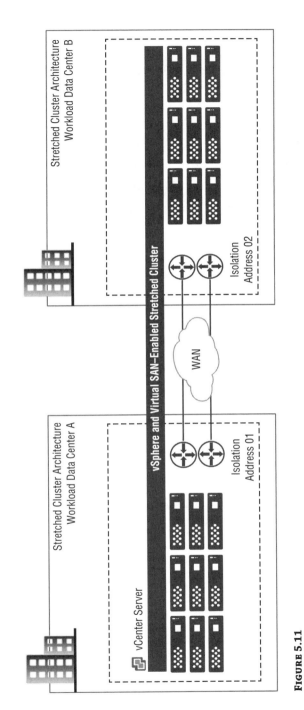

Figure 5.11
Stretched Cluster host isolation advanced settings

Finally, be sure to configure vSphere HA restart priorities as part of the overall design. Restart priorities help ensure that the most critical virtual machines will be started first when failed over to the alternate data center. The following sequence illustrates the priorities that can be applied to workloads; these allow vSphere HA to start virtual machines in a way that optimizes and respects application dependencies:

1. Agent virtual machines (a "tagged" virtual machine that gets powered on first in case of a host failure)
2. Virtual machines configured with a restart priority of High.
3. Virtual machines configured with a Medium restart priority.
4. Virtual machines configured with a Low restart priority.

The vSphere HA parameters should always be configured consistently across all Virtual SAN clusters in a stretched configuration, and therefore should be included in a design. This approach helps limit variability and simplifies operational management. Table 5.5 provides guidance for architects by providing the standard settings and options configured across the Virtual SAN cluster platform for vSphere HA values.

TABLE 5.5: Sample vSphere HA configuration for a Virtual SAN stretched cluster

PROPERTY	SETTING	CONFIGURATION
Cluster Feature	High Availability (HA)	Enabled
vSphere High Availability (HA)	Host Monitoring Status	Enabled
	Host Hardware Monitoring	Disabled (Default)
	Admission Control Policy	CPU: 50% Memory: 50%
	Virtual Machine Options ➤ Virtual Machine Restart Policy	Customer defined
	Virtual Machine Options ➤ Host Isolation Response	Power off and restart VMs
	VM Monitoring	Customer defined
	Datastore Heartbeating	N/A—"Use datastores only from the specified list." However, do not select any datastores from the list provided. This effectively disables datastore heartbeats.

TABLE 5.5: Sample vSphere HA configuration for a Virtual SAN stretched cluster *(CONTINUED)*

PROPERTY	SETTING	CONFIGURATION
Advanced Settings	`das.usedefaultisolationaddress`	False
	`das.isolationaddress0`	VSAN IP address on preferred site
	`das.isolationaddress1`	VSAN IP address on nonpreferred site

Stretched Cluster WAN Interconnect Design

As already highlighted, one of the key design factors associated with Virtual SAN stretched cluster design is the interconnect that provides connectivity between the two workload data-center locations. This cross-site connection and its associated bandwidth, latency, and availability are key components to every aspect of the design of a Virtual SAN stretched cluster architecture.

To determine the most appropriate cross-data-center WAN interconnect for a Virtual SAN stretched cluster design, the features of different WAN technologies must be evaluated against design requirements and constraints. The final design decision must be based on the technology and the business requirements that have been established by the customer. For a stretched cluster use case, the key factors in determining the most appropriate solution are the bandwidth and latency requirements for the platform, although other factors, such as availability, can provide additional design considerations.

Data-center WAN interconnects can broadly be defined as one of the following three options:

Private WAN   Is owned and managed by the user's business. These are expensive and potentially labor intensive to implement and maintain. They can also be difficult to reconfigure for dynamically changing business requirements. The primary advantages of using private WAN links include higher levels of security and transmission quality.

Leased WAN   Is maintained by a service provider, although the user's business might be required to purchase additional hardware to provide connectivity. The consumer typically pays for the allocated bandwidth, regardless of its utilization.

Shared WAN   Is a shared environment: the service provider is responsible for all operational maintenance, although the user's business might be required to purchase additional hardware to provide connectivity. While a shared WAN link is typically the least-expensive option to deploy and maintain, it also bears the highest security risks and the potential for resource conflicts, depending on configuration.

Evaluating WAN Platforms for Stretched Clusters

This section provides an evaluation of multiple WAN interconnect technologies, so that appropriate features can be mapped against the requirements for a Virtual SAN stretched cluster platform. Again, VMware provides no specific requirement or recommendation on this choice; it will be defined by the customer's specific requirements, key design factors, and services available to the data centers in question.

DARK FIBER

Dark fiber are strands of fiber-optic cables deployed underground, but they are not currently being used (see Figure 5.12). Because no light is transmitted across them, they are considered dark. Although any type of fiber (single or multimode) can technically be dark, the term *dark fiber* typically refers to 9μ single-mode fiber. Dark fiber can be privately owned, typically across a campus-style network, or might also be leased from a provider to create a metropolitan area network (MAN).

Dark fiber is a physical medium; that is, it has a direct connection between two switch points. However, other solutions can use dark fiber as their underlying platform, such as CWDM and DWDM, which are addressed later.

The maximum distance for dark fiber depends on various factors, including signal attenuation and switch optics. However, 10 km is generally considered the maximum distance for a standard dark-fiber link. In some special cases, it is possible to achieve far greater distances, of up to 80 km for 10 GbE and 40 km for 8 GB Fibre Channel traffic, but these types of links must be designed to meet stringent performance criteria and require specialized optics. Currently, dark fiber can support bandwidth of up to 100 Gb/s.

Because each fiber pair is dedicated to a pair of switch ports, different types of layer 2 traffic cannot be mixed. Ethernet and Fibre Channel traffic, for instance, cannot be mixed across the same fiber connections. To send both of these traffic types across dark fiber, at least two pairs of fiber strands must be used, one for each traffic type.

In summary, dark fiber has the following key design factors when evaluating the solution for a Virtual SAN stretched cluster architecture:

- 9μ single-mode fiber
- Viable for campuses and extended distances
- Used in pairs
- Can support any available bandwidth
- Typical distance is up to 10 km
- Up to 80 km possible for 10 GbE with specialized components and media
- Up to 40 km possible for 8 GB Fibre Channel with specialized components and media

WAVELENGTH DIVISION MULTIPLEXING

Wavelength division multiplexing (WDM) is a mechanism used to simultaneously transmit multiple streams of data on a single strand of fiber (or pair of fibers). Using WDM allows the environment to overcome the limitation imposed by dark fiber of requiring a separate fiber pair for each type of traffic to be transmitted. WDM has two types: dense wavelength-division multiplexing (DWDM) and coarse wavelength-division multiplexing (CWDM).

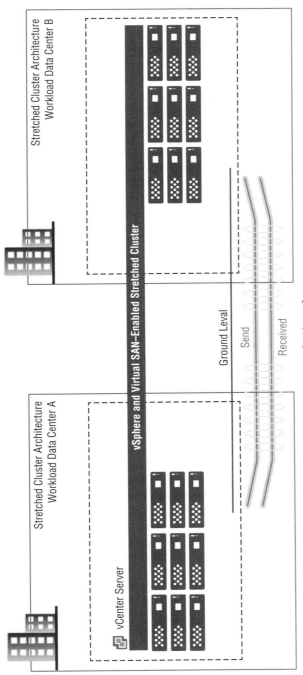

Figure 5.12
Dark fiber interconnect

DWDM, as its name suggests, supports more wavelengths, as they are spaced together more tightly. Figure 5.13 illustrates. Because the wavelength separation is so small, DWDM requires high-precision equipment that is typically costly to purchase.

This equipment provides filters to the light emitted from the data source, such as the switch, and then multiplexes it onto the fiber strands along with the light from other sources. At the far end of the fiber strand, another multiplexer separates the signals from the sources and forwards them to the appropriate destination equipment. Because the signals are also amplified by the DWDM system, they can travel much farther distances than with native dark fiber. With the right hardware and media components in place, DWDM could be extended to thousands of kilometers. The number of wavelengths that can be supported by DWDM varies by vendor, but 32, 64, and even 128 are possible.

DWDM has the following key design factors when evaluating the solution for a Virtual SAN stretched cluster architecture:

- Most typically used over short distances, such as 100 km to 200 km, but can extend further.
- Dark fiber must be available.
- Divides a single beam of light into discrete wavelengths (lambdas).
- Each signal can be carried at a different rate.
- Dedicated bandwidth for each multiplexed channel. Approximately 0.4 nm spacing.
- DWDM transponders can support multiple protocols and speeds (LAN, SAN, and other signals).

CWDM was introduced as a lower-cost alternative to DWDM. The functionality of the solution is almost identical to DWDM, except that the wavelengths (called *channels* in CWDM) are spaced farther apart. Fewer channels can fit on the fiber strands (a maximum of 16 channels per fiber pair). Also, CWDM does not require as many components as DWDM, and those components are significantly less costly. However, a CWDM solution does not amplify and clean the signal as DWDM does, so the distance limitations for CWDM are significantly lower. In most implementations, the maximum distance is typically less than 100 km, but multiple signals (such as Ethernet and Fibre Channel) can be carried, giving it a significant advantage over native dark fiber.

SONET/SDH Interconnect

Synchronous Optical Networking (*SONET*), more commonly known as *Synchronous Digital Hierarchy* (*SDH*) outside North America, supports longer network distances than the previously outlined optical transport technologies (see Figure 5.14). SONET/SDH is typically used for city-to-city or country-to-country communications.

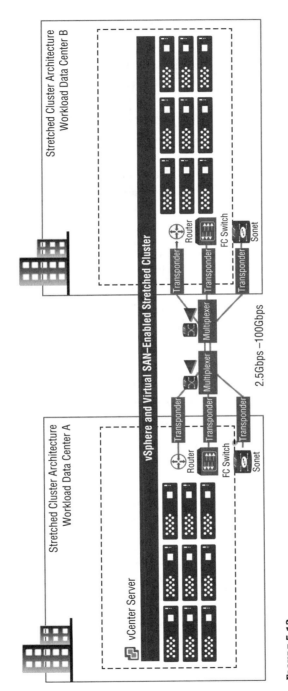

Figure 5.13
Dense wave division multiplexing (DWDM)

SONET/SDH provides the following benefits to a wide range of applications within the infrastructure:

- SONET/SDH are readily available in many areas.
- Spans longer network distances than CWDM or DWDM at a lower cost.
- Provides robust network management and troubleshooting capabilities. In SONET/SDH, the optical signal is converted to an electronic signal at all network access and regeneration points. Although this conversion introduces some latency, it allows much more robust monitoring.
- SONET/SDH provides a variety of protection systems, including $n + 1$, a more cost-effective protection system than $1 + 1$. DWDM and CWDM do not offer an $n + 1$ protection capability.

The $n + 1$ refers to one spare circuit providing redundancy for multiple active circuits. When the first active circuit fails, the standby circuit is enabled and used. In DWDM and CWDM, $1 + 1$ protection is a redundant set of fibers used to provide failover if the primary set fails.

In summary, SONET/SDH has the following key design considerations when evaluating the solution for a Virtual SAN stretched cluster architecture:

- SONET/SDH supports longer distances than the WDM mechanism.
- Most typically used over short and intermediate distances.
- Can be used where dark fiber is not available.
- Robust network management and troubleshooting.
- Significant installed infrastructure required.
- A variety of protection systems are typically offered.
- Can be combined with DWDM to increase capacity or redundancy.

Multiprotocol Label Switching

Multiprotocol Label Switching (MPLS), as illustrated in Figure 5.15, functions similarly to IP, but the infrastructure switches based on a predetermined path instead of calculating paths hop to hop. MPLS can also operate on a variety of layer 2 infrastructures.

For an MPLS or IP interconnect between data centers, the design changes from the fiber model. As the interconnect between the data centers is a layer 2.5 or layer 3 link, layer 2 connectivity is not available beyond the data-center boundary. However, as a Virtual SAN stretched cluster is the design goal, and therefore stretched layer 2 connectivity is a requirement in order to perform the same actions as a fiber link with cross-site layer 2 adjacency when using an IP-based WAN, a tunnel must be established between the two sites, and the Ethernet data encapsulated within that tunnel structure.

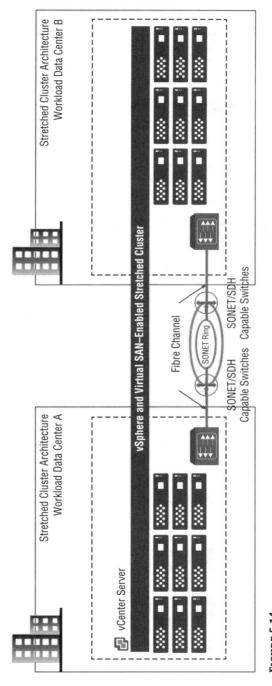

FIGURE 5.14
SONET or SDH

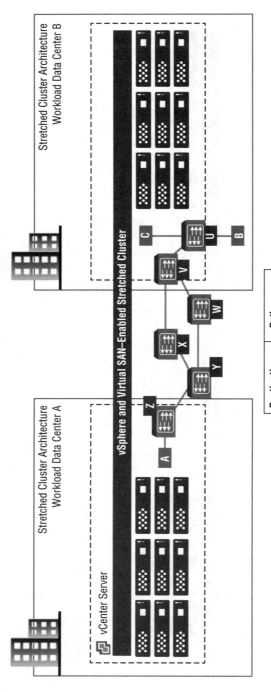

Figure 5.15
Multiprotocol Label Switching (MPLS)

In summary, MPLS has the following key design factors when evaluating the solution for a Virtual SAN stretched cluster architecture:

- Layer 2.5 protocol.
- Can be used over short or long distances.
- Can be used when dark fiber is not available.
- Links can be shared.
- Operates over a variety of layer 2 infrastructures.
- Uses predefined paths, reducing delays from dynamically calculated paths.
- MPLS performance (which affects delay, jitter, failover, and so on) is dependent on service provider service levels.

Deploying Stretched VLANs

One of the key requirements for a Virtual SAN stretched cluster is the ability to migrate a virtual machine to a different physical data center, while the virtual machine and its applications are still able to communicate and to be identified on the network, and running services can continue to serve end users without interruption.

For this to work, stretched VLANs are typically required. A *stretched VLAN* is a VLAN that spans multiple physical data centers. In a typical multisite data-center environment, locations are connected over a layer 3 WAN interconnect. This is the simplest configuration, which removes a lot of complex considerations from the environment. However, in a native layer 3 environment, workloads being migrated must change their IP address to match the addressing scheme at the other data center. Alternatively, the routing configuration for the entire VLAN subnet must be changed, meaning all resources on that VLAN must be transferred to the other site at the same time. This approach severely restricts the ability to migrate virtual machines from one data center to another, and does not provide the flexibility that a Virtual SAN stretched cluster requires.

Therefore, creating an environment in which workload migration over distance can occur requires the use of stretched VLANS, as illustrated in Figure 5.16. They can be extended beyond a single data-center site, allowing workloads to communicate as if they were local to one another.

Several mechanisms can be used to stretch VLANs across physical data-center sites, depending on the underlying WAN technology used in the environment.

As dark fiber, DWDM, and CWDM physically connect sites point to point, VLANs can be extended across the link as if they were located within the same physical data-center location (see Figure 5.17). There is no requirement for any additional configuration, assuming physical connectivity has been established.

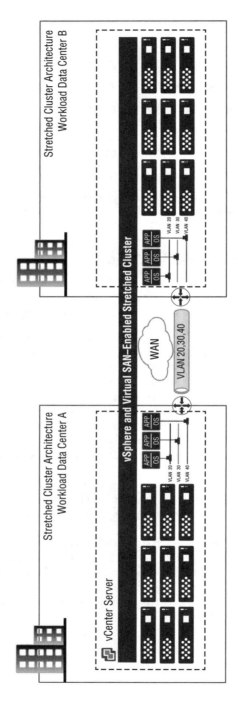

FIGURE 5.16
Stretched VLANs

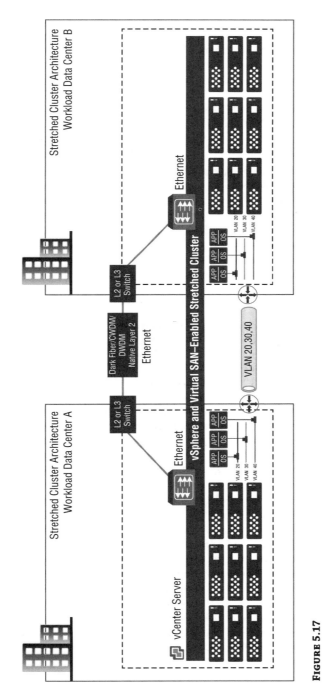

FIGURE 5.17
Stretched VLANs over dark fiber

However, extending VLANs across sites will likely require changes to the LAN configuration. Table 5.6 addresses design factors for extending VLANs across fiber-based data-center interconnects.

TABLE 5.6: Design factors for extending VLANs across fiber-based data-center interconnects

Layer 2 Connectivity	Distance	Fiber Requirements
Aggregation layer switches are connected to inter-data-center switches through layer 2 connections. Inter-data-center switches are aggregated into a virtual switch.	Limited to DWDM, CWDM, or dark fiber distances	Dark fiber: Two pairs of dark fiber at each site (more for additional bandwidth or traffic protocols)
		DWDM: Single fiber or single pair at each site
Spanning Tree Protocol (STP) should not be able to span sites. Block STP on WAN ports. Use Multiple Spanning Tree (MST) to create separate areas for each site.		CWDM: Single pair at each site, depending on bandwidth requirements

If the design uses an MPLS network between sites, either owned privately or leased, it can be used to tunnel the Ethernet frames. This can be configured by attaching both an MPLS virtual circuit label and MPLS tunnel ID label to the Ethernet frame (see Figure 5.18).

This type of encapsulation is referred to as Ethernet over MPLS (EoMPLS) or Virtual Private LAN Services (VPLS). EoMPLS is used for point-to-point configurations, while VPLS is used in point-to-multipoint scenarios or meshed environments.

If the only solution available to the customer is to obtain native IP between the data centers, there are several ways to extend a VLAN across the sites. The simplest option is to use the Layer 2 Tunneling Protocol version 3 (L2TPv3) to tunnel the Ethernet frames, as illustrated in Figure 5.19. In this approach, an L2TP header is attached to the Ethernet frame and encapsulates it in an IP packet. The Ethernet frame is then delivered to the remote site without having been seen by the connecting network.

Additionally, proprietary technologies such as OTV from Cisco can be adopted. This is the approach used in the sample Virtual SAN stretched cluster solution architecture provided later in this chapter.

While stretching VLANs and address space across two physical locations facilitates the Virtual SAN stretched cluster design, it also presents challenges, as providing these types of LAN extensions have a big impact on network design.

Unfortunately, it is not possible to simply allow layer 2 connectivity between data centers by using only layer 3 technologies, as this would have significant consequences on the traffic patterns between the two data centers, such as Spanning Tree Protocol (STP), Unicast floods, broadcasts, and ARP requests. Therefore, technology must be included that uses extended spanning-tree techniques to avoid loops and broadcast storms and that understands where an active IP address on the subnet exists at any given time.

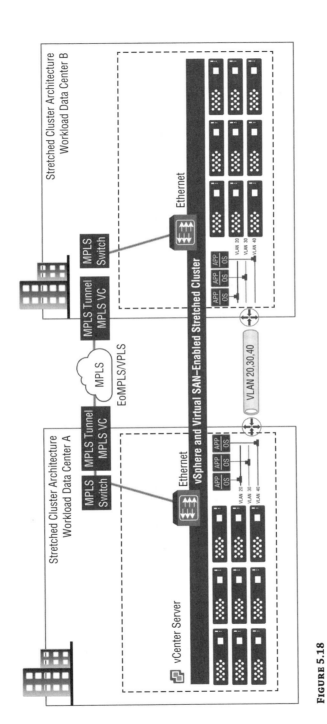

FIGURE 5.18
Stretched VLANs over MPLS

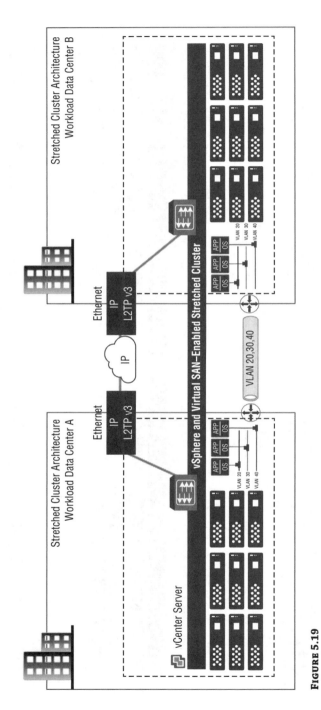

FIGURE 5.19
Stretched VLANs over L2TP version 3

For further information, refer to the network hardware documentation specific to the customer-planned implementation, or seek guidance from a network architect who is able to evaluate all of the requirements and constraints of a specific environment. However, when considering the deployment of a LAN extension across physical sites, it is important to address as key requirements both STP isolation and end-to-end loop prevention, to mitigate significant disruptions likely to be caused by remote data-center failure, or the propagation of unwanted network traffic behavior from one data center to the other.

WAN Interconnect High Availability

High availability for data center interconnect WAN links is significantly more complex than that used for LAN environments. Factors such as distance, cost, non-ownership of equipment, and connectivity have a much larger impact. The high availability model for WAN connectivity varies widely, depending on the type of WAN used. In general, a typical highly available solution can be classified as including either redundant components, redundant links, or both.

Redundant components are multiple devices, such as routers or multiplexers, at each data center. This provides redundancy in the event of component failure, and allows for nondisruptive maintenance and upgrades.

Redundant links are multiple connections between data centers. This can be achieved through multiple dedicated media, such as multiple strands of dark fiber, or through a combination of fiber and a shared medium, such as MPLS. A design can also use both methodologies to have a fully redundant data path for the highest level of availability.

Some network service providers may also provide the option of having a standby connection. This option often has a lower total cost of ownership, as you are charged only by the amount of data transmitted. This might be a cost-effective way of achieving a redundant link without having to pay for unrequired bandwidth.

Before making the appropriate design decision for providing WAN availability, understand the available options and what needs to be protected against. For instance, deploying redundant links or components does not provide any benefit if the single ISP providing the services has a large-scale outage.

Secure Communication

It may be a customer design requirement that all Virtual SAN traffic traversing the two data centers be secured through some form of *data inflight* encryption. The primary method used to secure data as it travels between data-center sites is *Internet Protocol Security*, or *IPsec*. IPsec is commonly used to scramble portions of the IP packet by encrypting the data flow at the edge of one site, and then decrypting it at the entry point to the second site. IPsec has two modes of operation:

- Transport mode
- Tunnel mode

Transport mode encrypts only the IP packet's payload and leaves the header in an unencrypted state so it can be read by other network devices. In *tunnel mode*, which is most commonly used for point-to-point or site-to-site VPNs, the entire IP packet, including its header, is encrypted, meaning that it must be encapsulated in another IP packet with an unencrypted header. Using this mechanism allows the source and destination information in the unencrypted header of the packet to be read by network devices, while the data remains secure.

Data Center Interconnect Design Considerations Summary

When evaluating the options to achieve the most viable data-center interconnect for a Virtual SAN stretched cluster design, key design factors must be considered. Table 5.7 provides a quick summary of the various interconnect options and the key design factors for each WAN type. It's important to be aware that many of these factors will differ based on location as well as additional service provider features and enhancements to the core technologies.

TABLE 5.7: Data-center interconnect key design factors

Design Factors	Description
Distance	WAN solutions are typically implemented at the physical layer, and often include a distance limitation. Although distances for DWDM are significant, this is probably not suitable for a cross-country solution. SONET, MPLS, and IP have essentially unlimited distances, as they span across multiple providers' infrastructures.
	Key Design Factors:
	Distance limitations for dark fiber, DWDM, and CWDM.
	SONET/SDH, MPLS, and IP have unlimited distances, but latency must be considered.
Speed (bandwidth and latency)	As the customer is the owner, private WAN solutions offer guaranteed bandwidth and latency levels, and the customer can control usage of the infrastructure and establish appropriate QoS policies. However, these leased solutions can often be more complex than other alternatives.
	Dedicated physical solutions, such as dark fiber, DWDM, and CWDM, provide a guaranteed amount of bandwidth, and a known level of latency because of the way the solution is provisioned. SONET is similar, as it generally uses underlying infrastructure that is provisioned physically.
	MPLS and IP use shared environments and are therefore subject to spikes in data transmission. These types of solutions allow businesses to lease a specific amount of bandwidth from a service provider. However, sometimes the full bandwidth may be unavailable, as it is typically overprovisioned by the provider.
	Key Design Factors:
	Privately owned solutions provide guaranteed bandwidth.
	Leased solutions can vary.
	Dark fiber, DWDM, CWDM, and SONET/SDH provide guaranteed bandwidth.
	MPLS and IP often provide variable bandwidth because they are shared.

TABLE 5.7: Data-center interconnect summary *(CONTINUED)*

DESIGN FACTORS	DESCRIPTION
Cost	Deploying a privately owned solution requires the purchase of the necessary equipment, and the media itself. Some of this equipment, such as DWDM optics and multiplexers, can come at a very high CapEx cost.
	Leased-line solutions may include some of the required equipment in the cost. However, typically, the customer must still purchase equipment to integrate into the provider's network.
	Key Design Factors:
	Privately owned requires the purchase of equipment.
	Leased solutions typically include some of the required equipment as part of the lease agreement.
Redundancy	With dark fiber, DWDM, and CWDM solutions, it can prove difficult to obtain connectivity across two physically separate and diverse paths. Having multiple fibers in the same bundle for the purposes of redundancy may increase the customer's exposure to the risk of a complete communications failure, due to a break in the media. One approach to protect against this type of communication failure is to use a different solution, such as MPLS or IP, as a backup. Leased-line solutions using multiple providers may offer additional redundancy in the event of a provider-wide outage.
	Key Design Factors:
	Redundant physical paths.
	Multiple service providers.
	Mixed solutions.

Table 5.8 contains a summary of the WAN technologies discussed, which could be used for our stretched cluster use case to connect multiple data centers. Note that the values provided in the following table are generalizations. The actual values will depend on various factors, such as the service provider's equipment, media, and so on.

TABLE 5.8: Data-center interconnect summary

DESIGN FACTOR	DARK FIBER	DWDM	CWDM	SONET/SDH	MPLS	IP (INTERNET)
Distance	40 km+	Up to 200 km[1]	Less than 100 km	1,000s of km	Unlimited	Unlimited
Latency	5 µs/km	Low	5 µs/km	Medium	Variable	Variable (high)

TABLE 5.8: Data-center interconnect summary *(CONTINUED)*

DESIGN FACTOR	DARK FIBER	DWDM	CWDM	SONET/ SDH	MPLS	IP (INTERNET)
Channels	N/A	100+	Up to 16	N/A	N/A	N/A
Bandwidth	100 Gb/s+	40 Gb/s (per channel)	10 Gb/s (per channel)	10 Gb/s	Variable	Variable
Reliability	1×10^{12}	High	1×10^{12}	High	High	High
Cost	Low/High	High	Medium	Medium	Low	Low
Type	Private Leased	Private Leased	Private Leased	Leased	Private Leased	Shared

[1] *With the appropriate equipment and media, DWDM can be extended beyond 200 km, but 200 km should be considered the typical limitation for most use cases.*

Stretched Cluster Solution Architecture Example

As you have seen, deploying a Virtual SAN stretched cluster across two data center locations carries two key requirements. First, there must be layer 2 adjacency between sites; they must be able to share the same VLANs. Second, the storage must be mirrored across both sites in real time. Without the ability to provide these two requirements, a Virtual SAN–enabled stretched cluster is not a viable design option.

By implementing a Virtual SAN stretched cluster across the two locations, the distributed datastore can be used to provide read and write access to the storage system at both sites simultaneously. This topology allows the vSphere cluster to span across both sites, providing mobility for workloads as well as enabling disaster-avoidance capabilities and a cross-site vSphere HA disaster-recovery solutions, all from a single platform.

In addition to the required vSphere components, the following example uses the Cisco OTV product to stretch the VLANs for use across both sites, essentially creating a single network. This enables virtual machines and their applications to retain their networking configuration at both locations, providing nondisruptive vMotion functionality across data centers.

Although all of these features create a more complex layered environment to support operationally, the management overhead to support the infrastructure is simplified by merging both physical locations into one logical data center.

In the following implementation example, VMware's Virtual SAN 6.2 enables the customer's virtualized stretch cluster requirements. However, as you have seen, the data center also requires multiple other components and technologies to use the design's disaster avoidance, continuous availability, and cross-site disaster recovery functionality. Figure 5.20 illustrates the logical architecture provided for the example.

After the architect carried out an extensive evaluation of the technology options available, the customer concluded that a 10 Gb/s DWDM WAN link for IP communication will provide the most appropriate solution for their architecture, based on design factors such as distance between data centers, latency, bandwidth, flexibility, and cost. Other decisions around hardware choice have aligned with the customer's *Cisco first* hardware strategy.

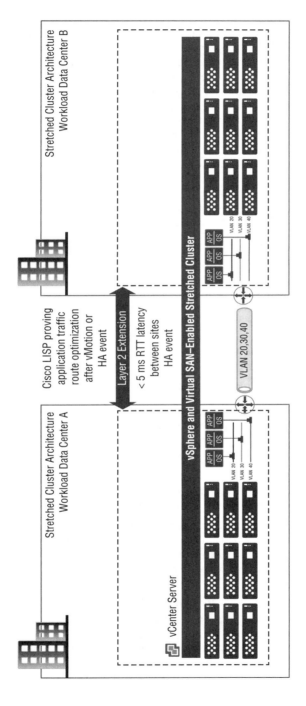

FIGURE 5.20
Use case example logical architecture

The end-to-end architecture leveraged for the Virtual SAN stretched cluster solution is depicted in Figure 5.21. This figure shows a high-level overview of the physical architecture used to achieve the solution, enabling the customer's enterprise applications to be protected across two data centers. Both data centers are using mirrored hyper-converged compute and network components from VMware and Cisco, to efficiently use the Virtual SAN stretched cluster to facilitate workload mobility and recovery of virtualized applications across sites, without end users experiencing any prolonged impact to application performance.

This architecture uses a typical modular separation of components at each of the physical data center locations. As illustrated in Figure 5.21, starting with compute, the example design provides multiple Cisco Unified Computing System C240 M4 rack-mount servers, connected to a redundant pair of 6296UP Fabric Interconnect devices. Each C-Series rack-mount server represents a single ESXi 6.0 host.

Each 6296UP Fabric Interconnect device is connected through a port channel to a pair of Nexus 7710 devices, representing a collapsed data-center core and aggregation layer. The pair of Nexus 7710s are then connected to the data-center WAN edge, to provide access to the layer 3 core of the network. In the layer 3 core of the network, the end-user and client connections accessing the specific data-center services and applications originate. The F5 Global Traffic Manager (GTM) devices are directly connected to the WAN edge, to provide global load-balancing and traffic management of client connections.

In this design, the two data-center locations are separated by 200 km and are connected through highly available, protected point-to-point DWDM circuits. The data-center interconnect solution presents various components that function alongside each other. The various technologies required to be considered for the implementation of the stretched cluster solution include, but are not limited to, the following:

LAN Extension   Given the availability of point-to-point circuits between the two sites and the hardware chosen for the customer's design, two options for LAN extension technologies have been considered. The first solution leverages the Cisco virtual PortChannel (vPC) capabilities of Nexus 7710 devices, to establish an end-to-end port channel between the Nexus 7710 pairs deployed in each data center. The second introduces OTV, a Cisco LAN extension technology, deployed across DWDM.

Routing   The data-center interconnect between locations is to be used for both sending LAN extension traffic and for routing communications between subnets that are not stretched. As outlined earlier in this chapter, satisfying this requirement has design implications that depend on the specific LAN extension technology deployed.

Storage and Compute Elasticity   Migrating workloads between sites brings challenges in terms of how these workloads impact the storage and compute solution. If this solution is aimed at facilitating disaster avoidance, or automated disaster recovery, sufficient compute, network, and storage resources must be available at both data centers.

Cisco vPC over DWDM and Dark Fiber

Cisco vPC provides distributed port channels across two separate devices, providing redundant and loop-free topologies, as illustrated in Figure 5.22. Although Cisco's vPC was designed and developed for intra-data-center use cases, its ability to bundle links belonging to separate devices into a single logical port channel provides one possible solution to stretch VLANs between separate physical sites, interconnected with DWDM and dark fiber.

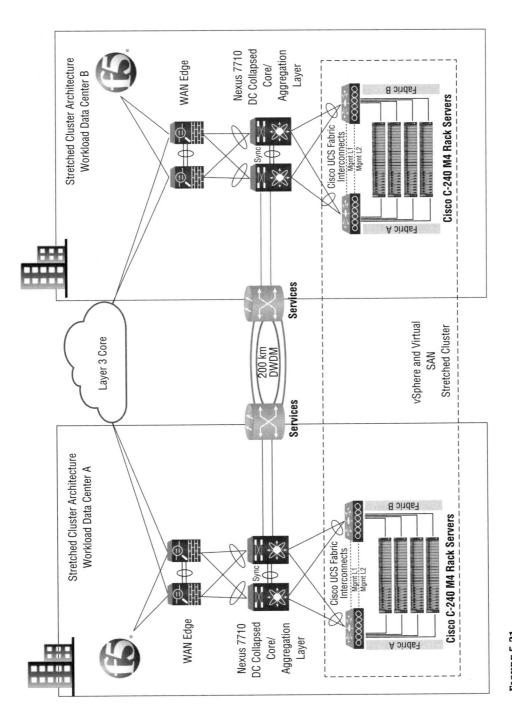

FIGURE 5.21
Physical architecture overview

The primary advantage of bundling physical point-to-point links interconnecting the two data centers is being able to provide stretched layer 2 domains, without creating layer 2 looped topologies. To achieve this, it is recommended to filter the spanning tree bridge protocol data units (BPDUs) across the port channel established between the two data centers, providing the ability to isolate the STP domains, one of the main technical challenges of any LAN extension technology. In essence, the idea is to replace STP with LACP as the control plane protocol.

One of the challenges of this architecture is the lack of ability to provide layer 2 and layer 3 communication across the same Cisco vPC bundled links. The reason for this is the lack of support for dynamic IGP peering establishment, across a Cisco vPC connection. One workaround is to leverage an extra pair of layer 3 links, specifically to be used for routed communication, as illustrated in Figure 5.23.

OTV over DWDM and Dark Fiber

Overlay Transport Virtualization (OTV) is an IP-based mechanism developed by Cisco to provide layer 2 extension capabilities over any sort of WAN-based transport infrastructure. Cisco's only requirement from the network infrastructure for OTV is IP connectivity between the data centers. This technology also provides an overlay, which enables layer 2 extension between separate layer 2 domains, while at the same time keeping these domains independent and preserving the fault isolation, resiliency, and load-balancing benefits of an IP-based interconnect.

Cisco OTV uses MAC routing, meaning a control plane protocol is used to exchange MAC location information between network devices, thus providing LAN extension capabilities. OTV also uses dynamic encapsulation for layer 2 traffic flows, which must be sent to the remote location. Every Ethernet frame is individually encapsulated into an IP packet, so they can be delivered across the transport network. Finally, OTV also provides a built-in multihoming capability with automatic detection, which is critical to improving high availability of the overall architecture. Figure 5.23 illustrates the deployment of Cisco OTV over the DWDM point-to-point connection, as defined in the example solution architecture.

To summarize the Cisco OTV over DWDM solution configuration:

- The cross-site links in the figure represent logical links to the OTV overlay and not physical connections.

- In this solution architecture example, the Nexus 7710s enforce the separation between SVI routing and Cisco OTV encapsulation for any given VLAN. This is a critical consideration for the customer's topology as the Nexus 7710s switches perform both functions. This separation can best be achieved through the use of virtual device contexts, a feature available on the Nexus 7710 platform.

- For this design, two virtual device contexts would be deployed:

    - A Cisco OTV virtual device context dedicated to perform OTV functions

    - A routing virtual device context to provide SVI routing support

- The use of OTV over DWDM or dark fiber provides several design advantages over the previously described Cisco vPC-based solution:

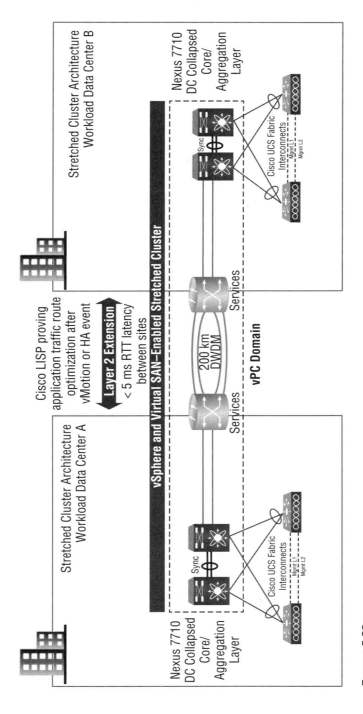

FIGURE 5.22
Cisco vPC domain

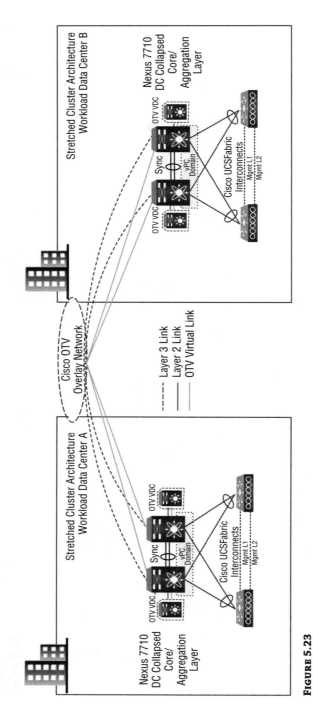

FIGURE 5.23
OTV deployment over DWDM and dark fiber

- Cisco OTV allows for the provisioning of layer 2 and layer 3 connectivity that can use the same DWDM or dark fiber connections. This is made possible because Cisco OTV encapsulates all traffic generated from the Cisco OTV virtual device context, as if it were normal IP traffic that can be exchanged between sites using a layer 3 routed connection.

- Cisco OTV's native failure domain isolation technology means there is no requirement to explicitly configure BPDU filtering to prevent the creation of a cross-site STP domain. In addition, ARP optimization is provided to limit the amount of ARP broadcast frames exchanged across the data-center interconnect.

- Cisco OTV's layer 2 data plane isolation means that storm-control configuration is simplified during deployment, due to the native suppression of unknown unicast frames.

- Cisco OTV's native multihoming LAN extension capability allows extending the service to additional remote data centers in a simple manner.

Cisco LISP Configuration Overview

The final critical Cisco technology required for this example solution is the Locator/ID Separation Protocol (LISP). The Cisco LISP mechanism performs traffic-route optimization after the cross-site vMotion or vSphere HA event has occurred, so that application traffic coming from end users outside the data center does not transverse the original source data center but instead reaches the application virtual machine directly. Detailed coverage of LISP is beyond the scope of this book, but in this type of design scenario, Cisco LISP provides key benefits to applications, including lower latency and faster application response times.

Stretched Cluster Failure Scenarios

It is important that, as the architect, you fully appreciate and can explain to the customer the considerations and expected behaviors should an outage occur. In addition, you should be able to provide guidance to the customer on operational testing procedures, which should be used to validate failure scenarios prior to the solution going into production.

Table 5.9 outlines the most common scenarios associated with stretched cluster environments, each of which should be validated as part of the infrastructure's verification plan.

TABLE 5.9: Virtual SAN stretched cluster failure scenarios

Test	Failure Scenario	Expected Result
Single-host failure	This test represents a single-host failure at one data center. This could be at either the preferred or nonpreferred site.	Virtual machines reboot locally, assuming available capacity. After rebooting, the virtual machines continue to be accessible on the same site, with greater than 50 percent of the components available. Virtual machines whose compute resides on other hosts in the cluster, but have replica components residing on the failed host, remain available, with greater than 50 percent of their components available.

TABLE 5.9: Virtual SAN stretched cluster failure scenarios *(CONTINUED)*

TEST	FAILURE SCENARIO	EXPECTED RESULT
Network partition	This test represents a failure of the interconnect between the workload data centers, and assumes that the vSphere HA host-isolation response value is configured as recommended by VMware.	The preferred site will form a cluster with the witness and therefore with the majority of the components. As a result, virtual machines can run only on the preferred site, where the majority exists. Virtual machines at the nonpreferred data center will no longer be able to access their storage. Therefore, these virtual machines will be powered on at the preferred data center, with greater than 50 percent of their components available. Virtual machines at the nonpreferred data center will remain powered on during the outage, although without access to their underlying storage. When normal operating conditions resume, the virtual machines will be stopped by the vSphere HA mechanism.
Full-workload data-center failure	This test simulates a full data-center outage at either the preferred or nonpreferred site. Being the preferred or nonpreferred site has no impact on this failure scenario.	The data-center failure is detected if five heartbeat misses are consecutive (5 seconds). Virtual machines located at the failed site will be restarted by the vSphere HA mechanism. Since both data centers will have a copy of the data, the second site can transparently take over with minimal downtime. vSphere administrators can continue to create new virtual machines, but they will be out of compliance, assuming FTT = 1 (with force provisioning being added by default). When the failed site comes back online, this is automatically detected, which begins a resynchronization of changed data. Once the resynchronization is complete, the vSphere administrator can use DRS to redistribute the virtual machines, based on affinity configuration.
Witness-site failure	This failure scenario represents a complete outage of the witness site, with neither workload data center having access.	There is no disruption to I/O traffic for virtual machines, which will continue to run as normal with no interruption, as the two workload data centers could still create a quorum. When the witness comes back online, all new metadata will be communicated, until the witness is updated with all new object data in the cluster. A new witness appliance can be provisioned if required, and connected to an existing cluster.

TABLE 5.9: Virtual SAN stretched cluster failure scenarios *(CONTINUED)*

TEST	FAILURE SCENARIO	EXPECTED RESULT
Workload-data-center to witness-network failure	This failure scenario represents an outage of the network to the witness location.	As per the previous test, the expectation is that this will not impact virtual machines running in the workload data centers, as there is at least one full replica of all data, and greater than 50 percent of components are available. Once network connectivity has resumed, the witness will come back online, and all new metadata will be communicated, until the witness is fully updated with all the new object data in the cluster.

Stretched Cluster Interoperability

A Virtual SAN stretched cluster is fully interoperable with vSphere Replication, Enhanced vMotion, and VMware Site Recovery Manager. These solutions can help provide additional solutions for disaster avoidance, disaster recovery, workload mobility, and data protection to the Virtual SAN stretched cluster platform, and therefore provide additional types of availability considerations for architects.

Support Limitations

Some vSphere features supported in a Virtual SAN single-site design are not supported by VMware in a stretched cluster environment:

- Fault Tolerance, both SMP and uniprocessor versions, is not supported in a Virtual SAN stretched cluster.
- Microsoft Windows Server Failover Clustering (WSFC), although supported in certain configurations in a single-site Virtual SAN enabled cluster, is not supported on a stretched cluster.
- The Number of Failures to Tolerate in a Virtual SAN stretched cluster can be configured with only a value of 1, as opposed to a single-site environment, which supports a maximum of 3.
- In a Virtual SAN stretched cluster, a maximum of three fault domains can be configured.
- A Virtual SAN stretched cluster requires Virtual SAN Enterprise licensing.

Chapter 6

Designing for Web-Scale Virtual SAN Platforms

The term *web-scale*, first introduced by Gartner in 2013, refers to an architectural approach employed by large-enterprise IT organizations and service providers to deliver the capabilities required by hyper-scale cloud providers such as Google, Amazon, Facebook, Netflix, and others, to build and operationalize an infrastructure platform at extreme levels of scale. The aim of this strategy is to not only achieve these extreme levels of scale, but also provide agility of operations through fixed processes and architectural standardization.

Web-scale is not a single technology. It's a methodology that pertains to architecting and managing data-center environments at any scale, employing a standardized and repeatable building-block design capable of delivering an infrastructure that can facilitate diverse business requirements.

The following points are typically considered the key requirements in delivering a web-scale infrastructure, either within a large-scale enterprise IT organization or a cloud service provider's infrastructure:

- Delivering a hyper-converged infrastructure (HCI) platform on $x86$ servers, with fully integrated compute and storage components.
- Providing the ability to distribute resources, cluster-wide, in order to deliver data and application services in a distributed manner.
- Facilitating a highly available and self-healing system infrastructure, capable of providing fault isolation and distributed recovery.
- Delivering API-driven automation through software-defined data-center concepts, and comprehensive analytics through low-level infrastructure monitoring.
- Providing the ability to host multiple application types simultaneously across the platform, while delivering key service requirements to workloads.

As highlighted in Chapter 4, "Policy-Driven Storage Design with Virtual SAN," Virtual SAN clusters have massive scalability, with a maximum of 64 nodes in vSphere 6, which can easily support tens of thousands of virtual machine workloads. When designing a web-scale Virtual SAN platform, you have two basic scaling strategies:

Scale-up Each host has more storage resources available, but a fewer number of overall nodes are employed.

Scale-out Hosts are scaled out by increasing the number of nodes, but the architecture will end up with a larger overall footprint.

In a web-scale architecture, it's common for the architect to think in terms of both a *scale-up* and *scale-out* strategy concurrently. The design may not be based around whether to have more smaller nodes per Virtual SAN cluster or fewer larger hosts, but is more likely to be based on a formulated building-block business architecture strategy, which combines scale-up and scale-out architectures alongside one another to provide a predictable and standardized platform.

Scale-up Architecture

A *scale-up strategy* in a Virtual SAN environment refers to increasing the amount of storage resources available on each host. This can be achieved by either increasing the number of capacity disks in each disk group, or increasing the number of disk groups on each Virtual SAN host. Adding capacity disks to an existing disk group is fully supported, as illustrated in Figure 6.1. However, an enterprise or service provider web-scale architecture will likely have a predefined building-block standard for disk group configuration—for instance, a 1:4 hybrid disk group configuration, with one flash device providing the write-buffering and read-caching mechanism to four mechanical capacity disks.

Virtual SAN supports disk groups composed of one endurance flash device and up to seven capacity disks, either mechanical or flash based, depending on the disk group model used in the design. In addition, each host in a Virtual SAN cluster can support up to five disk groups, each of which will contribute storage resources to the total capacity of the distributed Virtual SAN datastore.

As discussed in Chapter 4, by using multiple smaller disk groups rather than a single large disk group, the failure domain is reduced, and there are fewer components to rebuild should a capacity disk failure occur. This results in faster rebuild times. Using multiple smaller disk groups, as outlined in Figure 6.2, also improves performance, particularly in the hybrid model. Additional flash devices are used across the disk group architecture, making the ratio between flash and capacity storage smaller and allowing more of the current working data set to reside in read cache.

Defining operational configurations such as these is important. To achieve consistent performance across the distributed Virtual SAN datastore, a uniform disk group configuration is essential across all nodes in the cluster, as variability across a web-scale platform is typically unacceptable.

The number of storage I/O controllers per host can also be considered a design factor when using a scale-up strategy in a web-scale architecture. When disk groups are created on different storage controllers, the failure domain is reduced, and the controller queue is distributed across all storage controllers, typically providing superior performance.

Another design option may be to use SAS expanders rather than additional storage controllers. This storage technology allows the design to maximize the storage capability of a SAS storage controller card, by providing a means to go beyond its 8-, 12-, 16-, or 24-drive limit.

SAS expanders help place additional drives behind a single storage controller, which can typically be done at a lower cost than adding controllers. However, the performance and reliability of SAS expanders should be considered a design risk. SAS expanders are not typically recommended to be included in any Virtual SAN platform.

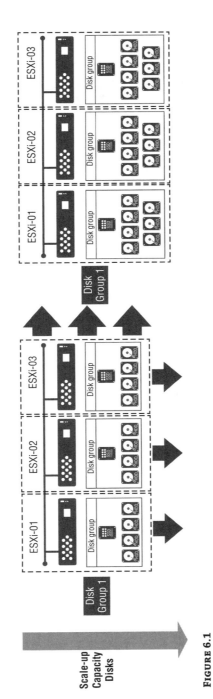

FIGURE 6.1
Disk group scale-up strategy (adding capacity disks)

FIGURE 6.2
Disk group scale-up strategy (adding disk groups)

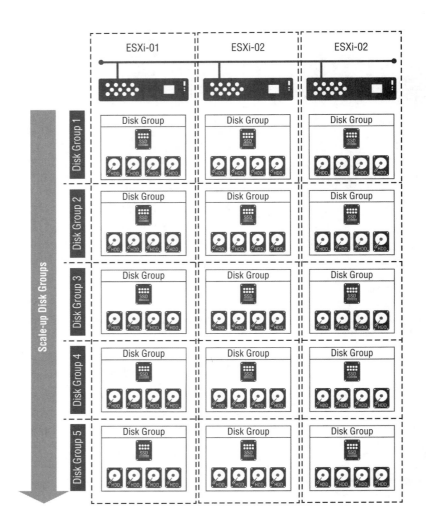

Scale-out Architecture

A *scale-out strategy* refers to the addition of new hosts into the Virtual SAN cluster to increase storage resources and compute capacity simultaneously. Although it is possible to scale out compute resources independently of storage, it is not possible to add nodes to only offer storage, unless DAS-based JBOD hardware is being used.

Virtual SAN supports the adding of nodes and disk groups during normal operations, without need for downtime. However, as with most physical remediation in the data center when making changes that require physical access and physical interaction with the hardware, these operations should normally be carried out during a soft maintenance window.

Figure 6.3 illustrates how Virtual SAN can scale to meet the needs of the most demanding enterprise or service provider environment. It can easily be designed and configured to support tens of thousands of virtual workloads, with single clusters as large as 64 nodes.

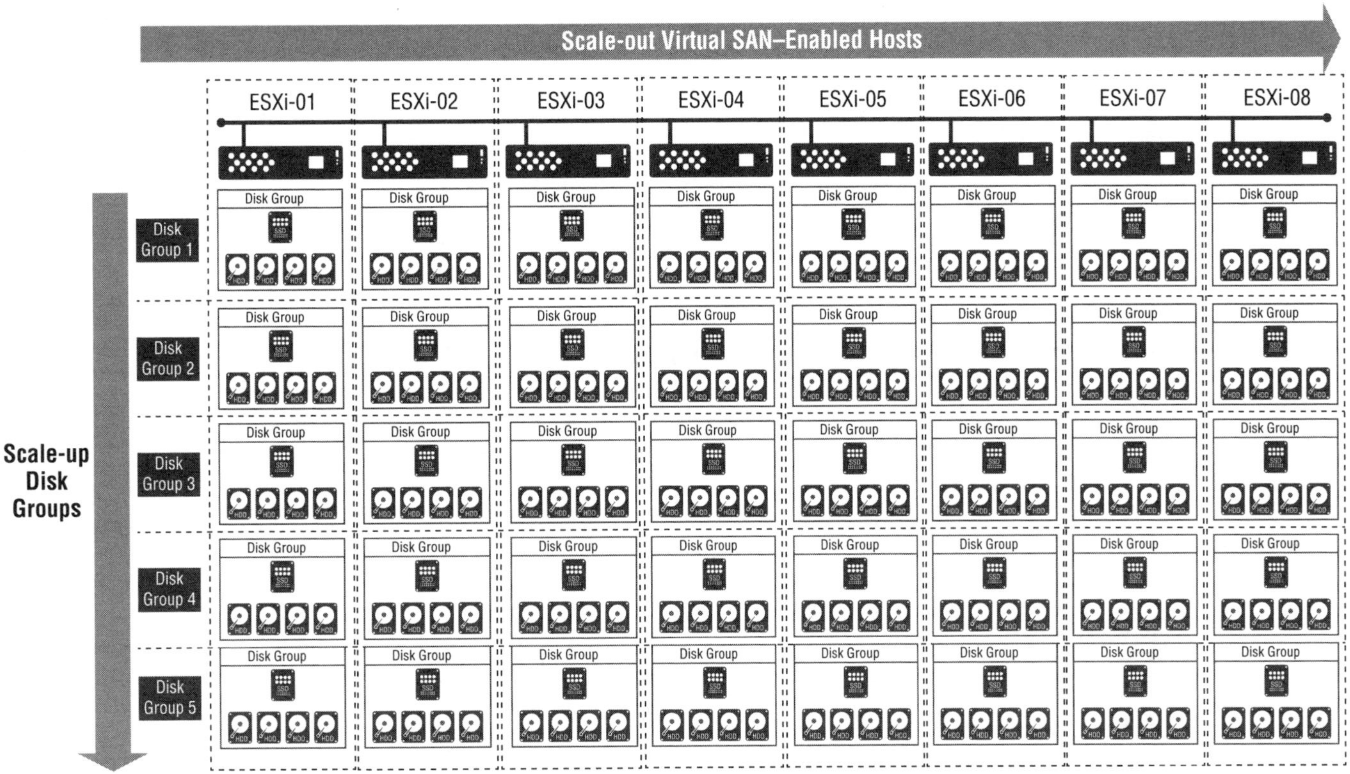

FIGURE 6.3
Virtual SAN–enabled vSphere cluster scaled up and out to eight hosts

Designing vSphere Host Clusters for Web-Scale

The Virtual SAN cluster is the boundary of shareable storage resources. Therefore, when planning the design of multiple large clusters, consider the following key design considerations:

Capacity Planning   Although it might be simpler to plan for growth with a smaller number of large clusters, the limitation on the number of hosts per cluster, and therefore the number of virtual machines, may warrant a *scale-out* approach to cluster design. For instance, employing sixteen 24-node clusters, as opposed to six 64-node clusters, may prove to be a more scalable building-block strategy.

Hardware Cost   Because a Virtual SAN cluster requires a defined amount of spare storage resources to accommodate failures, depending on the scale of the environment, having a large number of smaller clusters is likely to result in a higher hardware cost per virtual machine.

Security   In a multitenant or a multiple lines of business environment, isolating tenants or business groups into dedicated Virtual SAN clusters is one way to segment workloads, and control access through role-based access control (RBAC).

Performance   In a multitenant or a multiple lines of business environment, separating tenant workloads or specific lines of business applications onto dedicated Virtual SAN clusters provides a design mechanism to ensure that resources are consistently available for those consumers.

Building-Block Clusters and Scale-out Web-Scale Architecture

One simple and scalable approach to Virtual SAN cluster design is the *building-block approach*, which is used by many cloud service providers and large private cloud platforms hosted by enterprise customers. Each cluster is a standard container of resources that is provisioned consistently to provide a simple, scalable, and building-block approach to compute and storage. This methodology not only scales consistently across multiple data centers, but also eliminates variability, configuration drift, and the amount of operational effort involved with patch management and day-to-day operations. This approach is the simplest and most effective way to provide a flexible solution that can meet web-scale platform-scalability requirements for large enterprise IT organizations and cloud service provider infrastructures. This building-block approach standardizes the configuration of the Virtual SAN hosts, clusters, and even server cabinets to help provide a manageable and supportable infrastructure.

Standardizing not only the model, but also the physical and logical configuration of the Virtual SAN hosts and clusters, is critical to providing a manageable and supportable infrastructure in large-scale deployments by eliminating variability. An additional technology that should feature in the design to support this methodology is vSphere host profiles, which can be used to configure additional values consistently across hosts and Virtual SAN–enabled clusters, wherever applicable.

Scalability and Designing Physical Resources for Web-Scale

With any virtual infrastructure design that is required to scale extensively across hundreds or even thousands of hosts, provide petabytes of storage, and support large complex networks, extensibility is a key factor. For a successful web-scale design, scaling a large physical Virtual SAN platform while maintaining control, compliance, and security is critical. Taking a predefined building-block approach to this type of architecture from day one is paramount in planning for scalability.

In addition, the configuration and assembly process for each host should be standardized, with all components installed identically. Standardizing the configuration of the physical components is critical in providing a manageable, consistently performing, and supportable web-scale infrastructure. This standardization eliminates variability, which in turn reduces the amount of operational effort involved with patch management and helps provide a flexible building-block solution.

Although some aspects of the configuration and scaling are likely to be hardware-vendor dependent, this type of model should be part of any web-scale Virtual SAN platform design. The example illustrated in Figure 6.4 depicts just one possible building-block scenario.

In this example, each *web-scale pod* is made up of 96 rack-mount Virtual SAN hosts configured as four 24-node clusters that are split equally across six server cabinets. Each web-scale pod also houses two 48-port 10GbE "leaf" switches and two 1GbE IPMI management switches for out-of-band connectivity. Each pod is designed to provide multiple fault domains for Virtual SAN, as well as compute and network resources.

The number of web-scale pods in this example can be scaled out accordingly, based on design requirements, but also depending on hardware, software, and power limitations.

As illustrated in Figure 6.5, each resource pod, which provides 96 hosts, can be scaled out to form a true web-scale platform across multiple data-center availability zones and multiple physical data-center locations globally.

In this example, vSphere components in each web-scale pod are managed by a single vCenter Server instance. The compute, storage, and network resources available from each component layer of this building-block web-scale architecture are provided in Table 6.1.

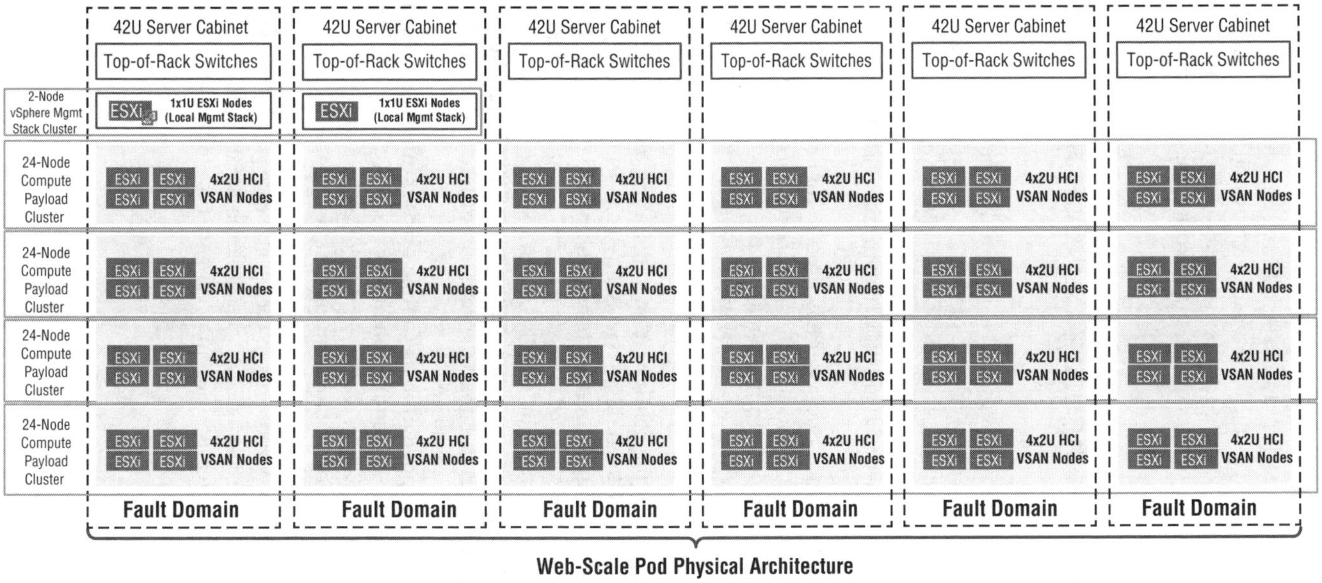

FIGURE 6.4
Web-scale pod logical architecture

TABLE 6.1: Example of capacity scalability of building-block web-scale architecture

Resource	Per Host	Per Cluster	Per Web-Scale Pod
Memory	512 GB DDR3	10.5 TB (24 nodes with 3 reserved for HA)	42 TB
CPU	2 × Intel E5 8-Core 3.1 GHz = 49.6 GHz	1,041.6 GHz (24 nodes with 3 reserved for HA)	4,166.4 GHz
Performance Storage clusters (tier 1) 1 × 400 GB MLC SSD (~15% of usable capacity) 5 × 1.2 TB 10k SAS	~20k IOPS read ~15k IOPS mixed 6 TB of raw storage	~480k IOPS read ~360k IOPS mixed 144 TB of raw storage	~1920k IOPS read ~1440k IOPS mixed 576 TB of raw storage
Standard storage clusters (tier 2) 1 × 400 GB MLC SSD (~10% of usable capacity) 7 × 2 TB 7.2k NL-SAS	~15k IOPS read ~10k IOPS mixed 14 TB of raw storage	~360k IOPS read ~240k IOPS mixed 336 TB of raw storage	~1440k IOPS read ~960k IOPS mixed 1.3 PB of raw storage
Capacity storage clusters (tier 3) 2 × 400 GB MLC SSD (~4% of usable capacity) 10 × 4 TB 7.2k NL-SAS	~10k IOPS read ~5k IOPS mixed 40 TB of raw storage	~240k IOPS read ~120k IOPS mixed 960 TB of raw storage	~960k IOPS read ~480k IOPS mixed 3.7 PB of raw storage
Network bandwidth	20 Gb/s	480 Gb/s	1,920 Gb/s (80 Gb/s MLAG to Spine)

NOTE In this example, IOPS is based on a mixed workload of 70 percent read and 80 percent mixed (random) I/O.

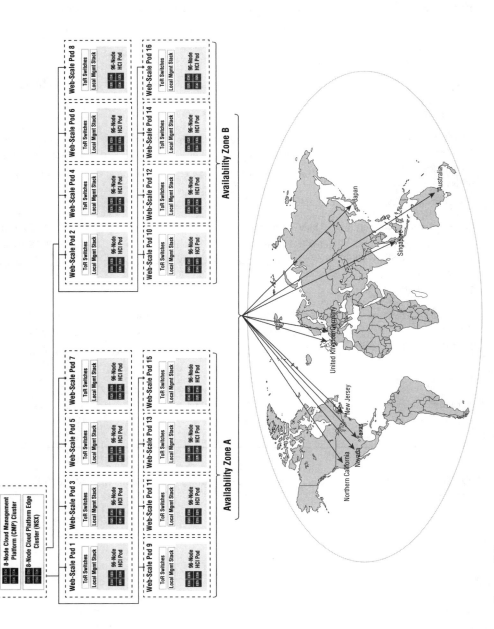

FIGURE 6.5
Web-scale pod scale-out data-center strategy

This example is only one possible web-scale platform architecture. There is no single design blueprint for delivering a Virtual SAN physical infrastructure platform that can provide this level of scalability. As with any Virtual SAN design, key factors will arise during the design phase that will play an important role in designing the web-scale building blocks. These are likely to include the following:

- Expectations of platform growth
- Hardware availability and lead times
- Physical hardware scalability limitations (such as management tools)
- Capital expenditure and hardware depreciation considerations
- Data-center power, space, zones, and cooling limitations

Leaf-Spine Web-Scale Architecture

The three-tier (core, aggregation, and access) network topology, which is optimized for transporting data into and out of a data center, is not typically best suited for the cross-rack internal transport required by a web-scale Virtual SAN platform.

The *leaf-spine architecture*, on the other hand, introduced in Chapter 4, uses a multirooted topology that actively manages the multiple paths between two end points by using equal cost multipathing (ECMP). In addition, the use of a high-port-count platform for the *spine* devices enables the deployment of a *folded-clos* (or fat-trees) design, without using additional switching components. The key characteristics of the leaf-spine topology as they relate to a web-scale Virtual SAN platform include the following:

- Multiple design options are available when using a variable-length spine with ECMP to have multiple available paths between leaf and spine switches.
- This architecture future-proofs the platform for higher performance and automated software-defined networking platforms, such as VMware NSX.

Figure 6.6 illustrates a leaf-spine architecture using a multirouted topology, which is aligned to the web-scale architecture example described in the previous section.

Other key design considerations for designing Virtual SAN in a highly scalable design include, but may not be limited to, these:

- Each Virtual SAN cluster should use a dedicated VLAN for isolated replication and workload traffic. This isolates Virtual SAN cluster traffic from any external interference and simplifies troubleshooting.
- In Virtual SAN 6, in order to enable Virtual SAN clusters to support 64 nodes, you must set three advanced configuration options on all hosts in the cluster:
    1. Configure the advanced settings for increased node support on each host in the cluster:

       ```
       esxcli system settings advanced set -o /VSAN/goto11 -i 1
       ```

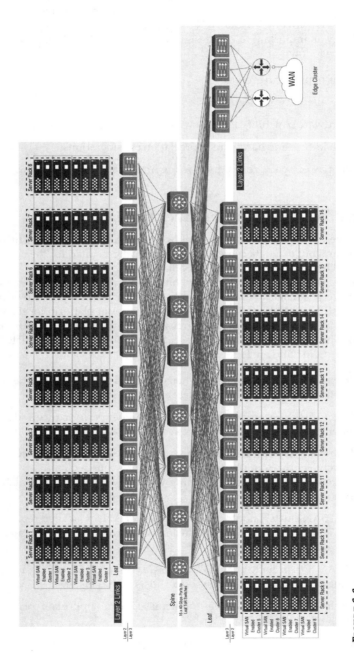

FIGURE 6.6
Web-scale leaf-spine architecture

2. Increase the TCP/IP heap size:

   `esxcli system settings advanced set -o /Net/TcpipHeapMax -i 1024`

3. Set the client limit to 65 to allow up to 64 hosts:

   `esxcli system settings advanced set -o /CMMDS/clientLimit 65`

All hosts must be restarted to allow these operational changes to take effect. In addition, you should review the latest VMware Knowledge Base articles for the latest configuration recommendations.

Furthermore, key scaling design factors that are pertinent in large, scalable, or web-scale environments include the following maximums, which take precedence over the vSphere core platform limits:

- 32 nodes per cluster in Virtual SAN 1 or 64 nodes per cluster in 6.0 to 6.2.
- Exactly one cache-tier flash device can be employed in each hybrid disk group.
- Between one and seven capacity-tier mechanical disks or capacity flash devices can be employed in each disk group.
- There is a maximum of five disk groups for each host.
- There is a maximum VMDK size of up to 2 TB in Virtual SAN 1 or 62 TB in Virtual SAN 6.0 to 6.2.
- There is a maximum of 100 virtual machines per host in Virtual SAN 1, or 200 virtual machines per host in 6.0 to 6.2.

In addition to these design factors, Table 6.2 highlights specific maximums that must also be adhered to when planning for a large-scale Virtual SAN deployment.

TABLE 6.2: Other Virtual SAN 6.0, 6.1, or 6.2 maximums

ATTRIBUTE	MAXIMUM VALUE
Virtual SAN datastores per cluster	1
Max virtual machines per host	200
Max components per host	9,000
Max virtual machines per cluster	6,400
Max HA-protected virtual machines per cluster	2,048

Chapter 7

Virtual SAN Use Case Library

Virtual SAN and Virtual Volumes both offer appropriate solutions for all vSphere-based use cases, as illustrated in Figure 7.1. Nevertheless, the design and implementation will almost certainly vary, based on the specific requirements and demands of the workload type. Therefore, it is critical that you address the customer-specific use cases as part of the design process from the beginning of the project, in order to ensure that the environment can meet the demands of their specific applications. For instance, designing a Virtual SAN hyper-converged infrastructure to support a business-critical, performance-dependent Oracle RAC implementation is likely to be significantly different from creating an architecture design to provide a disaster-recovery target for a development platform.

Typical Virtual SAN use cases include, but are definitely not limited to, the following:

- Tier 1 workloads (virtualized business-critical applications)
- Tier 2 and tier 3 workloads
- Virtual desktop infrastructure (VDI) and end-user applications
- Test, development, and staging workloads
- Backup and disaster-recovery target storage
- Isolated DMZ, perimeter zone, or edge clusters
- vSphere management or cloud management platform (CMP) clusters
- Two-node remote office / branch office (ROBO) solutions
- Virtual SAN–enabled stretched clusters

Virtual SAN–enabled clusters and Virtual Volumes, through their alignment with the software-defined storage model, can address most, if not all vSphere workload use cases for a design, when architected to meet the specific requirements of the customer's applications.

The use cases described in this chapter are presented progressively. Each use case provides, at a high level, the capabilities of the Virtual SAN platform, and describes how each component can be integrated to provide a solution for common business requirements. The use cases also demonstrate how the components of Virtual SAN can integrate with other vSphere-based technologies, and how this integration results in a highly available, secure, and manageable storage platform.

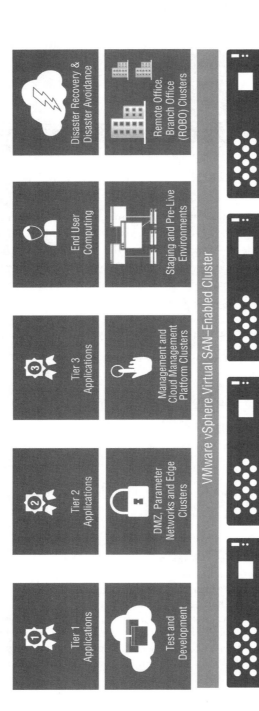

FIGURE 7.1
Virtual SAN use cases overview

The design, deployment, and management of any Virtual SAN environment will differ between organizations, and in some cases, within a virtual data center. Some organizations take advantage of their software-defined solution by migrating all applications onto the new platform, taking full advantage of the features made available through their Virtual SAN environment. However, other IT organizations take a more granular and cautious approach, and build the platform gradually, migrating applications only when their hardware life cycle requires a refresh, and using only the features required at the time of deployment.

Use Cases Overview

A Virtual SAN infrastructure can be an effective storage solution for almost any type of workload required in a modern, dynamic business environment. In these environments, the uptime of mission-critical applications and the ability to perform fast recovery from system failures are vital to meeting service-level agreements (SLAs).

A key requirement of each use case defined in this chapter is to ensure that the application assessment for each targeted workload is considered, as part of the design process, in order to guarantee that the host's disk group configuration can meet the performance and availability requirements of the applications.

It is common to see, particularly for enterprise IT organizations or cloud service providers, different types of application workload, such as DevOps, production, and business-critical applications, being separated onto dedicated Virtual SAN–enabled platforms. These dedicated clusters, often referred to as *island clusters*, as illustrated in Figure 7.2, can be used to isolate resources with specific performance, licensing, or security requirements.

As with any design, key design factors should be addressed through obtaining application requirements and dependencies, carrying out a full assessment of the targeted workloads, and obtaining verification of storage-policy prerequisites. However, it is also worth considering that it is deemed good practice for staging or pre-live environments to closely mirror the targeted production infrastructure, and therefore, storage policies should be aligned accordingly.

Many test and development environments have high consolidation ratios and transient workloads. Therefore, providing a compute and storage environment that can meet the requirements of a large number of virtual machines is one of the keys to providing a consistent and stable platform. It is also possible that availability and capacity are not as critical, as such an environment could, if required, be easily re-created. Alternatively, the workloads could be fully disposable and not be subject to any SLAs. An environment such as this is also likely to have a high turnover rate of virtual machine deployments, so automation is another key consideration, which we address further in Chapter 9, "Delivering a Storage-as-a-Service Design."

Virtual SAN can also provide a disaster-recovery solution target for replicated data, typically at a lower cost than traditional hardware-driven solutions. Additional vSphere features, such as vSphere Data Protection, vSphere Replication, and vCenter Site Recovery Manager, will also be required to allow the backup and replication mechanism needed for the solution design. For instance, such a solution will likely use vSphere Replication components, installed at both source and target locations, with Site Recovery Manager optionally providing the automation mechanism required to orchestrate the failover of virtual machine workloads, as illustrated in Figure 7.3.

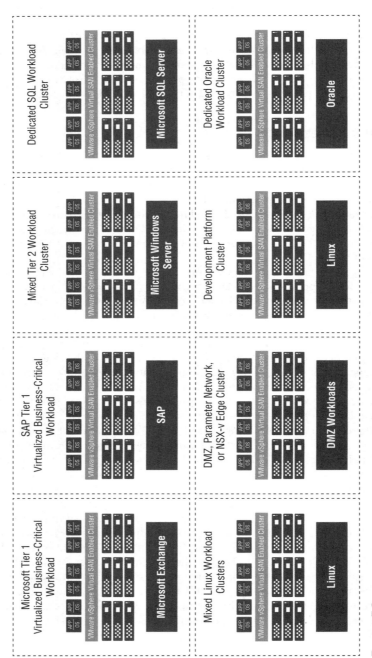

Figure 7.2
Virtual SAN island cluster design

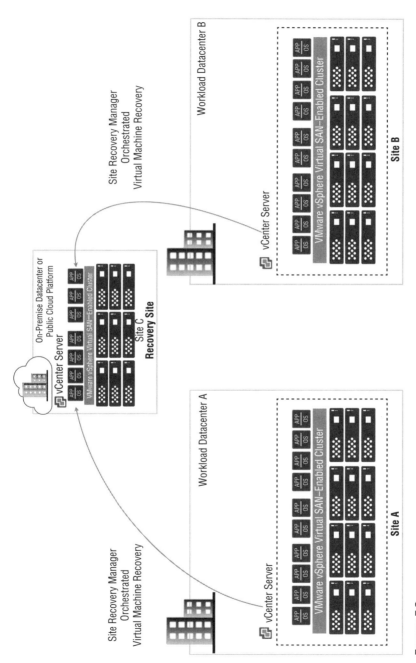

FIGURE 7.3
Disaster-recovery solution architecture example

The storage platform to be deployed at the disaster-recovery environment might be considered a capacity-based solution, which is not required to match the active site in terms of performance or availability. However, this will depend on the customer's requirements while failed over to the disaster recovery site, and whether the same SLAs are required to be met during the period of the incident. Therefore, performance of the target storage site is an important design consideration and should not be overlooked.

Another potential use case for a Virtual SAN platform is to provide an isolated DMZ, perimeter network, or edge cluster, as shown in Figure 7.4. It is common, and considered a good security practice, to create an isolated boundary for DMZ, perimeter network, or edge cluster workloads by implementing a dedicated island cluster for these publicly available or exposed application types.

By using Virtual SAN, the environment for this use case can be fully isolated into its own security zone at the compute, network, and storage layers. By using Virtual SAN, the DMZ, perimeter network, and edge cluster workloads can be easily isolated from other secured or firewalled applications, and corporate security policies can more easily be applied at the cluster level and on the underlying storage system.

In a Virtual SAN cluster design for DMZ, perimeter network, or edge cluster workloads, one key design factor that may need to be addressed is the migration of workloads to and from the hosts in this isolated cluster. This additional consideration arises because it is typical for a design such as this to use a fully secured and isolated network. Cold migrations could be the only option available for moving applications into and from this secured island cluster. Additional operational considerations may need to be applied because of the significant period of time required for such tasks to complete.

Two-Node Remote Office / Branch Office Design

An additional design configuration introduced in Virtual SAN 6.1 is a two-node remote office / branch office (ROBO) solution. The primary aim of this architecture is to offer these remote locations a simple, on-premises compute and shared storage solution, while minimizing up-front deployment costs and long-term operational overhead. Prior to support for this type of architecture, a three-node cluster was required to provide the minimum supported host configuration for Virtual SAN–enabled environments. However, with this specific ROBO architecture, it is now possible to design two-node clusters with an external witness site, assuming the architecture reflects the support boundaries set out by VMware.

As illustrated in Figure 7.5, the two-node Virtual SAN architecture builds on the concept of fault domains, previously addressed in Chapter 4, "Policy-Driven Storage Design with Virtual SAN." Each of the two VMware ESXi hosts located on the remote office premises represents a single fault domain. In Virtual SAN architecture, the objects that make up a virtual machine are typically stored in a redundant mirror across two fault domains, assuming the Number of Failures to Tolerate (FTT) is equal to 1. When one of the hosts goes offline, the virtual machines can continue to run, or be restarted, on the alternate node. To achieve this, a witness is required to act as a tiebreaker, to achieve a quorum and enable the surviving nodes in the cluster to restart the affected virtual machines.

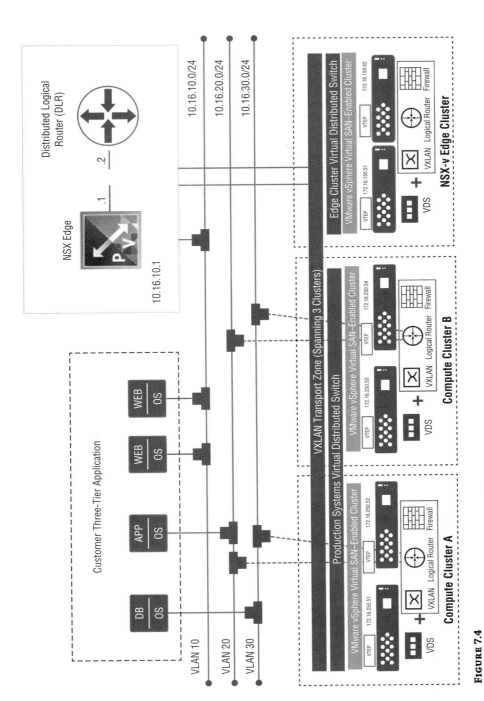

FIGURE 7.4
Isolated edge cluster design in an NSX implementation

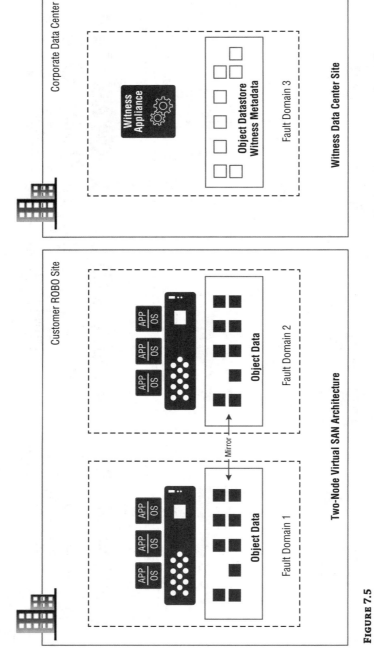

FIGURE 7.5
Remote office / branch office fault domain architecture

In a traditional Virtual SAN–enabled cluster, the witness objects are placed intelligently across the configured hosts. In contrast, in a two-node ROBO architecture (illustrated in Figure 7.6), the witness objects are located externally, on a different site. The witness virtual appliance node required for each remote-office instance can be located centrally, at the business's corporate data center, or may even be hosted on a public cloud platform. This architecture allows all ROBO sites and witness nodes to be managed centrally, through a single vCenter Server, allowing the centralization of operations and patch management.

This dedicated witness appliance is specifically configured to store just metadata, in fundamentally the same way as in the stretched cluster design described in Chapter 5, "Virtual SAN Stretched Cluster Design." This witness appliance provides the quorum services required if a host failure occurs, as depicted in Figure 7.7.

This ROBO solution would typically be suited to supporting a small number of virtual machines at each ROBO site, while also providing a highly available solution at each location. Providing this type of vSphere High Availability–enabled two-node cluster would not be possible if the design used local VMFS storage in each of the remote two-node clusters, or without some sort of shared storage device being used.

By using a dedicated virtual appliance to provide witness services, this architecture eliminates the need to deploy a third vSphere host on the ROBO site, reducing overall costs without sacrificing the availability benefits of shared storage. As with the stretched cluster architecture, the witness appliance is a specially modified nested ESXi host specifically designed to store only witness objects and cluster metadata. Also, like the stretched Virtual SAN cluster, the witness appliance does not contribute to the compute and storage capacity of the solution, and cannot be used to host virtual machines. The use of a witness appliance in a Virtual SAN–enabled configuration is supported by VMware only for this type of two-node architecture, and with a Virtual SAN stretched cluster design, as detailed in Chapter 5.

Like the stretched cluster deployment, the nested ESXi Virtual SAN witness appliance is automatically deployed with both flash and mechanical disks embedded. One of the appliance's VMDKs is tagged as a flash device during provisioning. No manual configuration is required by the Virtual SAN administrator. In addition, there is no requirement for a physical flash device in the vSphere host that is hosting the witness, and if required, all of the appliance's virtual disks can be thin provisioned.

To store the required metadata, the witness requires 16 MB of storage capacity for each witness component stored, with one witness component per object. Therefore, unlike a stretched cluster architecture, most deployments will host only a small number of virtual machines at each remote office, so the *tiny* configuration of two vCPUs with 8 GB of assigned memory should typically be more than sufficient and allow support for up to 750 components.

Because the witness appliance does not host virtual machines and therefore does not have to service virtual machine read and write requests, the network connectivity requirements between the remote office and the corporate data center or cloud platform are minimal. Typically, a WAN interconnect with 1.5 Mb/s of available bandwidth and latency as high as 500 ms round-trip time (RTT) is sufficient to provide network communication between the two-node cluster and the witness appliance. However, just as in a traditional Virtual SAN deployment, multicast must be enabled for communication between the hosts in the two-node ROBO cluster. There is no requirement for multicast to be enabled for WAN communication to the witness appliance.

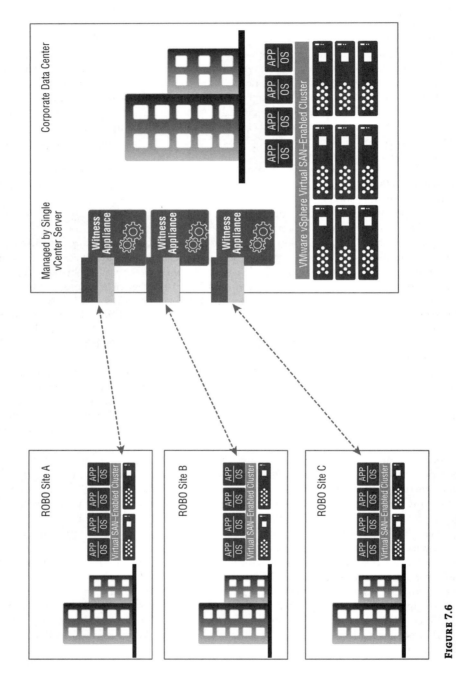

FIGURE 7.6
Two-node ROBO solution architecture overview

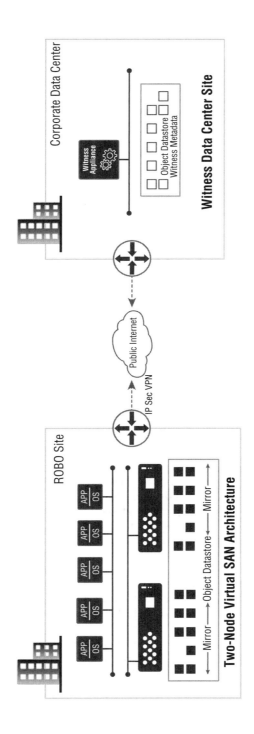

FIGURE 7.7
Witness object metadata architecture

The Virtual SAN storage policy–based management (SPBM) capability, the FTT, is pertinent to this ROBO architecture. As discussed in Chapter 4, this capability delivers the mirrored configuration, which provides the $n + 1$ redundancy for virtual machines. In a Virtual SAN two-node architecture, an FTT of 1 is explicitly required, as there are exactly three configured fault domains, as illustrated in Figure 7.5. With this configuration applied, a mirrored copy of each virtual machine is created and automatically maintained on the two separate physical nodes. It is this mechanism that allows one host within this two-node architecture to fail, with users retaining full access to the application without disruption, or the workload being restarted within a few minutes by the vSphere HA process. The level of application availability ultimately depends on whether it was residing on the failed node, or configured at the application layer for high availability and deployed and load-balanced across both nodes in the cluster.

Another option that provides application availability across the two-node cluster is the vSphere Fault Tolerance feature. This feature, which is also compatible with this Virtual SAN architecture, can provide continuous availability to workloads with up to four vCPUs in the event of a host failure.

As previously noted, vSphere High Availability is a critical component of this architecture and these use cases, as shared storage is tightly integrated with the process of restarting virtual machines after a host outage. With vSphere HA enabled, if a host fails, the virtual machines affected by the outage reboot on other hosts in the cluster, minimizing downtime. However, it is important to understand and to advise accordingly that in a two-node architecture, to ensure that enough CPU and memory resources are available to restart all impacted virtual machines, effectively running 100 percent of the workload on a single host, the vSphere HA Admission Control Policy must be configured to reserve 50 percent of both memory and CPU resources, irrespective of the amount of available storage. As a result, in this configuration, only 50 percent of the compute resources available to the two-node cluster will be free to run workloads at the remote office site.

Horizon and Virtual Desktop Infrastructure

One use case that sees particularly strong adoption of the Virtual SAN platform is a VDI, delivered through the VMware Horizon portfolio, as illustrated in Figure 7.8.

The VMware Horizon platform provides cloud resource flexibility and agility to desktop computing, by delivering end-user desktops from a highly available pool of resources. VMware Horizon enables the end user to access desktops and applications across multiple devices, such as Microsoft Windows, Mac OS X, and Linux desktop computers, and iOS and Android tablets and phones, in any location where a connection is available.

Virtual SAN provides a scale-out repeatable building-block storage infrastructure, which aligns nicely with the virtual desktop predictive design deployment model. In addition, virtual desktop storage traditionally came at a high cost, because of the high speed and low latency typically required from this use case. Significant savings can be realized through the low cost per I/O that is achievable by using Virtual SAN in an optimized VDI environment.

When designing and implementing Virtual SAN for the Horizon use case, several specific design factors need to be taken into account as part of the architecture. For instance, most VDI platforms require a high-performance storage design, in order to reduce the latency associated with serving large numbers of desktops and applications concurrently. However, availability may not be as pertinent a design factor as it would be with virtual server workloads, because in many desktop platforms, the virtual desktops are by design disposable, and destroyed after a single use.

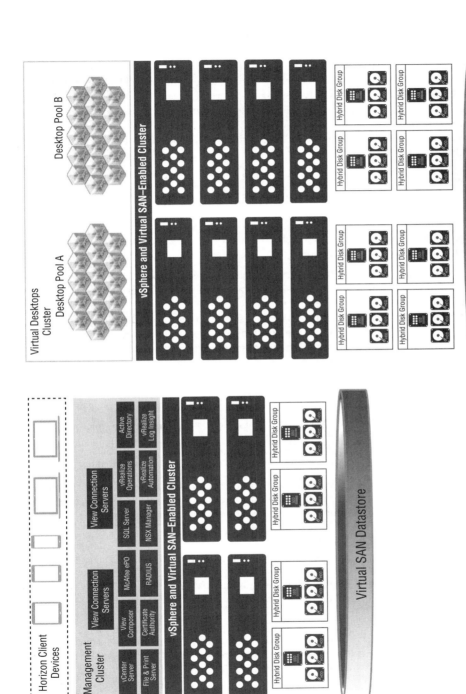

FIGURE 7.8
Virtual SAN and VDI architecture

This workload type is referred to as a *nonpersistent linked clone*. Although it is also important to consider application availability, which is typically managed outside the desktop, the nonpersistent linked clones themselves are frequently refreshed, resulting in all locally stored user data being lost. In addition, this workload type provides a predictable storage footprint that can be tuned in an all-flash configuration to provide highly efficient performance with minimal storage capacity utilization.

For these reasons, when designing a storage policy for this workload type, a typical configuration might use the following:

- Availability: Number of Failures to Tolerate = 0
- Sizing: (Number of VMs × (Delta Disk Max Growth + Unreserved Memory size)) + Replica Size
- Object Space Reservation: = x% (where x is based on typical disk usage of the desktop workload)

A Horizon virtual desktop design might also use other workload disk types or a variety of classes of desktop that require storage persistency, and will therefore need to be protected by a different FTT policy.

For instance, *dedicated linked clones* typically have similar design considerations to those of nonpersistent linked clones, in that they require desktops to be recomposed regularly in order to prevent linked-clone storage growth. As with nonpersistent linked clones, this desktop type requires separately maintained application and profile management, in order to preserve user-specific configurations. A dedicated linked clone workload typically uses a storage policy such as this:

- Availability: Number of Failures to Tolerate = 1
- Sizing: ((Number of VMs × (Delta Disk Max Growth + Unreserved Memory size) × (FTT + 1)) + Replica Size
- Object Space Reservation: = x% (where x is based on typical disk usage of the VM)

Finally, *full clone desktops* require a significantly larger disk footprint from Virtual SAN, and typically use traditional desktop application management and local profiles. A storage policy for full clone workloads might commonly be expected to provide the following:

- Availability: Number of Failures to Tolerate = 1
- Sizing: (VMDK Size + Unreserved memory size) × (FTT + 1)

Designing a Horizon-based solution that uses Virtual SAN requires additional design considerations beyond those required for a traditional server-based infrastructure. These include, but may not be limited to, the following:

- Virtual SAN is Horizon View Composer *aware*, and creates Virtual SAN policies when the pool is first created. In addition, a refresh, recompose, or rebalance operation in Horizon will enumerate and re-create the storage policy if required.
- Virtual SAN creates different policies based on the pool type, although the desktop storage policies are not deleted when the associated pool is removed.
- Virtual SAN does not support the use of space-efficient virtual disks.

- Ensure that the design includes a plan for appropriate delta disk growth, which aligns with the virtual desktop refresh cycles.

As with any Virtual SAN design, for a Horizon-based solution, you must be sure to do the following:

- Understand the customer's business drivers for their virtual desktop infrastructure.
- Understand the integration of Horizon with Virtual SAN.
- Identify the required information to architect a Horizon Virtual SAN solution.
- Conduct a sizing exercise for the Horizon Virtual SAN cluster, to ensure that the solution is properly sized, in terms of desktop performance and scale.
- Advise the customer on the selection of VSAN Ready Nodes, EVO:RAIL, or build-your-own hardware.
- Architect a Horizon Virtual SAN solution that is highly available and easily managed.

Virtual SAN File Services

Virtual SAN, when used with third-party software, can provide file or block disk services as part of a solution, as illustrated in Figure 7.9. This may prove useful when a design includes requirements to add file services, such as Server Message Block (SMB) or NFS, on top of Virtual SAN—for instance, VDI user home directories or ROBO file shares. These third-party solutions use SPBM and the underlining Virtual SAN capabilities to provide abstracted pools of disk resources that can be used to meet the storage architecture requirements for multiple types of business applications.

Although it is beyond the scope of this book to address these third-party solutions in detail, two products that could be considered to meet these specific use case requirements include NexentaConnect for Virtual SAN and EMC CloudArray.

Solution Architecture Example: Building a Cloud Management Platform with Virtual SAN

The final section of this chapter offers an overview of a typical Virtual SAN use case: providing storage for a dedicated cloud management platform (CMP) cluster. This overview shows you how Virtual SAN can be integrated into an infrastructure design and demonstrates features, functionality, and configurations. The example design helps explain the architecture and the reasons for specific configurations and design decisions.

Each customer design is typically unique, based around their specific infrastructure and application requirements. The following provides a high-level overview only, and not the detailed, low-level design typically required before implementation can begin.

Introduction and Conceptual Design

In recent years, with the growth in new management tools and services required to maintain complex self-service automation components, a dedicated, out-of-band cluster has significant advantages in providing a fully centralized ecosystem for the operational administration and maintenance of a cloud management platform.

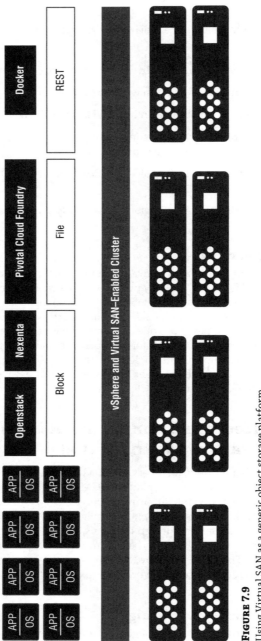

FIGURE 7.9
Using Virtual SAN as a generic object storage platform

In this use case example, the cloud management platform cluster hosts all of the components and services required to maintain the IT organization's cloud infrastructure, and is separated from other end-user application workloads. Separating the CMP infrastructure components from end-user workloads allows for better segmentation of resources and improves the manageability and security of the environment.

One of the traditional obstacles blocking the implementation of a dedicated cloud management platform cluster has been cost. It could often be far too costly to create an out-of-band infrastructure dedicated to the cloud and other management systems. However, with VMware Virtual SAN, enterprise IT organizations and cloud service providers can now eliminate the requirement for costly shared storage systems dedicated to hosting management components.

Designing a cluster to support an out-of-band cloud management infrastructure is similar to designing any other stand-alone Virtual SAN–enabled vSphere environment. However, it is important to design the management environment so that no dependencies exist between it and the production workloads. Creating dependencies between the management platform and the production resources limits the isolation of the administration components, and therefore risks creating a circle of dependency between management and production systems.

In this use case example, the cloud management platform cluster will host the virtual machines and appliances that provide cloud management infrastructure services to the entire vSphere environment and beyond.

A typical cloud management platform cluster can include, but is not limited to, the following components:

- vCenter Server and Platform Services Controller (PSC) components
- vRealize Automation cloud platform components
- Dedicated SQL database server for management components
- vCenter Support Assistant
- vSphere Update Manager (VUM)
- NSX Manager
- vRealize Operations Manager
- vRealize Log Insight nodes
- Third-party software, such as antivirus and other hygiene-management components
- Third-party management and monitoring software, such as hyper-converged server vendor management tools
- Active Directory Domain Services components, for authentication and federated authentication services

Running the cloud management components within a mixed production cluster can make troubleshooting time-consuming and can make tracking down management virtual machines difficult in a disaster-recovery scenario. Providing the enterprise infrastructure with a dedicated cloud management cluster and management component separation, as illustrated in Figure 7.10, has the following key benefits:

- It separates the management components from the resources they are managing.
- It facilitates quicker troubleshooting and problem resolution, as management components are strictly contained in a relatively small and manageable cluster.
- It isolates resources between workloads running in the production environment, and the actual systems used to manage the infrastructure, avoiding resource contention.

The following host, storage, and networking considerations will apply to the design of a dedicated cloud management platform:

- If possible, avoid booting from local disks; this makes them unavailable for use as part of a Virtual SAN disk group.
- Ensure that the design has a highly available vSphere cluster configuration, with redundancy at each component layer.
- Provide a minimum of $n + 1$ resilience of all physical components, including ensuring high availability of virtual and physical network switching.
- Take advantage of the shared storage provided by Virtual SAN to facilitate vSphere High Availability (HA), vSphere vMotion, and vSphere Distributed Resource Scheduler (DRS).
- Ensuring simplicity is a primary goal of the management environment. Designing a simple and static environment minimizes the risks that can be caused by misconfiguration or human error, which in turn reduces recovery time objectives (RTOs).
- Three nodes provide the minimum number of hosts required to support a Virtual SAN–enabled cluster. In this cloud management platform, the minimum three-node cluster might provide sufficient resources for the management components and maintain $n+1$ operational availability, but it wouldn't provide sufficient accessibility to storage resources during maintenance operations. This example therefore uses a four-node cluster. It will also be possible to scale out hosts later if the management cluster becomes resource-constrained, or new management components need to be deployed.
- Where possible, physically separate the cloud management cluster and switches into dedicated racks, to help distinguish between management and production workloads. You should also aim to design the out-of-band cloud management infrastructure in such a way that any type of incident or outage affecting the production system does not affect the management cluster. Likewise, issues with the management cluster cannot affect the production workloads.

Customer Design Requirements and Constraints

Requirements are the key demands on the design and must be satisfied if the design is to be successful. A *constraint* is anything that restricts the Virtual SAN or vSphere design options in a way that risks not satisfying a business requirement.

The customer has defined the following key requirements and constraints for the cloud management platform cluster design. These details were gathered during a series of workshops.

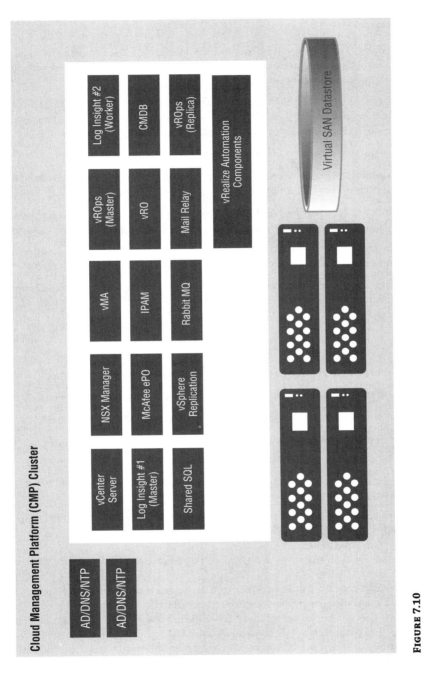

FIGURE 7.10
Architectural overview of enterprise cloud management cluster

Requirements

- The proposed management system must be designed for scalability and elasticity. Components must be easily upgraded, migrated, added, and removed without any impact on the applications running on the cloud management platform.

- Management components should be separated from production workloads on independent and isolated management hardware. Unpredictable events affecting availability should not have an impact on consumers of production systems.

- N + 1 resiliency should be factored in to all components in the design.

- The availability of services required by the management platform is defined as 99.9 percent, allowing a maximum of 8.76 hours per year of downtime.

- The ability to centrally log vSphere, and other management platform component events, in a highly available manner for log retention purposes, is a key security requirement.

- The ability to proactively monitor all vSphere components and the overall solution status, in terms of utilization and performance metrics, is required by the design.

- All management components should be deployed with security as the primary factor for all configuration choices. The security design should aim to conform with all vendor best practices.

Constraints

- Cisco and Intel have been preselected as the compute platform of choice.

- Cisco physical switches have been preselected as the network platform of choice.

- Shared storage is to be provided exclusively through VMware Virtual SAN.

- Network connectivity to the hosts and core data center is to be via multiple 10GbE connections.

Host Compute Design

This section discusses the design and implementation details of the compute components of the example design, to ensure that a stable and consistent environment is achieved within the cloud management platform cluster.

For this example design, compute is provided by four Cisco VSAN Ready Nodes, configured as a single vSphere cluster, using both vSphere HA and DRS. Although it is possible to meet the customer's compute and storage resource requirements by using a three-node cluster, it is not possible to meet the requirements set out for availability during periods of maintenance. Therefore, four Cisco C-Series servers will be configured with dual Intel Xeon CPUs, 256 GB of RAM, a single pass-through RAID controller, 1 × 400 GB SSD, and 7 × 900 GB 10K SAS disks. Each cluster node will run the ESXi 6 hypervisor on an internally installed, industrial-grade 32 GB USB device. This configuration is detailed further in the figures and tables that follow.

Ensure that you use the same hardware specifications for all hosts in the cluster (where possible), and apply consistent configuration settings to the hardware. This reduces the amount of operational effort involved with patch management and provides a flexible basis for change and growth.

This is made easier in this example design as the customer will be procuring Cisco UCS Virtual SAN Ready Nodes to provide the cloud management platform cluster. A *Virtual SAN Ready Node* is a validated server configuration that is tested and certified by the hardware vendor for Virtual SAN deployment, and then jointly recommended by the server OEM and VMware.

Various options are available to provide uplink connectivity from the Cisco C240 M3 rack-mount servers to the core switches. For instance, Cisco UCS Fabric Interconnects could be used to enhance manageability of the hosts, with similar functionality to that available to the B-Series blade system. However, to simplify the management design, limit costs, and most important, to not place the cloud management platform design in the same data path as the production systems, creating a circle of dependency, the Cisco C-Series management hardware will be connected directly to the layer 3–enabled Nexus 5548UP switches, using 10 Gb/s FCoE connections, via the Cisco 1225-VIC PCIe converged network adapter (CNA) cards, installed in each of the rack-mount chassis.

Figure 7.11 shows the physical architecture of the Cisco UCS C240 M3 solution for the cloud management platform cluster.

This detailed architecture and configuration for the Cisco UCS C240 rack-mount servers with VMware Virtual SAN solution consists of the components listed in Table 7.1. In the sample design, each ESXi host in the cloud management cluster has the following hardware specifications.

TABLE 7.1: ESXi host hardware specifications

ATTRIBUTE	SPECIFICATION
VMware vSphere hypervisor ESXi for Cisco UCS C240 M3 Series rack server	ESXi 6 Update 2	ISO Build 3620759
Cisco UCS[1]	4 Cisco UCS C240 M3 rack servers (*x*86 servers), each with
	• 2 Intel Xeon processor E5-2660B CPUs
	• 24 8 GB 1,600 MHz DDR3 RDIMMs, PC3-12800, dual rank, 1.35V (256 GB)
	• 7 Seagate 900 GB SAS disks (10k)
	• 1 SAMSUNG 400 GB SAS SSDs
	• 1 Cisco 9300-8i 12G SAS HBA
	• 1 Cisco UCS VIC 1225 CNA
	• 2 Cisco Flexible Flash (FlexFlash) cards
Cisco C-Series firmware & IMC	2.0(9e)
Processors	Dual Intel Xeon Processor E5-2660B
	(2.20 GHz E5-2660 v2/95W 10C/25 MB Cache/DDR3 1866 MHz)
Virtual SAN storage (per node)	Enterprise Performance SAS SSD: 1 × 400 GB
	Magnetic disks: 7 × 900 GB 10k SAS (SFF)

TABLE 7.1: ESXi host hardware specifications *(CONTINUED)*

ATTRIBUTE	SPECIFICATION
Disk controller	Cisco 9300-8i 12G SAS HBA
Disk controller queue depth	1024
Boot device	32 GB industrial-grade, wide-temperature USB/SD flash drive
Networking	Cisco UCS VIC 1225 CNA Configured with: 2 × 10 Gigabit Ethernet Connections (802.1Q trunk port connectivity participating in the VLANs outlined in the network design)
Memory	256 GB

[1] *Each host is configured with one of three available disk groups, each composed of one 400 GB SSD and seven 900 GB 10k SAS disks. This configuration uses 8 of the 24 slots available on the Cisco UCS C240 M3 rack server.*

Table 7.2 lists the high-level CPU and memory requirements for the management hosts and the aggregate resources available from all four hosts.

TABLE 7.2: Host resources

ATTRIBUTE	SINGLE-HOST SPECIFICATION	TOTAL AVAILABLE CAPACITY (4 HOSTS)
Number of CPUs (sockets)	2	8
Number of cores per CPU	8	N/A
GHz per CPU core	2.20 GHz	N/A
Total CPU GHz per CPU	17.6.4 GHz	N/A
Total CPU GHz available	35.2 GHz	140.8 GHz
Proposed maximum host CPU utilization	80%	80%
Available CPU GHz per host	28.16 GHz	112.64 GHz
Total physical memory	256 GB	1,024 GB
Proposed maximum host RAM utilization	80%	80%
Available RAM for use	153.6 GB	614.4 GB

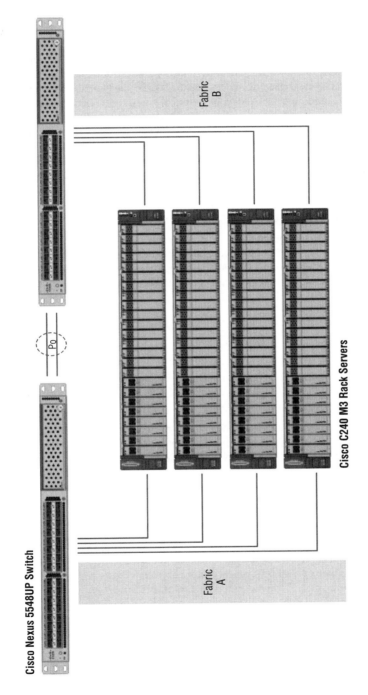

FIGURE 7.11
Virtual SAN with Cisco UCS environment physical connectivity details

Boot Device Architecture

In this example design configuration, VMware ESXi is to be booted from the on-board Cisco FlexFlash SD cards. Two Cisco FlexFlash SD cards, in a RAID 1 configuration, will be used to help ensure reliability through redundancy in the cloud management environment.

It is important to remember that persistent disks in the host server that are used by Virtual SAN cannot be used as a boot device. In a typical Virtual SAN environment, you would configure hosts to boot from USB, SD, or other nonpersistent storage in order to maximize the number of persistent disk slots available for disk groups. It is also important to remember that when the ESXi installation device is a USB device or SD card, a local scratch partition is not created on the installation media automatically during the deployment, and therefore logs are not retained locally by default after a host reboot. For more information on recommendations when booting hosts from nonpersistent storage, please refer to Chapter 4.

Cluster Configuration

The vSphere cloud management hosts and their resources will be pooled together into a single cluster, which will contain the aggregate CPU, memory, network, and storage resources available for allocation to virtual machines. To maximize resource utilization and to meet availability requirements, the cluster is to be configured with the following vSphere availability and load-balancing technologies:

- vSphere vMotion
- vSphere Distributed Resource Scheduler
- vSphere High Availability

vSphere High Availability

The cloud management platform cluster will employ vSphere High Availability (HA) to automatically recover virtual machines, should a host or specific virtual machine fail. In a Virtual SAN environment, vSphere HA behavior is slightly different from the traditional mechanism. External datastore heartbeats are typically not used and therefore become irrelevant. In addition, the HA agent uses the Virtual SAN network to communicate instead of the management network. However, the management gateway is still used by the host to detect if it has become isolated. For more information about vSphere HA behavior in a Virtual SAN–enabled cluster, please refer to Chapter 4.

In this example cloud management platform cluster design, illustrated in Figure 7.12, vSphere HA will employ the percentage-based admission control policy, in an $n + 1$ fashion, instead of defining the number of host failures a cluster can tolerate or specifying specific failover hosts.

As a four-node cluster, the percentage of tolerable failure to maintain $n + 1$ resilience is 25 percent of the total available compute resource. This equates to suffering a single host failure and still being able to maintain services at an adequate level. For additional availability, you could add an extra host for an $n + 2$ cluster, although this is not a customer requirement for this design.

Within the scope of this example design, the vSphere HA parameters shown in Table 7.3 are to be configured. Changes to the values defined in this design must be made only with approval through the IT organization's change management procedure framework.

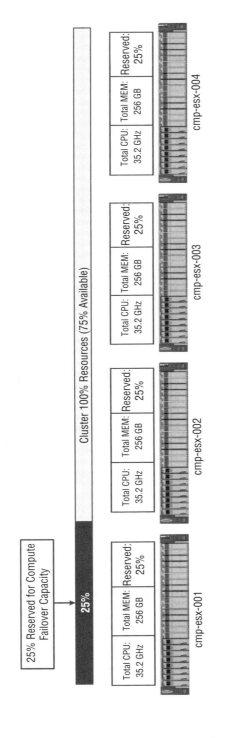

Figure 7.12
Percentage-based admission control

TABLE 7.3: vSphere HA example design values

Attribute	Configuration
Cluster name	CMP-01
Number of ESXi hosts	4
Host monitoring	Enabled
Admission control response	Prevent virtual machines from being powered on if they violate availability.
Admission control policy	Enabled—percentage of resources reserved: 25% CPU 25% memory $N+1$ for four-host cluster
Default virtual machine restart priority	Medium (majority of VMs) Modify as necessary at VM level for High (critical VMs) / Disabled (noncritical VMs)
Host isolation response	Power off, then fail over (Virtual SAN best practice)[2]
Virtual machine monitoring	Disabled
Virtual machine monitoring sensitivity	Medium
Heartbeat datastores	N/A (see Chapter 4 for further details)
Maintenance mode (Virtual SAN data migration)	Always plan for accessibility: Ensure accessibility / full data migration

[2] *Host isolation response refers to the action that vSphere HA takes when the host becomes isolated. The best practice in a Virtual SAN environment is different than in a traditional storage HA design. For information about the specific Virtual SAN host behavior during a host isolation event, please refer to Chapter 4.*

vSphere Distributed Resource Scheduler

vSphere DRS intelligently allocates available compute resources among the virtual machines, based on a predefined set of rules, and migrates virtual machines to reflect business or application requirements. The use of vSphere DRS rules can also allow for the separation of virtual machines, ensuring there is limited impact on the application or service if a host failure or isolation event occurs. A number of anti-affinity rules will be implemented as part of this example design.

vSphere DRS collects resource usage information from all hosts and virtual machines in the cluster, and will live-migrate virtual machines through vSphere vMotion in order to improve resource utilization across the cluster. In addition, vSphere DRS will perform initial placement actions when a virtual machine is first powered on, to ensure it's running on the most appropriate host.

Although it is not essential that you configure vSphere DRS on a cloud management cluster, where resource contention is not expected, it is recommended as a mechanism for balancing workloads across hosts to provide optimal performance, particularly as the cluster grows.

In smaller clusters, like that being defined here, it is desirable to set the vSphere DRS migration threshold to avoid automatic vMotion actions, which may have only a short-term performance benefit. Therefore, for this cluster, the recommendation for this design, where resource contention is not expected, is to configure the DRS migration threshold level to Conservative.

Based on the defined customer requirements and the design factors outlined, the values listed in Table 7.4 should be used for the vSphere DRS parameters for this example design. As with the vSphere HA configuration, these settings should not be modified without approval being sought through the appropriate change management framework.

TABLE 7.4: vSphere DRS example design values

ATTRIBUTE	CONFIGURATION
Cluster name	CMP-01
Number of ESXi hosts	4
DRS	Enabled
Automation level	Fully Automated
Migration threshold	Conservative
VMware DPM	Not Supported
Enhanced vMotion compatibility	Enabled

Separating servers and other critical roles that hold redundancy at the application layer is recommended to avoid service downtime. For this to be effective, it is necessary to apply vSphere DRS anti-affinity rules on the applicable cloud management components. These rules specify the relationship between groups of virtual machines, so that they remain separated from each other at the host level. As a result, if one vSphere host fails, the impact to the service is limited or not seen at all by end users. However, a design should apply affinity and anti-affinity rules sparingly, as they add overhead to the vSphere DRS algorithm and limit virtual machine migration options. The rules listed in Table 7.5 are to be applied to components hosted on this example cloud management platform cluster.

TABLE 7.5: Anti-affinity rule guidelines for cloud management cluster applications

SERVICE	RULE	DESCRIPTION
vRealize Automation appliances	Anti-affinity	Keep load-balanced services separated.
vRealize Automation .NET components	Anti-affinity	Keep servers separated.
Active Directory / DNS servers	Anti-affinity	Keep Active Directory / DNS servers separated.
vRealize Log Insight	Anti-affinity	Keep clustered Log Insight appliance nodes from running on same host.
vRealize Operations Manager	Anti-affinity	Keep master and replica nodes separate.

Network-Layer Design

This network layer design example addresses all communications between the cloud management virtual machines, at the logical and the physical network level, as well as the platform interactions associated with the infrastructure, such as Virtual SAN, vMotion, and management. The key design qualities typically associated with networking include performance, availability, and security, all of which are addressed in this example.

The network configuration for the cloud management cluster include, but are not limited to, the following design best practices:

- Network connectivity, with virtual switch port groups and 802.1Q VLAN tagging, is to be used to segment traffic into VLANs, to address security and traffic load prioritization. ESXi management, virtual machine, vMotion, and Virtual SAN storage traffic will all be separated and tagged by the virtual switch for QoS.

- The network design will provide isolation of layer 2 traffic that does not need to be routed outside the cloud management platform, such as Virtual SAN replication and vMotion traffic.

- The design will provide network redundancy, with at least two 10GbE active physical adapters from each host server.

- Where possible, the design will provide redundancy across different physical adapters to protect against host PCIe slot failure.

- Each of the four ESXi cloud platform management hosts are to be configured on the same, management-only, vSphere Distributed Switch (VDS).
- The physical network adapter cards are to be connected to redundant physical switches in such a way as to tolerate a single adapter or physical switch failure.

Management vSphere Virtual Switch

When designing the cloud management network infrastructure, a key design factor is whether to adopt the vSphere standard switch or the vSphere Distributed Switch (VDS). As outlined in Chapter 4, the main benefit of the vSphere standard switch is its simple configuration. However, if the design is to benefit from features such as Network I/O Control (NIOC), Link Aggregation Control Protocol (LACP), and NetFlow, these features are offered only by the VDS. Furthermore, if the platform is to include network virtualization, provided by VMware NSX, then the VDS is typically the only design option.

In this example, a single VDD with 10 Gb/s network uplinks is to be used. This configuration will ensure that Virtual SAN replication and synchronization activities, which will be imposed on the network, are prioritized appropriately, and will also provide the design with the ability to manage contention through NIOC and QoS, should it be required during periods of high utilization.

Network I/O Control Configuration

In this example, the two 10 Gb/s uplink interfaces, carrying multiple traffic flows from each host, are to be configured from the dedicated cloud management vSphere Distributed Switch (dvSwitch-CMP). Network I/O Control will be used to monitor the virtual network. Whenever it identifies congestion, it will automatically transfer resources to the highest-priority traffic type, as defined by the shares value in the NIOC policy.

As illustrated in Figure 7.13, the four management hosts are to be configured with a single dvUplink group, which sees both of the two active 10 Gb/s Ethernet adapters traversing traffic flows across the dedicated VDS, which has been configured to carry all cloud management platform network traffic.

Each of the physical network switch ports, connected to the host's network adapters, are to be configured as trunk ports. Figure 7.13 also illustrates the port groups that are to be used to segment traffic logically by VLAN, with VLAN tagging applied to traffic at the virtual switch level. Uplinks will be configured in an active/active configuration, with the choice of load-balancing algorithm depending on how the physical switches have been configured, as outlined in Table 7.6. Both the virtual and physical switches are to be configured to pass traffic specifically for VLANs being used by the cloud management platform, as opposed to allowing all VLANs to be trunked.

Table 7.6 provides the configuration that is to be applied to the cloud management VDS configuration.

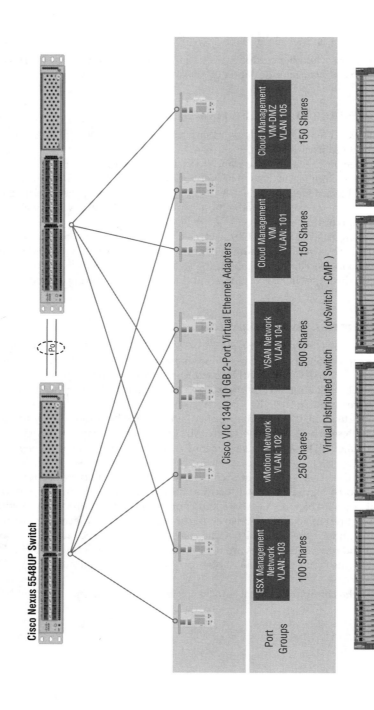

FIGURE 7.13
Network I/O Control

TABLE 7.6: vSphere Distributed Switch configuration

PARAMETER	PORT GROUP	CONFIGURATION SETTING
Load balancing	All	Route based on originating port ID. Depending on physical switch type, configuration, and 802.3ad (static or dynamic) support for downstream links, configure the management host dvSwitch with an appropriate load-balancing policy. See Chapter 4 for more details.
Failover detection	All	Link status
Notify switches	All	Enabled
Failback	All	No
Port binding	All	Static

As outlined here, the two 10 Gb/s network interfaces will carry all ingress and egress Ethernet traffic on all configured VLANs. The user-defined network resources should be configured port group by port group, as shown in Table 7.7.

TABLE 7.7: Example CMP Network I/O Control policy

PORT GROUP	VLAN ID	BROADCAST DOMAIN	NO. SHARES	QOS TAG	FUNCTION DESCRIPTION
Cloud-Mgmt-VM	101	172.16.101.0/24	150	2	Management VM traffic
vMotion Network	102	172.16.102.0/24	250	3	Layer 2 vMotion traffic (not routed)
ESX Management Network	103	172.16.103.0/24	100	1	ESXi management traffic
VSAN Network	104	172.16.104.0/24	500	5	ESXi VMware storage Virtual SAN traffic (not routed)
Cloud-Mgmt-VM-DMZ	105	172.16.105.0/24	150	2	Management DMZ virtual machine traffic

In addition, an appropriate QoS (802.1P) tag, as noted in the table, will also be associated with all outgoing packets, allowing the compatible upstream switches to recognize and apply the QoS configuration.

By taking this approach to the design of the VDS and segmenting vMotion, virtual machine, and Virtual SAN traffic into separate VLAN-backed port groups, and using shares and the QoS mechanism, it will be possible to sustain the required level of performance for each traffic type, even during periods of network contention.

Virtual SAN VMkernel Ports

To provide connectivity to the shared Virtual SAN datastore, a dedicated VMkernel port is to be configured on each cloud management host. The VMkernel port is to be configured on the dedicated Virtual SAN port group, and will be used to pass traffic to the storage kernel. In this example design, Virtual SAN traffic will traverse only the designated layer 2 segment and will not be passed outside the cloud management platform.

Jumbo Frames

The Virtual SAN port group will be configured to use jumbo frames, which are to be carried end to end throughout the Virtual SAN platform. As Virtual SAN traffic is to be configured over the designated layer 2 segment only, and not be routed outside the dedicated cloud management platform switches, jumbo frames need to be configured on this traffic type only within the cloud management platform network itself.

Virtual SAN Multicasting

The dedicated Virtual SAN VLAN, used for the Virtual SAN traffic, will be configured with multicast enabled within its layer 2 network segment. This configuration is to be applied to the Cisco Nexus 5548UP switches as part of the Virtual SAN VLAN multicast policy, with IGMP snooping enabled.

Physical Switch Configuration Overview

The cloud management platform physical network design uses a dedicated pair of Cisco Nexus 5548UP switches to isolate specific traffic types to traverse only within the cloud management platform environment. The Cisco Nexus 5548UP switches will require the optional layer 3 daughter board to be installed and configured appropriately. Figure 7.14 illustrates, at a high level, the cloud management platform environment and upstream connectivity to the production systems through the northbound aggregation layer network.

Storage-Layer Design

The four-node cloud management cluster must be configured with shared storage in order to take advantage of vSphere's vMotion and High Availability features. This design provides shared storage to the cloud management platform by using Virtual SAN, without creating any dependency on the primary production storage system, which in this example design is provided by a Virtual Volumes–enabled EMC VMAX3 400K system.

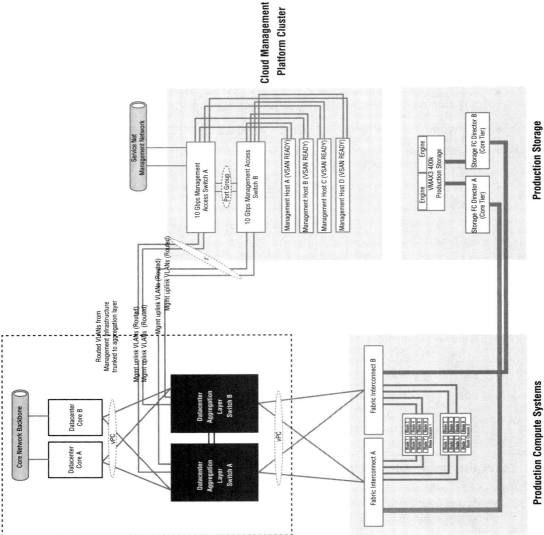

FIGURE 7.14
High-level physical network design

In the past, providing dedicated out-of-band storage for management-only clusters could be cost-prohibitive. It was often necessary to purchase an additional independent storage array purely for the management platform, one that could provide the required performance and availability for I/O-intensive and highly available virtual machines. In addition, a design might have required the provisioning of an isolated storage fabric, increasing costs even further. However, by using Virtual SAN, these costs can be significantly reduced, making this type of dedicated out-of-band management environment a viable and affordable option for most medium and large organizations.

In this example design, the cloud management platform will use a Virtual SAN 6.2 hybrid configuration to deliver true out-of-band storage to the dedicated management infrastructure. This will provide the following key design benefits to the platform:

- The system is inherently fault-tolerant and based on a distributed RAIN architecture, with no single point of failure, and failures are handled without downtime.
- The entire system is tightly integrated with and automated by vCenter Server.
- The locally attached storage is aggregated from the vSphere hosts in the cluster to form the distributed Virtual SAN datastore.
- Virtual SAN is flash-optimized, with the sole purpose of the SSD being for I/O acceleration.
- The design is to be based on virtual machine–centric operations, and policy-driven management principles.
- The Virtual SAN platform is fully integrated with vSphere, the vCenter Server Web Client, and the vRealize suite of applications.

In addition to these features, this example design applies the following Virtual SAN best practices to the cloud management platform architecture:

- A four-node cluster is used, with all nodes contributing storage, instead of the minimum configuration of three nodes.
- The cluster's storage is to be entirely balanced, with identical host configurations across the platform.
- The design will employ a redundant RAID 1 SD card configuration on all hosts as a boot device and will use remote logging.

STORAGE DESIGN SPECIFICATIONS

This section addresses the storage specifications for the cloud management platform cluster. These values, relating to virtual machine size and storage policy configuration, are based on assumptions. Table 7.8 provides the source metrics, which can then be used for datastore sizing estimations in this example design.

SOLUTION ARCHITECTURE EXAMPLE: BUILDING A CLOUD MANAGEMENT PLATFORM WITH VIRTUAL SAN

TABLE 7.8: Cloud management platform virtual machine requirements

VARIABLE	VALUE
Estimated number of virtual machines	40
Average size of virtual disk(s) per VM	120 GB (~60% consumed)
Average memory size per VM	6 GB
Safety margin	20% (to avoid warning alerts) and to allow for snapshots
Maximum Number of Failures to Tolerate capability	1

Management Virtual Machine Storage Requirement Calculation: Based on a Number of Failures to Tolerate Capability of 1, the raw storage requirement for the Virtual SAN datastore is no greater than 12 TB ((40 x 120GB) + (40 x 6GB)) + 20% = (4800GB + 240GB + 100%)= 6048GB ((No. VMs x AvgSize) + (No.VMs x Avg.Mem)) + Growth

CLUSTER ARCHITECTURE

As indicated in the preceding formula, the design requires 12 usable terabytes of storage capacity, to meet the initial storage requirements for cloud management virtual machines. The hosts in the cloud management cluster will have access to only the single Virtual SAN datastore; to ensure it remains fully out of band, no external storage is to be made available.

Figure 7.15 shows the high-level, logical storage configuration for this example design.

To meet these minimum storage requirements, the design will employ four identically specified Cisco UCS C240 M3 rack-mount servers. The selected host server holds 24 × 2.5-inch small form factor (SFF) drives, in a single rack-mount chassis. The 24-drive chassis gives the ability to create at least three fully populated Virtual SAN disk groups (7 + 1) in each host, for future growth and expansion. To make room for the use of all 24 disk slots, the design specifies an industrial-grade SD card device for the ESXi hypervisor installation (redundantly configured in RAID 1, which is supported by the chosen hardware).

To meet the storage requirement of 12 TB of usable datastore capacity, the initial Virtual SAN disk group is to be composed of 7 × 900 GB 10k SAS drives, as illustrated in Figure 7.16. As shown, the flash device to be used for the hybrid Virtual SAN read-caching and write-buffering mechanism is a 400 GB SAS SSD drive. The choice of 400 GB SSDs for the flash tier is based on VMware's recommendation for the flash capacity to be at least 10 percent of the anticipated consumed storage capacity, before the FTT is considered.

NOTE In this design, a redundant SD card configuration is being used as the ESXi install location, which does not allow for the creation of the scratch partition during the initial setup process. As part of the deployment, configure a local .locker directory on the shared datastore for each host, to act as a scratch partition. In addition, be sure to configure each host to log to the remote syslog system.

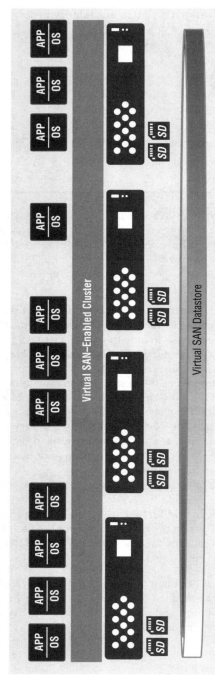
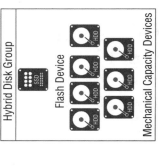
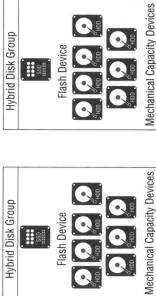
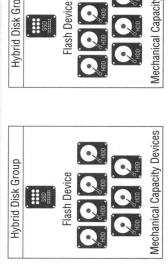

Figure 7.15
Virtual SAN Storage Configuration

FIGURE 7.16
Virtual SAN hybrid disk group configuration

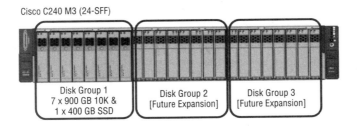

STORAGE CONTROLLER

VMware Virtual SAN supports storage controllers in two modes:

- Pass-through (or JBOD) mode
- RAID 0 mode

An important consideration when choosing a storage controller for VMware Virtual SAN is whether it supports pass-through mode, RAID 0 mode, or both. For the selected Cisco hardware, both modes are supported, as listed in the VMware Virtual SAN Hardware Compatibility Guide.

When Virtual SAN is implemented with pass-through controllers, Virtual SAN accesses the drives directly, and a RAID configuration is not necessary. When Virtual SAN is implemented with controllers that do not support pass-through mode, a virtual RAID 0 drive must be created for each physical disk that Virtual SAN will use.

For this reason, this design will use pass-through mode with Cisco UCS-RAID-9300-8i 12G SAS HBA providing direct connectivity to all the drives in the host. This controller was chosen as it allows for true pass-through mode, in order to present the drives directly to the hypervisor for Virtual SAN to use. In addition, this controller achieves higher performance compared to other controllers, because of its queue depth of 1,024.

CAPACITY CALCULATIONS

This design includes a requirement to configure a minimum of 12 TB of usable storage across the four-node cloud management cluster. The following calculations are based on the requirements dictated by the design.

Raw Cluster Capacity:

```
Hosts x NumDskGrpPerHst x NumDskPerDskGrp x SzHDD = RawClusterCapacity
Management Design Formula: 4 x 1 x 7 x 900 GB = 25,200GB
```

Based on this formula, the planned VSAN Ready Nodes cluster configuration, outlined previously in this example design, is able to provide more than enough usable capacity to meet the sizing requirements set out by the customer.

Storage Policy Configuration

Virtual SAN employs an SPBM framework in which storage policies are used to guarantee that a virtual machine has access to appropriate storage resources, based on its assigned policy. The administrator can define on a per virtual machine basis what type of capabilities the workload receives from within the policy-driven control plane. The policies are then enforced by the hosts, which ensure that the virtual machine receives exactly the capabilities that the administrator has defined.

It is considered good practice to create storage policies based on business needs relating to performance, capacity, and availability. This example uses two policies that will meet the requirements of all the cloud management component workloads. These policies address performance, capacity, and the availability requirements of the targeted management component. For more information on SPBM, please refer to Chapter 4.

For this example design, all of the virtual machines running in the environment have similar requirements. Therefore, the initial implementation is to use just two storage policies, a Performance-Based Specification and Availability-Based Specification, as outlined in Table 7.9.

The following two storage policies should be considered as a starting point in order to meet the initial requirements in this example design. However, a far more granular approach may be required during the life cycle of the environment. For instance, the design could also take into account additional virtual machine requirements, such as storage read, write, quality of service, disposability, redundancy, and performance, to create far more granular policies that might be performance-based, capacity-based, or balanced.

Table 7.9: Example design storage policy specification

Policy	Performance-Based Specification	Availability-Based Specification
Number of disk stripes per object	2	1
Flash-memory read-cache reservation	0%	0%
Number of failures to tolerate	1	1
Forced provisioning	Enabled	Enabled
Object-space reservation	0%	0%

Installing vCenter Server on a Virtual SAN Datastore

A unique consideration to this specific use case is the challenge associated with installing a vCenter Server virtual machine on a management cluster Virtual SAN datastore. This creates a catch-22 situation. The aim is to install vCenter Server on a virtual machine that resides on a Virtual SAN distributed datastore. However, you first need the datastore available to build the virtual machines and install the vCenter Server components. And you need vCenter Server to deploy a Virtual SAN–enabled cluster and configure the distributed datastore. Or do you? Fortunately, you have two relatively simple options to achieve this goal.

The first option requires a spare host in the environment, with sufficient local storage to deploy and configure the vCenter Server virtual machine and other required management components. This can be achieved either by using the traditional vSphere C# client, with a direct connection to the host, or by using the ESXi embedded host client. The vCenter Server, now residing on the spare host, can then be used to deploy and configure the Virtual SAN distributed datastore across the cloud management cluster. After vCenter Server has been migrated from the temporary host's local datastore to the Virtual SAN distributed datastore, the temporary host can be removed and decommissioned, or reconfigured to be joined to the Virtual SAN cluster. For instance, with the example design addressed here, the initial configuration of the Virtual SAN cluster would be configured with just three hosts, with the final host being added later. This first option is illustrated in Figure 7.17.

The second option, referred to as *bootstrap*, takes advantage of ESXCLI to configure a Virtual SAN datastore on a single node (see Figure 7.18). It then involves installing and configuring the vCenter Server virtual machine and other required management components on the single host's Virtual SAN datastore, using the traditional vSphere C# client with a direct connection to the host, or by using the ESXi embedded host client. These host connections allow you to see, but not manage, the single-host-configured Virtual SAN datastore. Finally, the remaining hosts are added to the cluster.

To perform a *bootstrap* installation of vCenter Server, you must first modify the default storage policy, allowing the administrator to sidestep the availability rules that typically govern the deployment of virtual machines on a Virtual SAN datastore.

The process for accomplishing a bootstrap deployment of vCenter and other required management components, to get the Virtual SAN enabled cluster up and running, is outlined in the following section.

Deploy ESXi onto a locally installed SD/USB device as you would typically do, based on the vSphere Installation and Setup Guide. This host will be used as the stand-alone Virtual SAN node. Follow these steps:

1. Change the default Virtual SAN force provisioning storage policy from Disabled to Enabled on the objects shown, by running the following two commands:

   ```
   esxcli vsan policy setdefault -c vdisk -p "((\"hostFailuresToTolerate\" i1)
   (\"forceProvisioning\" i1))"
   esxcli vsan policy setdefault -c vmnamespace -p "((\"hostFailuresToTolerate\" i1)
   (\"forceProvisioning\" i1))"
   ```

2. Confirm that you have successfully modified the Virtual SAN default storage policy from Disabled to Enabled by executing the following command:

   ```
   esxcli vsan policy getdefault
   ```

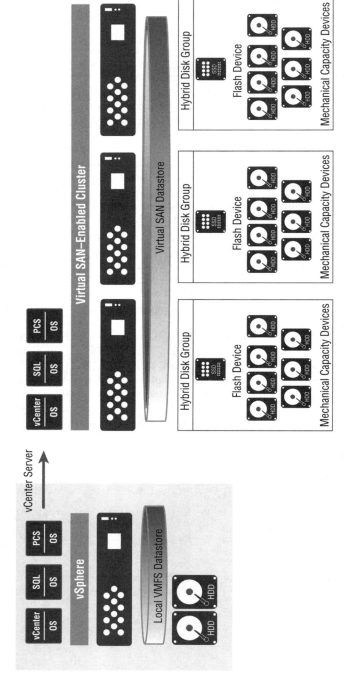

FIGURE 7.17
vCenter Server migration option

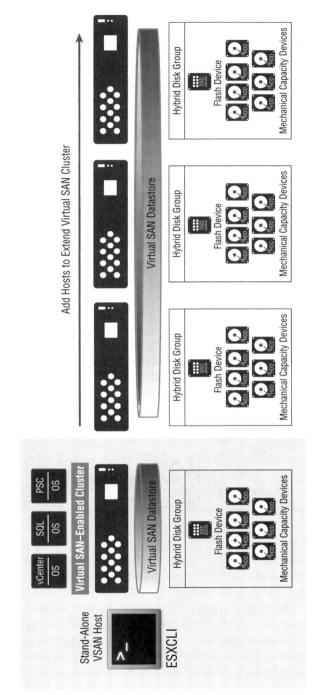

FIGURE 7.18
vCenter Server bootstrap option

Original Default Policy

```
~ # esxcli vsan policy getdefault
Policy Class  Policy Value
------------  ------------
cluster       (("hostFailuresToTolerate" i1))
vdisk         (("hostFailuresToTolerate" i1))
vmnamespace   (("hostFailuresToTolerate" i1))
vmswap        (("hostFailuresToTolerate" i1) ("forceProvisioning" i1))
~ #
```

Modified Default Policy

```
~ # esxcli vsan policy getdefault
Policy Class  Policy Value
------------  ------------
cluster       (("hostFailuresToTolerate" i1))
vdisk         (("hostFailuresToTolerate" i1) ("forceProvisioning" i1))
vmnamespace   (("hostFailuresToTolerate" i1) ("forceProvisioning" i1))
vmswap        (("hostFailuresToTolerate" i1) ("forceProvisioning" i1))
~ #
```

3. Identify which disks within the single host will be used to configure the initial disk group for the Virtual SAN datastore. Identify the required disks with the following command:

   `esxcli storage core device list`

   To further confirm which identifiers refer to SSD or magnetic disks, extend the command as follows, and record the UIDs of the disks you want to use:

   `esxcli storage core device list -d <disk ID>`

```
Is SSD: true
Is Offline: false
Is Perennially Reserved: false
Queue Full Sample Size: 0
Queue Full Threshold: 0
Thin Provisioning Status: unknown
Attached Filters:
VAAI Status: unsupported
Other UIDs: vml.0000000000766d686261313a303a30
```

4. Configure the Virtual SAN cluster and provide it with a UUID, which can be obtained from a site such as www.uuidgenerator.net. For instance, to create the Virtual SAN cluster with a UUID of 1b86d43c-6520-11e4-b116-123b93f75cba, use the following command:

   `esxcli vsan cluster join -u 1b86d43c-6520-11e4-b116-123b93f75cba`

5. View information about the newly created Virtual SAN cluster by executing the following command:

   `esxcli vsan cluster get`

```
~ # esxcli vsan cluster get
Cluster Information
   Enabled: true
   Current Local Time: 2014-11-05T03:09:05Z
   Local Node UUID: 5399f376-7db4-9261-1bde-000c2991b581
   Local Node State: MASTER
   Local Node Health State: HEALTHY
   Sub-Cluster Master UUID: 5399f376-7db4-9261-1bde-000c2991b581
   Sub-Cluster Backup UUID:
   Sub-Cluster UUID: 1b86d43c-6520-11e4-b116-123b93f75cba
   Sub-Cluster Membership Entry Revision: 0
   Sub-Cluster Member UUIDs: 5399f376-7db4-9261-1bde-000c2991b581
   Sub-Cluster Membership UUID: b4945954-f824-b5e5-ecba-000c2991b581
~ #
```

6. Add the SSD and magnetic disks, recorded earlier in this procedure, to the new Virtual SAN cluster. The -s option identifies an SSD disk, and the -d option identifies all magnetic disks. The following example is configured with three magnetic disks (HDD) and the required flash SSD disk:

```
esxcli vsan cluster add -d <HDD 1 ID> -d <HDD 2 ID> -d <HDD 3 ID> -s <SSD ID>
```

7. After the disks are added, you can view the disks contributing to the Virtual SAN by executing the following command:

```
esxcli vsan storage list
```

You can also verify the creation of the Virtual SAN datastore on the stand-alone host by logging in to the host via the traditional Windows C# client with a direct connection, or the ESXi embedded host client with root credentials. You will be able to view the Virtual SAN datastore already mounted on the single ESXi hosts.

After completing the installation of vCenter Server, database and other critical components on the stand-alone Virtual SAN host, you can add the initial host and the remaining hosts to vCenter Server by using the vSphere Web Client. Other outstanding vSphere Web Client tasks include configuring the Virtual SAN VMkernel interfaces and attaching the appropriate storage policy to the management component virtual machines. In addition, remember to revert the default force provisioning storage policy to Disabled.

In environments where the vCenter Server already exists, and you are just creating a new management cluster, you can simply migrate the vCenter components to the new Virtual SAN datastore, using standard operating procedures.

Cloud Management Platform Security Design

As with any new IT implementation, security is critical. Any security vulnerability or risk exposed by this new vSphere cloud management platform could seriously impair the availability of the systems in question. This section outlines a high-level security and operational approach for the example design, which should be applied consistently across the environment during the implementation phase.

No cloud service grants users access to the vSphere infrastructure directly. They use the cloud platform application interfaces for connectivity to services. Where possible, Microsoft Active Directory or other directory services are used to govern this access. Access is granted only when required to perform a specific and authorized job function.

The following security considerations should apply to the dedicated cloud management platform:

- Management components will have banner messages notifying users of monitoring, lack of privacy, and civil and criminal responsibilities for malicious or damaging behavior, regardless of intent.
- Default or well-known accounts within the cloud platform management components will be removed, as they provide an attacker with an advantage in attempting to compromise the device.
- Cloud management components will be configured to require strong passwords, to prevent an attacker from deciphering the password and gaining unauthorized access.
- Cloud platform management ports will be configured with a relatively short connection time-out period, to minimize the risk of session hijacking.
- Antivirus programs, backups, and regular patching will be applied on the systems hosting the cloud platform applications.
- Cloud management component databases will be highly secured. The vSphere and cloud platform databases hold information about system configuration, history, performance statistics, permissions and roles, and could provide an attacker with a wealth of useful information.

CLOUD MANAGEMENT VIRTUAL MACHINE SECURITY

To protect the cloud management component virtual machines from unauthorized access or a malicious attack, it is necessary to set up a baseline hardening configuration. Table 7.10 lists some of the hardening parameters to be configured on each management virtual machine on the cloud management platform.

These standards provide a baseline and are not comprehensive. They may need to be modified based on organization-specific internal processes. For instance, for governmental certification and accreditation programs, an organization may need to certify that the new environment has applied an appropriate risk management methodology to the workloads and the infrastructure supporting them.

NOTE For more guidance on vSphere hardware, refer to the publicly available VMware Hardening Guides, which provide detailed guidance for those who are required to deploy and operate products in secure environments.

TABLE 7.10: Cloud platform virtual machine security baseline

CONFIGURATION	DESCRIPTION

Prevent virtual disk shrinking	By default, administrative and nonadministrative users within a virtual machine have the capability to initiate the shrinking of a virtual disk. A denial of service can occur if this is done repeatedly, as it could cause the disk to become unavailable while the shrinking takes place. Mitigate this risk and the vulnerability inherent with this configuration by ensuring that the shrinking of virtual disks is restricted.
	`isolation.tools.diskWiper.disable=TRUE`
	`isolation.tools.diskShrink.disable=TRUE`

Ensure unauthorized devices are not connected to virtual machines	Virtual devices such as serial and parallel ports, CD/DVD drives, and USB ports. are available in virtual machines but are rarely used.
	Enabled virtual devices such as these provide additional attack vectors that should be secured where possible. Therefore, this example design aims to ensure that unused virtual devices are disabled, to eliminate potential attack risks.
	`floppyX.present=FALSE`
	`serialX.present=FALSE`
	`parallelX.present=FALSE`
	`usb.present=FALSE`
	`ideX:Y.present=FALSE`

Prevent unauthorized removal, connection, and modification of devices	Nonadministrative users within virtual machines can connect and/or disconnect devices such as CD-ROM drives and network adapters. In addition, they can modify device settings and configurations.
	For instance, nonadministrative users can reconnect a disconnected CD-ROM drive and access information left mounted in it, or can disconnect and change network adapter settings, and disrupt service to the virtual machine.
	Therefore, this example design ensures that this functionality is disabled, in order to eliminate these potential attack risks.
	`isolation.device.connectable.disable=TRUE`
	`isolation.device.edit.disable=TRUE`

System Logging

System message logging is to be configured to direct syslog messages to external syslog appliances, provided by vRealize Log Insight. It is recommended to configure logging to the syslog servers from all hardware, including vSphere hosts, UCS servers, and network components. In addition, in order to facilitate an accurate syslog configuration, consistent Network Time Protocol (NTP) parameters across the entire infrastructure are highly recommended. This synchronizes all the clocks in each component in the environment and therefore eases the task of reconciling logs, debugging, and tracing information.

Compute and Storage-Layer Security Hardening

Compute and storage-layer hardening addresses security standards that apply to all Cisco C240 M3 servers. To provide a baseline level of security, in the VSAN Ready compute and storage hardware stack, the configurations shown in Table 7.11 will be addressed during the implementation phase.

Table 7.11: Cisco C-Series hardening baseline

Configuration	Description
Centralized logging	A risk exists that operational or security-related alerts and events can be missed when logs are not centrally managed. The centralization of logs improves administration as well as security investigation capabilities. By configuring host server hardware to use a centralized logging server, aggregate analysis and searches become possible, which in turn provides visibility into events affecting multiple hosts.
NTP is enabled	By not using a centralized NTP source, the risk exists that the correlation and auditing of logs will be difficult and inaccurate. All systems within the design will be configured to use the same time NTP source.
Change the default Cisco IMC password	The Cisco Integrated Management Controller (IMC) provides an interface to manage Cisco UCS rack-mount servers. This is a local password that will be changed to help mitigate against the guessing or cracking of credentials.
Turn on IP blocking	The risk exists that without blocking enabled, an attacker would have unlimited attempts to try to guess a password, or perform a brute-force attack, in order to gain access to the IMC WebGUI interface.

Network-Layer Hardening

To configure a security baseline for the network access layer of the CMP environment, this example design will apply the additional hardening to the Cisco Nexus 5548UP devices described in Table 7.12.

TABLE 7.12: Cisco Nexus 5548UP hardening baseline

CONFIGURATION	DESCRIPTION
Configure remote syslog	A risk exists that operational or security-related alerts and events can be missed when logs are not centrally managed. The centralization of logs improves administration as well as security investigation capabilities. By configuring network devices to use a centralized logging server, aggregate analysis and searches become possible, which in turn provides visibility into events affecting multiple platform components.
Ensure the use of strong passwords	Passwords must be of sufficient length and meet complexity requirements, to not only meet policy and regulatory requirements, but to also help mitigate the guessing or cracking of credentials.
Use banner messages	Banner messages are to be used to notify anyone connecting to the device that authorization is required and activities are monitored. A risk exists that without a banner message explicitly warning the user of monitoring and a lack of privacy, the carrying out of legal follow-through against attackers may be impacted.
Enable SSH	SSH is to be exclusively used to provide a secure and encrypted channel for communication with remote terminals. A risk exists that by not using SSH, insecure sessions will be established and sensitive information may be exposed and compromised.
Configure NTP	All systems should be configured to use the same time source. By not using a centralized, consistent time source, the risk exists that correlation and auditing of logs will be difficult and inaccurate.
Enable IP Source Guard	A risk exists for the impersonation of hosts by the assumption of a legitimate host's IP address. Source IP address filtering on a layer 2 port is provided by IP Source Guard.

Change and Configuration Management

Change and configuration management processes are important when maintaining any computer system. Anything that can impact the SLA becomes in-scope for change and configuration management. The following should be documented to maintain, support, and administer the cloud management platform infrastructure on a daily basis:

- A list of people who can approve changes to each component
- A list of appropriate business decision makers for shared systems
- Approvals for all configuration changes
- Default and other component settings
- Information from monitoring configuration drift
- Procedures for the regular backup of all configuration settings

Management Patch and Update Practices

Scheduled and emergency patches and updates protect systems from security vulnerabilities and provide performance stability. vSphere components should be maintained, and devices patched and updated in line with the organization's internal policies and VMware recommendations. For the patching and update plan, do the following:

- Document the version of each hardware and software component within the environment.
- Document the risk if patches are delayed or not installed in a timely manner.
- Identify ways to reduce the risk when patches cannot be installed.
- Follow change management procedures for documentation and internal approvals.
- Establish regular patch cycles for high- and low-priority patches (for example, weekly and monthly).
- Establish and test processes for emergency out-of-cycle patching.
- Repatch virtual machines if they have been restored from a snapshot prior to a scheduled patch date.

In addition, the following are some basic operational security requirements in an enterprise IT organization or service provider that should be applied to the cloud management platform:

- Keep the system updated with all required patches and service packs.
- Implement standard Windows system protection such as antivirus.
- Limit the number of users who can access the systems.
- Give the allowed users only the permissions that they require for their job role.
- Monitor the Microsoft Windows event logs.
- Use the Microsoft Windows firewall.

Summary

This chapter has provided an architectural example for a cloud management platform cluster exclusively using Virtual SAN storage across the hyper-converged infrastructure platform, bringing host compute, network, and storage into a single building-block solution. Although this design example does not provide a fully comprehensive architecture, it does, to a greater or lesser extent, meet the design requirements outlined at the outset.

Chapter 8

Policy-Driven Storage Design with Virtual Volumes

VMware-based server and desktop virtualization has been extremely successful at addressing a wide range of challenges facing IT organizations since 1998 by maximizing efficiency, automation, and operations. However, in more recent years, a lack of tighter integration between shared storage and virtualized applications has somewhat limited progress and slowed the pace.

To help address this gap between applications and storage systems, VMware introduced a technology that targets shared storage systems and allows them, through far closer integration with the vSphere layer, to meet the challenges of the software-defined storage era. Unlike Virtual SAN, *Virtual Volumes* (*VVOLs*) uses centralized storage systems provided by third-party storage vendors, which allow vSphere storage administrators to unlock a whole range of possibilities to improve application alignment with array-based storage resources.

As addressed in Chapter 2, "Classic Storage Models and Constructs," in classic storage mechanisms, storage arrays predominantly integrate with vSphere at the datastore level, using VMware's VMFS. However, the aim of Virtual Volumes is to address the primary challenges associated with the classic storage model by not requiring large numbers of fixed-size uniform LUNs. In addition, Virtual Volumes help address the lack of granular control over storage policies, which have LUN-centric storage configurations and rigid service levels based on LUN capability, and not virtual machine workload requirements.

These challenges with classic storage systems have often led to massive overprovisioning of storage resources, making it difficult to forecast the right capacity and performance use over time, and also making it difficult to change policies once allocated based on the workload's predefined requirements. With this next-generation shared storage model, vSphere can use Virtual Volumes' new framework for managing virtual machine disks to enable array-based operations at the virtual disk level, facilitating far more advanced integrated capabilities than were previously possible.

For Virtual Volumes, we return to the centralized shared storage model illustrated in Figure 8.1. However, the presentation of the storage, and the relationship between the vSphere components and the storage array, have changed significantly.

Introduced with vSphere 6, Virtual Volumes enables SAN and NAS storage systems to be managed at a virtual machine level, and enables array-based data services and storage array capabilities through VASA 2.0. These features, and this new level of integration with shared storage systems, enables vSphere storage administrators to take a virtual machine–centric and application-centric approach to workload provisioning, allowing them to provide the performance capabilities and data services at the granularity of an individual virtual disk.

The Virtual Volumes mechanism represents a significant change in the way virtual machines interact with shared-storage systems, and can potentially eliminate the need for IT organizations to have to provision and manage large numbers of LUNs or volumes per vSphere cluster. Virtual Volumes introduces two major changes in its approach:

- The implementation of VMware's Storage Policy–Based Management (SPBM) framework to simplify storage management and automate storage provisioning
- Granularity at the virtual machine level, by introducing a one-to-one mapping of virtual machines to storage volumes

Virtual Volumes enables storage operations at the level of virtual machines or virtual disks, through the same SPBM mechanism used by Virtual SAN. However, a virtual volume is not a LUN (in the context of Chapter 2, as a classic storage system logical construct). Likewise, a virtual volume is not a NAS mount point, and neither is it an object store (in the same sense as AWS S3 or others). Virtual Volumes defines a new type of virtual disk container, implemented on shared storage systems, which are independent of the underlying physical subsystem. This virtual disk container becomes the primary unit of virtual machine data management, and therefore eliminates the need for preallocated LUNs or volumes.

Introduction to Virtual Volumes Technology

Virtual Volumes enables storage vendor hardware to use a new set of APIs, called *vSphere APIs for Storage Awareness 2.0* (*VASA 2.0*), which allows direct bidirectional communication between vSphere 6–based systems and the storage array. This provides vSphere storage administrators, and therefore applications and workloads, with significant additional features that can be provided directly to the virtual machines.

The connection between the vSphere host and Virtual Volumes is performed through an abstraction layer known as a *protocol endpoint* (PE), which provides the storage administrator the freedom to use several protocols at once, such as Fibre Channel, iSCSI, or NFS.

To use Virtual Volumes in the data centers, storage administrators have several requirements that must be met. In addition to requiring vSphere, which must be at least at version 6, the array vendor must support Virtual Volumes via the APIs provided by VMware (VASA 2.0). Storage administrators also must meet any other vendor-specific requirements—for instance, in the case of NetApp, the Virtual Storage Console (VSC) is required.

Typically, in a virtual data center, storage arrays provide various performance and availability capabilities that can be aligned with the requirements of different applications. However, until now, it has been up to storage administrators to ensure that this alignment occurs. Without Virtual Volumes, storage capabilities cannot be applied directly to individual virtual machines or virtual disks in a shared storage design, and can be applied at only a datastore or LUN level, which typically contains numerous virtual machines. However, with the addition of Virtual Volumes functionality into the storage array, VMware vSphere can use VASA 2.0, providing vSphere operational teams the ability to assign storage profiles on a per virtual machine or virtual disk basis, and select the specific storage capabilities required explicitly for that application.

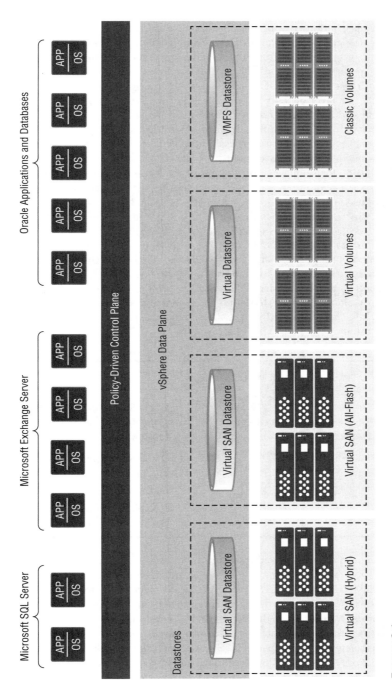

FIGURE 8.1
Next-generation storage model

With the classic storage approach, in which a SCSI LUN is presented to the host or cluster, the underlying storage system has little or no knowledge of the hypervisor, filesystem, guest operating system, or application. In this type of design, it is left up to the hypervisor and vCenter Server to map objects and files, such as VMDKs, to their corresponding extents, pages, and logical block addresses (LBAs) that are understood by the storage system. In addition, in the case of a NAS solution, a layer of abstraction is placed over the underlying block storage to handle file management and the associated file-to-LBA mapping activity.

However, with Virtual Volumes, instead of simply presenting a LUN, the vSphere host can now manage data placement and access on the storage system, and can also gain insight into which LBAs correspond to various entities, such as a VMDK, VMX, log, clone, swap, or other object. With this additional insight, storage systems can provide far more granular and native functions, such as cloning, replication, and snapshots at the virtual machine or virtual disk level, as opposed to simply working on a LUN-by-LUN basis.

This architecture also allows Virtual Volumes to move more intelligence and operations from the hypervisor down into the storage system. With Virtual Volumes, instead of the storage system simply presenting a SCSI LUN or NFS mount point and providing limited (VASA 1.0) or no visibility into the underlying storage array, the storage profile configures a set of rules that can define the service level assigned to individual virtual machines, based on the different capabilities published by the storage array. This storage profile can then be selected during the virtual machine provisioning process to define placement and available features for the workload. This framework helps to address administrative challenges related to vSphere storage and infrastructure, as set out in Table 8.1.

TABLE 8.1:   vSphere operational priorities

vSphere Storage Administrators' Primary Operational Issues	vSphere Infrastructure Administrators' Primary Operational Issues
Capacity management	On-demand storage provisioning for workloads
Access control	Availability of appropriate data services at the virtual machine level
Achieving application SLA requirements	
Data security and integrity	SLA compliance checks, throughout the virtual machine life cycle

The level of integration into the storage array, offered through the SPBM mechanism, also makes common tasks associated with storage management, such as LUN provisioning, less of an administrative overhead. By using the SPBM mechanism to provision virtual volumes that meet the required service levels during the virtual machine provisioning process, storage administrators should see a reduction in their operational overhead. Administrators won't have to pre-provision LUNs and datastores, so costs can be reduced by eliminating these tasks. These concepts also extend to NAS-based NFS storage access via appropriate protocol endpoints.

Figure 8.2 provides a high-level comparison between the traditional VMware VMFS LUN-based datastore architecture and a Virtual Volumes storage architecture.

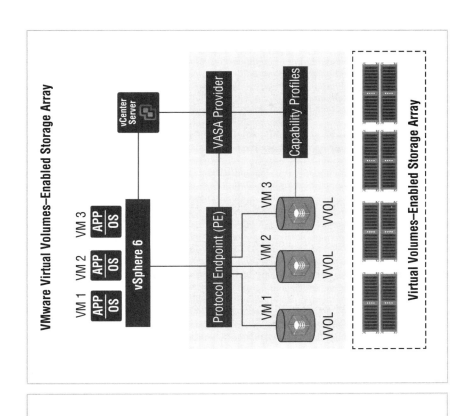

FIGURE 8.2
Comparing the classic storage architecture with Virtual Volumes

In addition, Virtual Volumes enables the underlying storage system to become more closely integrated with the vSphere host and its associated management tools, including supported cloud management platforms (CMPs) and classes or categories of service for performance, availability, capacity, or economics. This architectural approach to a shared storage model also enables third-party storage vendors to improve the storage-capacity consumption for IT organizations and service providers by doing the following:

- Moving from a per LUN to a per virtual machine provisioning-based service level, which can be applied to either virtual machines or individual virtual disks.

- Enabling storage arrays to present their unique capabilities, via the SPBM mechanism. These may include features such as availability, deduplication, compression, or encryption.

- Providing storage arrays the control over data, at the level of individual virtual disks.

- Simplifying the virtual machine provisioning process, and providing virtual disk access to the storage array through preconfigured protocol endpoints.

Virtual Volumes Component Technology Architecture

The Virtual Volumes architecture is made up of new storage constructs, concepts, and terminology. This section presents each, so that you can become more familiar with the technology, better understand the Virtual Volumes storage architecture, and see how the components interact with one another. Figure 8.3 illustrates, at a high level, each of the technology components to be discussed next.

Virtual Volumes Object Architecture

As previously noted, Virtual Volumes, like Virtual SAN, represents a different type of storage mechanism. Virtual machine objects are stored natively in containers on the storage system.

The virtual volume itself is the container, which encapsulates the virtual machine files, virtual disks, and their derivatives. A single virtual machine consists of several VVOLs, including one for configuration data, one for each of the virtual disk files, one for the virtual machine's swap file, and additional ones for the memory and data from any virtual machine snapshots that currently exist. The containers, which store the Virtual Volumes objects, are then mapped to their derivative source files, such as the VMDK or virtual machine swap file.

Virtual volumes—which are automatically created for virtual machine operations such as creation, powering on, cloning, and snapshotting—have five types of objects. Each maps to its corresponding virtual machine file, as described in Table 8.2.

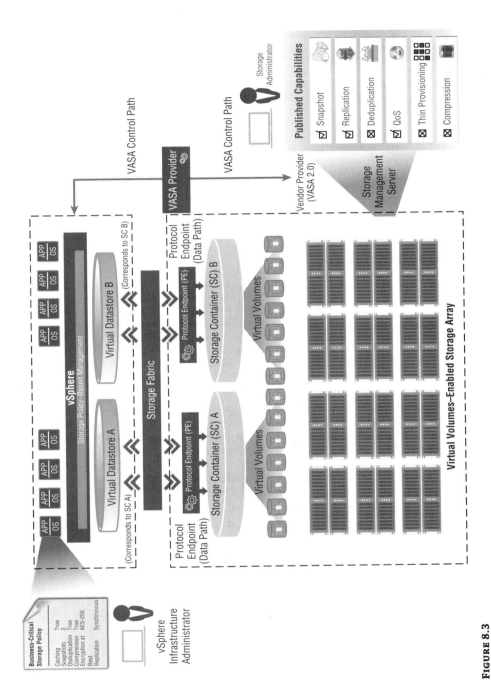

FIGURE 8.3
vSphere Virtual Volumes component architecture

TABLE 8.2: Virtual Volumes object types

OBJECT TYPE	MAPPED DERIVATIVE
Config-VVOL	VM home namespace container, which stores configuration-related files, such as the logs and VMX file
Data-VVOL	Maps to virtual disks (VMDKs)
Memory-VVOL	Snapshot files
Swap-VVOL	Virtual machine memory swap files
Other-VVOL	Generic object type, defined for hardware or software vendor-specific solutions

Management Plane

The Virtual Volumes *management plane* is defined as the management and control path for communication, which takes place between the vSphere components and storage array.

VASA 2.0 Specification

Virtual volumes are built on VMware technologies first introduced in vSphere 5, including vStorage API for Array Integration (VAAI), and VASA. As addressed in Chapter 2, with VAAI, the vSphere host can off-load some common functions to storage systems that support features, such as copy, clone, and zeroing. In addition, we addressed the VASA mechanism, which is used to provide visibility, insight, and awareness between the hypervisor and its associated vCenter management, as well as the storage system. This functionality allows a storage array from any vendor to be able to publish to vSphere its various capabilities, such as for storage capacity, availability, performance, configuration, or other features.

Virtual Volumes is part of the VASA 2.0 specification, which defines a newer, more integrated standard for abstracting storage at the virtual machine level. VASA 2.0 includes a much wider range of APIs, allowing vSphere to query storage capabilities, which can then be used by the SPBM mechanism to make decisions about virtual disk placement.

The VASA 2.0 specification, which was introduced with vSphere 6, uses a bidirectional communication mechanism, whereby the hypervisor and management tools can tell the storage array of its configuration and activities. In addition, the shared storage array can set out how virtual volumes can be placed, to provide the required capabilities to each virtual machine through its own set of technology features. With this new level of integrated storage technologies, each virtual machine is provisioned with a separate set of virtual volumes within the storage system's *storage container*.

VASA Provider

The *VASA provider*, also sometimes referred to as the *storage provider*, is presented by the storage vendor's system or appliance, using the VASA APIs. The VASA provider is a two-way communication mechanism that uses VASA for reporting information, configuration, and other insight, up to vSphere hosts, vCenter Server, or other management tools in the control path. In addition, the VASA provider receives VASA information from the vSphere stack on how to configure the storage system components, such as the storage containers.

As illustrated in Figure 8.4, the VASA provider passes this information about the storage topology, its capabilities, and current status, to the vCenter Server and vSphere hosts.

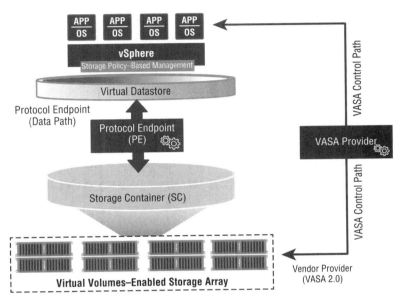

FIGURE 8.4
VASA control path

Also consider the VASA provider as the out-of-band management interface between the storage system supporting and presenting virtual volumes and the vSphere management components, including the vSphere hosts, vCenter Server, and other management tools. The VASA provider typically resides within the storage array's firmware, although it can also be external, on a physical server or virtual appliance, as defined by the storage array's hardware or software vendor.

Data Plane

The *data plane* is made up of several constructs within the Virtual Volumes architecture. These include storage containers, protocol endpoints, and binding operations. We address each of these technologies and mechanisms individually over the following few sections of this chapter.

Storage Container

Instead of using LUNs carved from a pool of raw storage on the array, Virtual Volumes uses *storage containers* (*SCs*), a logical representation of a pool of raw storage capacity in which created virtual volumes can reside. The SC is created on the shared array, using the vCenter Server Web

Client, which works alongside the array's VASA provider to provision the raw disks into a single logical entity. The storage container's capabilities can then be advertised via the VASA provider.

Even though the storage container cannot be thought of as a LUN, with a VASA provider and *protocol endpoint* in place, the storage container becomes visible to the vSphere hosts as a virtual volume datastore, referred to as a *virtual datastore*. The storage container must have a one-to-one relationship with the virtual datastore, similar to that of the distributed Virtual SAN datastore, addressed previously in Chapter 4, "Policy-Driven Storage Design with Virtual SAN." Once created, the virtual datastore can be presented to the required hosts or cluster, at which time the provisioning of virtual machines can begin, allowing virtual volumes to be created within the storage container. This entity must be created as a datastore within vSphere in order to allow all the traditional features of vSphere, such as High Availability (HA), to interact with the new virtual datastore construct.

As noted previously, a storage container cannot be thought of as a LUN, but instead must be considered as a new entity type for the placement of Virtual Volumes–backed virtual machines. Table 8.3 highlights the distinct differences between the storage container entity and the traditional classic concept of a LUN.

TABLE 8.3: Comparison of storage container and classic Volumes/LUNs

STORAGE CONTAINERS	CLASSIC VOLUMES/LUN
Size can be based on the entire array's capacity.	LUNs have fixed sizes and are typically provisioned in large numbers.
The maximum number of storage containers depends only on the array's vendor-defined capabilities.	LUNs require a VMFS filesystem.
The size of the storage container can be extended, just like any other pool of raw storage.	Can apply only a homogeneous capability on all virtual machine disks (VMDKs) provisioned in that LUN.
Storage containers can distinguish between heterogeneous capabilities for different virtual machines (Virtual Volumes) provisioned within that entity.	LUNs are managed by in-band SCSI filesystem commands.

The technologies used by the storage manufacturer define the features that can be provided by the array's storage containers, which in turn, dictates the storage capabilities that are presented to vSphere. For instance, these capabilities may include features such as level of storage performance (tier 1, tier 2, or tier 3), backups, snapshots, deduplication, and encryption at rest. As a result, we can either use storage containers as logical partitions to apply storage needs and requirements, or provide multiple capabilities into a single storage container.

In addition, there is no direct one-to-one mapping between a storage container and a protocol endpoint. As we will address in the next section, a protocol endpoint can manage multiple storage containers, or multiple protocol endpoints can be used to manage a single storage container.

Furthermore, although the implementation of storage containers will vary from storage array vendor to storage array vendor, there is no requirement to stick to a single, uniform set of capabilities for a Virtual Volume storage container. For instance, a storage container does not need to be built exclusively on SAS disks or flash devices, but may instead be formed from a heterogeneous pool of physical storage resources. A storage container is, in effect, an aggregate of a storage pool, containing raw physical disks, and includes capabilities which go to make up that pool, such as the specific RAID level configured, or mix of physical disks, illustrated in Figure 8.5. However, how the storage container's capabilities (such as snapshots, replication, deduplication, or encryption at rest) are applied to the vSphere platform varies from vendor to vendor.

FIGURE 8.5
Storage container architecture

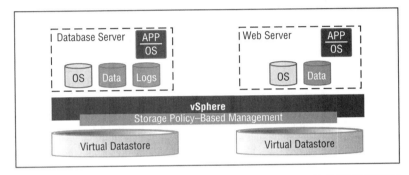

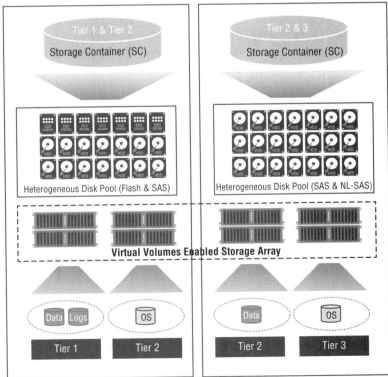

Finally, storage containers are not visible via the in-band data path. The VASA provider manages storage containers and reports their existence to the vCenter Server and vSphere hosts via the out-of-band control path, through the VASA provider's URL. The storage container discovery and provisioning process is illustrated in Figure 8.6.

Protocol Endpoints

The *protocol endpoint* represents the I/O access point for a virtual volume. Protocol endpoints are like LUNs or mount points, as they can be mounted or discovered by multiple hosts. In the Virtual Volumes architecture, vSphere hosts do not have direct access to virtual volumes on the storage array. Instead, hosts use the protocol endpoint logical I/O proxy to communicate with virtual volumes, by replacing the traditional LUNs and mount points, allowing the protocol endpoint to serve as the data path between hosts and the virtual machines' respective VVOLs.

More specifically, all paths to the virtual volumes are administered by protocol endpoints. These protocol endpoints provide access to the specific virtual machine objects, such as the VMDK or VMX file, that are stored within their virtual volumes. The protocol endpoint acts like a pass-through device, handling all I/O requests from the virtual machine to multiple VVOLs, by using protocol endpoint multipathing. A single or multiple protocol endpoint can serve a vSphere cluster of hosts and a storage array. In short, a protocol endpoint and storage container combined create the virtual datastore within the vSphere storage stack.

As illustrated in Figure 8.7, the protocol endpoint enables the vSphere hypervisor to see and access VMDKs, as well as other objects stored in virtual volumes.

The operation of the protocol endpoint depends on the storage protocol being used. NFS version 3, iSCSI, Fibre Channel, and FCoE are all fully supported. For NFS storage, the protocol endpoint is simply an NFS mount point, and the virtual disks are files beneath that mount point. However, for SAN-based block protocols, the protocol endpoint is a proxy LUN, which acts as a multiplexor that allows each host to address thousands of underlying VVOLs, each corresponding to a virtual disk or other object type, and with a unique identifier. Storage arrays that support multiple storage I/O paths or storage protocols such as Fibre Channel, iSCSI, and NFS, can have multiple protocol endpoints that point to the same storage container. The Virtual Volumes protocol endpoint construct can be discovered and rediscovered, using the regular LUN scanning commands traditionally used in vSphere.

It's important for the architect to appreciate that for storage I/O operations, the protocol endpoint is simply a pass-through mechanism that does not store the VMDK or other vSphere virtual machine component data.

Virtual Volumes protocol endpoints use multipathing policies in the same way as they are used in classic shared storage environments, with the same policy options being available to a protocol endpoint as a traditional LUN. Therefore, if a path failover occurs, it is applicable to all virtual volumes bound on that protocol endpoint, with the native ESXi multipathing plug-in being modified to ensure that it does not treat an internal Virtual Volumes error condition as a path failure. In addition, vSphere will ensure that older multipathing plug-ins cannot claim protocol endpoint virtual volumes. The protocol endpoint discovery and provisioning process is illustrated in Figure 8.8.

FIGURE 8.6
Storage container provisioning process

FIGURE 8.7
Protocol endpoint architecture

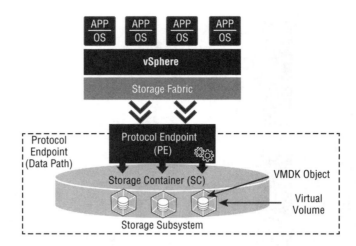

When a virtual volume is created, it is not immediately accessible for I/O. To access virtual volumes, vSphere first needs to issue a *bind* operation to the VASA provider, which creates an I/O access point for a virtual volume on a protocol endpoint chosen by the VASA provider.

Binding Operations

Binding operations, illustrated in Figure 8.9, coordinate data-path mechanisms that occur between VASA providers and vSphere hosts to provide access to the virtual volume. Bindings are created by the VASA provider in coordination with the array, by request from a vSphere host. This binding is then used between the host and the array, for access to the virtual volume.

There are three types of binding mechanisms that perform different operations within the virtual volume's life cycle:

Binding   Allows the array to create I/O channels for a virtual volume.

Unbind   Destroys the I/O channel between the virtual volume and a given vSphere host to the protocol endpoint. In an unbind operation, the VASA provider deletes binding at the request of the host. Typically, the virtual machine must be powered off to perform an unbind operation.

Rebind   Provides the ability to change the I/O channel (protocol endpoint) for a given virtual volume, such as moving a virtual volume to a different protocol endpoint. The VASA provider may also initiate a rebind operation if a storage migration (svMotion) event occurs across storage containers, or for load-balancing purposes across multiple protocol endpoints. A VASA provider may also choose to rebind a VVOL to a protocol endpoint for several other reasons, to provide vendor-specific capabilities.

In the binding mechanism workflow, the vSphere hosts reference a virtual volume via a particular binding ID, and make a call to the VASA provider by using the `bindVirtualVolume` API with the given `vvolID`, which is the VVOL ID referencing a particular VVOL. The VASA provider will then return with a protocol endpoint ID, as well as a secondary ID. In the case of SCSI, this is a second-level LUN ID, whereas in the case of NAS, this is an object for I/O. These values make up the `objectID`.

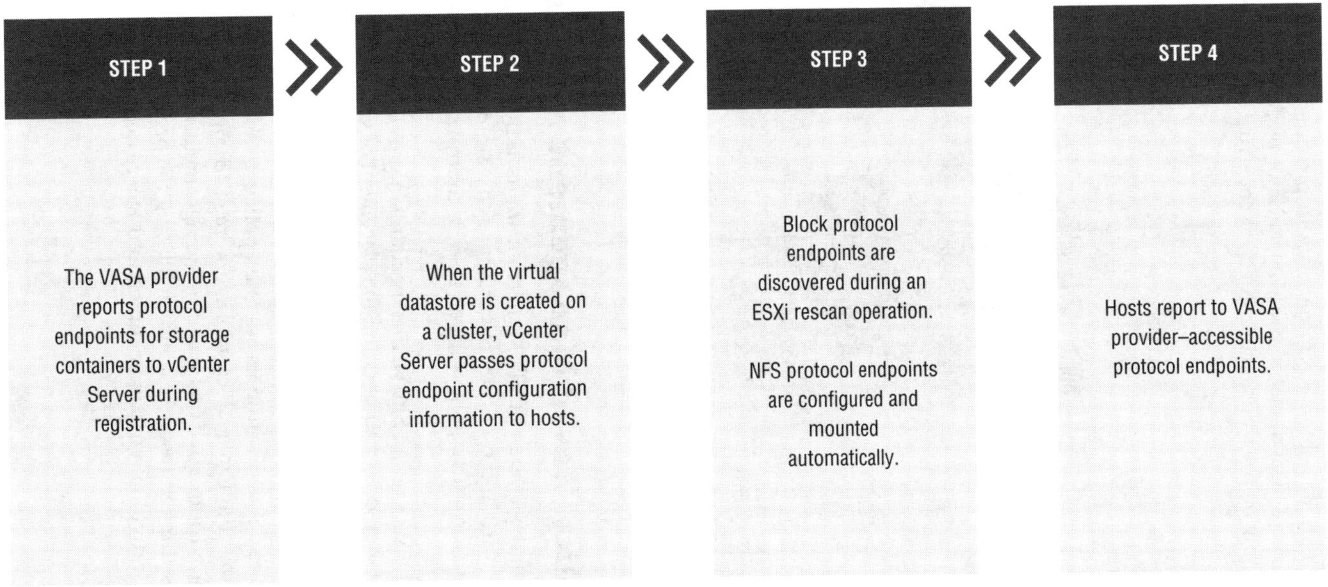

STEP 1	»	STEP 2	»	STEP 3	»	STEP 4
The VASA provider reports protocol endpoints for storage containers to vCenter Server during registration.		When the virtual datastore is created on a cluster, vCenter Server passes protocol endpoint configuration information to hosts.		Block protocol endpoints are discovered during an ESXi rescan operation. NFS protocol endpoints are configured and mounted automatically.		Hosts report to VASA provider-accessible protocol endpoints.

Figure 8.8
Protocol endpoint provisioning process

Once this information is obtained from the VASA provider, in the case of a SCSI block device, it will be used to open a SCSI device in the SCSI stack, or perform a File Open command on the NFS client, depending on the transport protocol, to construct the data path on the vSphere host to the virtual volume.

FIGURE 8.9
Binding operations

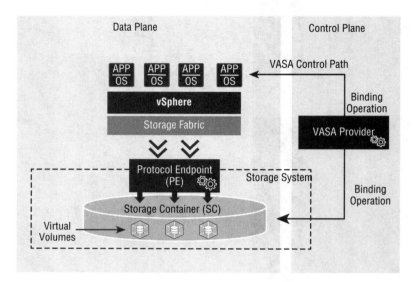

Storage Policy–Based Management with Virtual Volumes

Virtual Volumes employs the same SPBM mechanism used by Virtual SAN, and therefore takes us even further forward in the journey toward our software-defined storage vision. SPBM provides a common platform for a policy-driven storage control plane across the entire vSphere layer (Figure 8.10).

As with Virtual SAN's use of SPBM, this mechanism provides vSphere administrators with a quick and simple method for automating the placement of virtual machines, and offers the additional granularity sought by application owners by providing workload capabilities at the individual disk level. With SPBM, when the virtual machine is provisioned, a storage policy is used to automate the selection of the most appropriate virtual datastore that fulfills the workload requirements as defined within its assigned policy.

This simplified mechanism enables vSphere storage administrators to overcome several challenges related to provisioning virtual machine storage, including capacity planning, managing diverse SLAs across limited resources, and facilitating the provisioning of appropriately defined storage policies to individual virtual machines, and where necessary, on a virtual disk-by-disk basis (Figure 8.11).

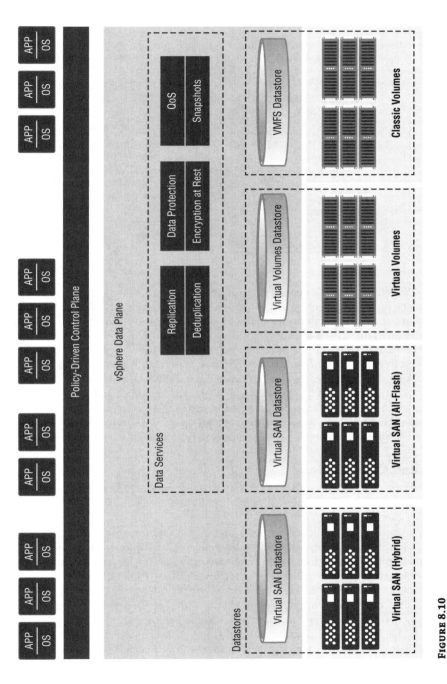

Figure 8.10
Common management platform for policy-driven storage

Figure 8.11
Storage policy example

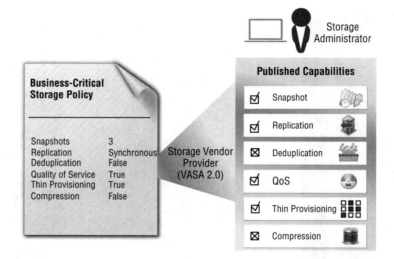

The SPBM framework also uses a programmable API, allowing the consumption of the control plane and storage policy mechanism via scripting or cloud automation tools. It is this feature that provides the framework for a self-service consumption model and storage administration and operations from a CMP.

In addition to optimizing the provisioning process within a vSphere environment, SPBM can be used by other VMware or third-party tools to automate storage management. VMware vRealize Automation, the vSphere API, PowerCLI, and VMware Integrated Open Stack (VIO) are all applications that can use the vSphere SPBM API to automate storage management operations within the software-defined storage infrastructure, as illustrated in Figure 8.12. These cloud automation concepts, as well as delivering storage as a service (STaaS), are examined more closely in Chapter 9, "Delivering a Storage-as-a-Service Design."

Published Capabilities

The list of storage capabilities that a specific storage array can deliver are system-defined and vendor specific and as we have already highlighted, are published by the storage array's VASA provider service, via VASA 2.0.

The VASA provider publishes storage system capability information from the array to vSphere, in the form of the specific attributes delivered by the available physical storage. These can include capabilities such as RAID level, thin provisioning, drive type, snapshots, encryption at rest, compression, deduplication, and many more. These capabilities are unique to a specific storage system, and once published, can be used by the SPBM-based storage policy mechanism. These policies, once associated with a virtual machine, are used to define the required levels of performance, capacity, availability, or other published storage services necessary for that application workload or its individual virtual disks.

In a Virtual Volumes storage policy, you can reference two types of storage capabilities:

- Vendor-specific storage capabilities
- User-defined metadata tags

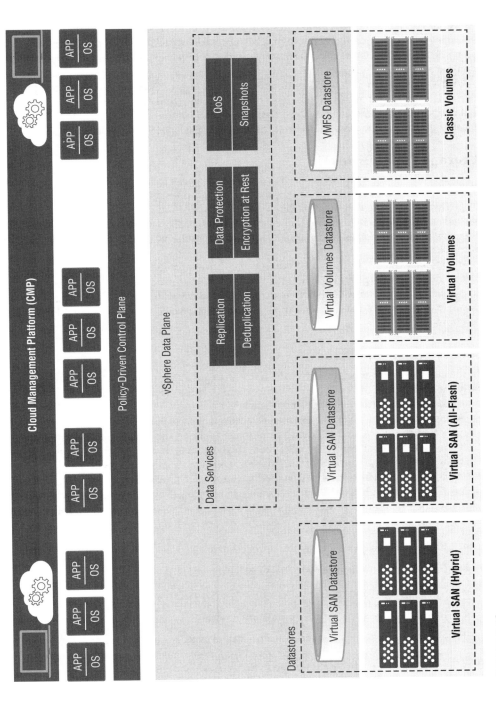

FIGURE 8.12
Storage policy–driven cloud platform

Vendor-specific storage capabilities outline the quality of service and other features that a storage system can deliver. vSphere may also provide capabilities that are common to all datastores. However, the storage capabilities provided and published by a storage system are system-defined and vendor specific.

A user-defined metadata tag is an optional tag that is defined manually by a vSphere storage administrator and that can then be associated with a virtual datastore. The user-defined metadata tags allow storage policies to be created with additional criteria, in conjunction with, and further extending the capabilities that the storage array's VASA provider publishes.

Storage Capabilities

The VASA 2.0 framework, which dictates the level of integration and the feature sets that are supported between storage arrays and virtual volumes, will continue to be developed by VMware. This development of the VASA 2.0 specification will be accompanied by individual storage vendors enabling these new capabilities and additional functionalities within their hardware platforms, and publishing them out to the VMware stack through the VASA provider.

The following will highlight common capabilities, provided by storage vendors and published via the VASA provider, to enable Virtual Volumes functionality. However, it is impossible to provide a comprehensive list of vendor-supported capabilities and their specific features, because these are system-defined and highly vendor specific, as well as being developed on an ongoing basis.

Array-Based Snapshots

In the classic storage model with VMFS datastores, array-based snapshots could be taken at only the entire LUN level. Therefore, restoring individual virtual machines required additional operational procedures to be put into place.

A vSphere Virtual Volumes–enabled environment can support two types of virtual machine snapshot mechanisms: managed and unmanaged. The traditional vSphere virtual machine snapshot, which is performed by vCenter at the host level, is referred to as a *managed* snapshot. It's managed entirely by vSphere components. *Unmanaged* snapshots, with an enabled storage array capability, can be entirely off-loaded to the storage hardware.

With a supported Virtual Volumes–enabled array, *unmanaged* snapshots can be taken on individual virtual machines, providing the ability to off-load vSphere snapshots entirely onto the array. Other benefits provided by this capability include the following:

- Facilitating much simpler restore procedures
- Providing a quiesced copy for backup or archival purposes
- Creating a test and rollback environment for applications
- Instantly provisioning new application images

This feature also allows vSphere administrators to fully and transparently manage these array-based snapshots through the vSphere Web Client, making them as simple to perform as if they were using the traditional managed vSphere snapshot mechanism.

Array-Based Thin Provisioning

When using classic VMFS datastores, the vSphere storage administrator can elect to perform thin provisioning through vSphere, or on the storage array, or with due consideration, can use both mechanisms together. However, on a Virtual Volumes–enabled storage system, array-based thin provisioning replaces vSphere thin provisioning, so this design factor no longer has to be addressed by the architect. All virtual machines deployed through vSphere, which are configured for thin provisioning, will automatically be thinly provisioned on the virtual datastore at the array level.

Space Reclamation

On a VMFS datastore, the storage array has no means of discovering when a virtual machine has been deleted or migrated off a LUN. Space reclamation in this type of environment is a long, resource-intensive, and often manual process, initiated by storage administrators using command-line tools. With Virtual Volumes, space can be reclaimed immediately, as the storage array will be completely aware of all virtual machine operations taking place. As a result, the space allocated to a deleted virtual machine can be reclaimed instantaneously, automatically, and without impact to the environment.

Deduplication

A VMFS datastore is unable to provide any sort of virtual machine–based deduplication of data, although some classic storage arrays might provide it at the array-block level. However, with an enabled Virtual Volumes capability, which supports array-side deduplication, it is possible to apply deduplication to specific running virtual machine workloads. This provides the ability to target specific workloads that would benefit the most from this technology, without impacting other higher-priority applications that you might want to protect from its associated performance overhead.

Storage Capabilities Summary

The preceding features are just a small number of example capabilities, which are of course subject to support by array vendors. By performing these capabilities on the array, or by off-loading them from the vSphere compute host to the storage hardware, they can be performed much faster and more efficiently. This approach to storage operations helps increase host resources and simplifies both vSphere and storage-associated tasks and procedures.

Benefits of Designing for Virtual Volumes

You've had an overview of the features, functionalities, and capabilities associated with Virtual Volumes that might typically be used in a design. Now we will finalize this discussion by addressing the key design factors and use cases associated with Virtual Volumes. You'll learn why they might lead to a recommendation for a specific environment over the use of traditional VMFS block-level or NFS datastores.

Enhanced Performance

With Virtual Volumes allowing vSphere to off-load far more tasks to the storage array than were previously supported through VAAI, workloads can perform faster and more efficiently with direct access to the storage layer. This tightly integrated framework not only allows storage arrays to complete common tasks off-loaded from vSphere, such as snapshots and thin provisioning, but also enables vendors to publish unique capabilities to vSphere, providing hardware-vendor-specific storage technologies that can be integrated into the vSphere software stack.

Greater Application Control

Prior to the availability of Virtual Volumes, with classic VMFS datastores the storage array had little or no visibility into the virtual machine disk. As a result, storage array features, such as QoS and snapshots, could be performed at only the LUN level. However, with Virtual Volumes, storage array capabilities, which are supported with VVOLs, can now be implemented on the individual virtual machine or individual virtual disk, through the SPBM framework, to provide a significantly higher level of control and efficiency of shared storage. Applications can benefit from this granular level of interaction with business-critical and cost-sensitive storage resources.

Operational Simplification

The SPBM framework simplifies common storage management tasks through its use of storage policies. By using a common set of storage policies to automatically provision virtual volumes and assign appropriate capabilities during the virtual machine creation process, no pre-provisioning of datastores is required by storage administrators, reducing operational overhead and costs by eliminating unnecessary tasks.

Reduced Wasted Capacity

Virtual volumes use only the space required by the virtual machines on the storage array. Therefore, there is no longer a requirement to allocate large chunks of capacity to LUNs, typically in predefined sizes. In addition, the thin-provisioning and deduplication capabilities can, if available, further reduce the amount of disk capacity consumed by virtual machines. Other array-enabled capabilities, such as automatic space reclamation when a virtual machine is deleted or moved, also ensure that no space is wasted and the storage array maintains as small a footprint as possible, further reducing costs.

Virtual Volumes Key Design Requirements

Virtual Volumes requires several key prerequisites to be addressed as part of a vSphere storage design. To include a Virtual Volumes storage architecture as part of a design, the following key requirements must be met:

- The vSphere core platform must be at least at version 6.
- The storage vendor must also support Virtual Volumes, through the APIs provided by VMware in VASA 2.0.
- The design might also have to meet other storage-vendor-specific requirements, such as NetApp, which requires its VSC to be included as part of the storage implementation.

vSphere Storage Feature Interoperability

Virtual Volumes, like Virtual SAN, is fully integrated and fully interoperable with most vSphere features. However, again like Virtual SAN, the Storage Distributed Resource Scheduler (SDRS) becomes a redundant feature because of the storage provisioning mechanism. Key supported features include, linked clones, vMotion, host profiles, SvMotion (across different virtual datastores or arrays), Distributed Resource Scheduler (DRS), Enhanced vMotion, the vSphere SDK and vCenter APIs, VDPA and VDP, View, the View Storage Accelerator, vRealize Operations, and vRealize Automation.

VAAI and Virtual Volumes

In Chapter 3, "Fabric Connectivity and Storage I/O Architecture," we addressed the concept of VAAI, or the vStorage API for Array Integration. The VAAI primitives enable certain storage tasks, such as block zeroing, to be off-loaded from the vSphere platform to the storage-array hardware. The VAAI primitives, along with their Virtual Volumes interoperability status, are addressed here.

Just as when addressing the classic storage model, the individual array vendors must also support the specific primitive for use with Virtual Volumes in order for it to work. For instance, atomic test-and-set (ATS) can be supported, as there is still a requirement to provide clustered filesystem semantics and locking for the configuration of the Virtual Volumes VM Home Object type. Support for this primitive is detected based on ATS support for a protocol endpoint LUN, to which VVOLs are bound.

XCOPY (cloning and linked clones) can also be supported, as vSphere has the ability, through API calls, to instruct the array to clone a Virtual Volumes object, on the behalf of vSphere.

However, block zeroing, whose primary purpose is to initialize thick disks provisioned on VMFS datastores, is not used for Virtual Volumes. The reason is that in a Virtual Volume VMFS-based object (such as the VM Home object, which has a default size of 4 GB for housing several small files including disk descriptors, virtual machine configuration files, stats, and log data), the off-loading of the block zeroing of such a small amount of capacity would bring little value.

Other storage array primitives may also be supported, but it is important to establish these with the individual storage array vendor. These technologies require vendor support to be interoperable with both classic VMFS-based storage and Virtual Volumes.

Virtual Volumes Summary

With Virtual Volumes and Virtual SAN, we have truly entered the era of software-defined storage, and as such, can take full advantage of the vSphere SPBM mechanism.

Virtual Volumes can significantly simplify vSphere operations through storage policies, reducing storage complexity and improving the overall efficiency of storage resource management through the offloading of common vSphere tasks to the storage array. The key design factor to be considered by the architect, relating to the implementation of a Virtual Volumes–enabled storage environment, is that the capabilities offered by the array, which are published to vSphere components, will vary from storage vendor to storage vendor. Therefore, the mileage you gain is dependent on the vendor chosen and their implementation of this technology in their storage hardware.

Both Virtual SAN and Virtual Volumes are policy-driven solutions that allow vSphere storage administrators to configure the requirements for a virtual machine or individual virtual disk, in terms of policies or storage service levels that can then be applied to the individual objects in question. Once assigned, the SPBM framework will take these requirements and ensure that the assigned virtual machines, and underlying virtual disks that are hosted on the Virtual SAN or Virtual Volumes datastore, are always in compliance with the configured policy capabilities.

Most Virtual SAN and Virtual Volumes daily operational tasks can be performed through the simple management interface offered by the vSphere Web Client, and are tightly integrated with existing vSphere technologies such as replication, backups, vMotion, DRS, and HA. However, in the following chapter, we will take this journey to its next logical step, and address the provisioning and management of storage resources through the use of a cloud automation and management platform. The aim is to further reduce the operational overhead associated with daily storage tasks.

Chapter 9

Delivering a Storage-as-a-Service Design

One of the key goals for many IT organizations and cloud service providers is to deliver self-service access to resources via a cloud management platform (CMP). vSphere storage administrators aim to automate storage infrastructure management in order to accelerate provisioning times, and to simplify operations in order to improve efficiency. In addition, for many industries, a key requirement of providing storage as a service (STaaS) is to remove the barriers that developers and end users encounter when requiring resources, by making the underlying storage infrastructure all but invisible during the provisioning process.

When requesting additional storage in the data center, most organizations must address various questions and considerations before those storage resources can be provisioned. Some of these queries might need to be addressed by various teams within the IT organization, and interaction between all involved may be required, as illustrated in Figure 9.1. Because of this complex and prolonged process, deploying new storage resources may involve long delays in provisioning times. This in turn can lead to bottlenecks in service provisioning, affecting service-level agreements (SLAs) provided to end customers.

If you break down this process, you can quickly understand why the provisioning of storage resources into an environment might not be as simple as first thought. As outlined in Figure 9.1, in this manual scenario, when the cloud services operator requests new storage resources, several questions must be answered. After that happens, several configuration tasks are required on various hardware components, such as the switches and array, which might be performed by different teams. This makes the challenge of provisioning physical storage resources into such an environment a lengthy and protracted one, as illustrated in Figure 9.2.

The primary aim of a STaaS cloud platform offering is to automate a substantial part of this process, and reduce the requirements for significant interaction between administrators in the IT organization. In addition, by automating daily tasks with software, operational teams are relieved of performing repetitive manual duties. From this, additional benefits can be realized, such as providing shorter delivery-based SLAs and improving quality, while at the same time standardizing storage operations and reporting mechanisms, as shown in Figure 9.3.

Furthermore, with a storage-vendor plug-in for vRealize Orchestrator, a cloud services architect can deliver full storage life-cycle management into the software-defined data center. Instead of opening a storage-provisioning request, answering the required questions and waiting, a STaaS design can integrate the process of provisioning storage resources directly into vRealize Automation catalogs, with approval processes and automatic provisioning occurring in the background.

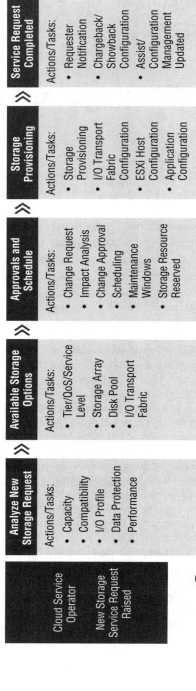

Figure 9.1
Manual storage provisioning process

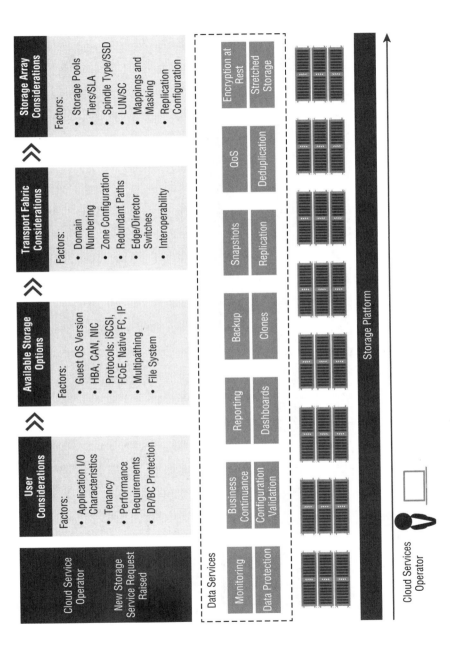

FIGURE 9.2
Complex storage provisioning process

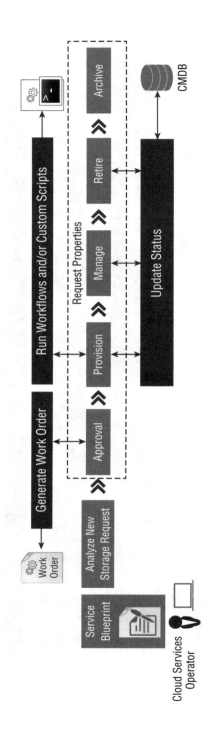

FIGURE 9.3
Example of a storage-as-a-service request workflow

STaaS Service Definition

Storage as a service can mean different things to different people and platforms. For instance, Amazon's Simple Storage Service (S3) can be considered a public STaaS platform; it provides a multitenanted storage environment that's typically seen as an alternative for small and medium-sized businesses that lack the CapEx budget and technical personnel to implement and maintain their own storage infrastructure. Public STaaS platforms are also promoted as a way for businesses to address risks in a disaster-recovery scenario, provide long-term archiving solutions for data retention, and address additional business continuity and availability of business-critical data. In addition, public cloud storage can provide a service model in which data is maintained, managed, and backed up remotely, and then made available to end users over the public Internet.

However, in the context of this chapter, we're referring to STaaS in how it relates to VMware's software-defined storage model, and the software-defined data-center vision, for private and hybrid cloud service offerings. This type of service offering is typically provided by either an internal IT organization within an enterprise, or by an external cloud service provider. This external cloud provides either a fully managed service or partly managed service to the various lines of business within its customer's organization.

As outlined in previous chapters, one of the primary challenges facing IT organizations is providing the skills and operational resources to undertake frequent storage-related administrative tasks, which are both essential and continual. In addition, administrative problems, such as lack of skills, knowledge, human error, or slow reaction times can create an ongoing bottleneck. Many of these problems can be addressed by providing a sophisticated cloud platform solution to automate these daily storage-related tasks. This can be achieved by employing a workflow engine with scripts and a cloud management platform that is capable of providing not only the framework to interact with end users, but also the capability to manage, store, and execute scripts and workflows in response to predefined events.

The primary aim of delivering a STaaS design is to enable self-provisioning through a simple self-service portal, which has the ability to interact with the back-end storage constructs required to deliver the fully automated provisioning of storage resources.

This chapter shows how you can use the VMware vRealize Automation cloud management platform to provide the basic building blocks for delivering STaaS. However, it is also true that before you can attain the desired simplicity, you must first, to some extent, increase the level of complexity.

In order for the cloud management platform to deliver self-service of storage resources, you must first define the storage service that is to be made available through the service catalog. For instance, storage services such as enabling disaster recovery, providing quality of service, or provisioning datastores might all be included within a STaaS design. Figure 9.4 illustrates what a cloud storage catalog may look like when viewed by a cloud service consumer within the vRealize Automation web interface.

As outlined later in this chapter, these services are enabled through software-defined storage workflows, via vRealize Orchestrator, within the storage management and orchestration layers of the cloud management platform.

FIGURE 9.4
vRealize Automation storage service catalog example

Cloud Platforms Overview

A cloud platform cannot be delivered just by deploying a single product, despite what some vendors might tell you. The aim of cloud computing is to provide infrastructure technology resources via a private or public portal and to offer on-demand services, such as infrastructure as a service (IaaS), platform as a service (PaaS), or software as a service (SaaS).

The cloud provides shared or dedicated resources, such as applications or data platforms. It also can be provided to end users as a service over a private or public network infrastructure. Figure 9.5 illustrates one view of traditional cloud service offerings, which are often seen and commonly delivered *as a service*.

Cloud computing services can be classified into the three categories shown in Figure 9.6.

Infrastructure as a service (*IaaS*) enables delivering compute resources, such as virtual machines, directly to cloud consumers. The cloud consumers then use those virtual machines to deploy guest operating systems, applications, middleware products, and databases. Typically, cloud consumers find this more within their comfort zone than deploying physical hardware. An IaaS platform does not provide cloud consumers with any control over the underlying hardware resources. Changes to the platform at the infrastructure level must be requested via the customer's cloud consumer portal.

Platform as a service (*PaaS*) provides the resources required, on top of the IaaS cloud infrastructure, to deliver consumer-developed applications, libraries, services, or other tools to the cloud consumer. On a PaaS platform, much like an IaaS platform, the service consumers cannot control the underlying infrastructure, but can only request changes from their cloud service provider via their portal.

Software as a service (*SaaS*) delivers provisioned applications, which reside on the cloud platform, directly to end users. The applications are typically accessible from multiple device types that comply with the appropriate corporate security policies, or the security component of the cloud service provider's SLA. Here again, the end user or cloud consumer cannot control the underlying hardware, or even software, but can make requests for changes to their cloud platform provider via their consumer portal.

The aim of a cloud computing strategy is to enable IT organizations or service providers to build an *IT as a service* solution for businesses and individuals. This solution is capable of providing resources simply, efficiently, and at a lower cost than traditional mechanisms.

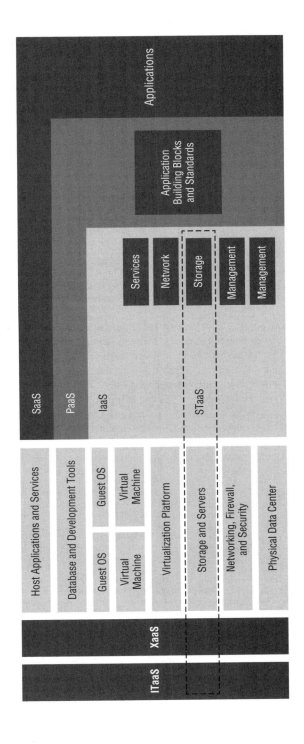

FIGURE 9.5
IT optimization computing components, delivered as a service

FIGURE 9.6 Common cloud computing services

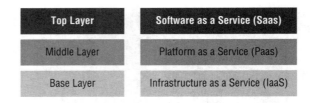

Cloud services are typically built using one or more of the following three cloud computing models:

- Private cloud
- Public cloud
- Hybrid cloud

A *private cloud* platform is managed by its own IT organization, which controls its own compute, storage, and networking resources. Alternatively, a *public cloud* infrastructure is provided by and managed by a third-party cloud service provider, which typically offers other IT organizations or consumers access to their shared platform of compute, storage, and network resources.

In terms of storage resources specifically, private cloud services provide a dedicated environment, protected behind an organization's firewall. These storage services are typically used by IT organizations that require customization and more control over their business data. Public cloud storage services, such as Amazon's S3, provide a multitenanted storage environment over the Internet, and are most often used for unstructured data.

The *hybrid cloud* model, shown in Figure 9.7, is the extension of a private cloud into a public cloud entity. This is the cloud computing model gaining the most traction in the market. For most internal IT organizations, a private cloud cannot exist in complete isolation from other systems. Therefore, either to a a lesser or greater extent, all private cloud platforms extend horizontally into cloud service provider data centers.

FIGURE 9.7 Hybrid cloud platform

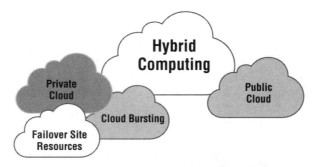

Hybrid cloud storage combines these two models and includes at least one private cloud and one public cloud infrastructure. An IT organization might, for instance, store actively used and structured data in a private cloud, in on-premises data-center storage devices, and place unstructured or infrequently accessed archival data on a public cloud platform.

IT as a service (*ITaaS*) is another cloud services model: IT organizations enable their lines of business or individual business units to run much like consumers of a cloud service provider. The IT organization functions as a separate entity and creates products and services within its own enterprise business unit, offering them to internal customers. The primary aim of offering ITaaS to an enterprise is to deliver the following:

- Increased availability
- Increased sustainability
- Increased efficiency
- Agile and rapid services
- Lower CapEx and OpEx
- Less downtime

Providing STaaS via a private cloud model may represent one of the component offerings that make up an entire ITaaS solution, offered to the different lines of business within an organization. Such a solution would allow individual IT users from differing business units to provision storage and other resources to facilitate their own departmental IT requirements. However, building such a solution requires, as we have already stated, multiple tools that can be integrated and customized to meet the specific needs of the storage services being delivered.

Cloud Management Platform Architectural Overview

This type of CMP requires numerous software components running alongside one another, within the cloud stack, in order to deliver the cloud storage catalog to users. As the architect, you should fully understand which software components are required, and which component pieces are employed for which task, to allow the cloud platform to provision and modify storage resources.

Figure 9.8 illustrates a logical architectural overview of the component parts included in a STaaS platform delivered through vRealize Automation. Each component within the VMware and storage vendor software stack is required to be integrated, in order to provide end users self-service access to the automated storage services offered through the service catalog.

The remainder of this section covers each of these software components, in a top-down approach, and provides insight into each of the required layers of the cloud stack.

vRealize Automation Cloud Management Platform

In this example solution architecture, service delivery and the cloud management layer are provided by VMware's vRealize Automation platform, illustrated in Figure 9.9. This platform is used to create, present, and execute the storage service catalog to end users.

vRealize Automation provides a secure, web-based user interface used by appropriately authorized users, administrators, and developers to request new IT services or resources. vRealize Automation ensures that its cloud service consumers can request only those resources that comply with their IT compliance, security, or business policies. Through the web-based portal, services such as desktop applications, infrastructure, or *anything as a service* (*XaaS*) can be requested via a shared service catalog, and can be approved and provisioned by users with limited or no knowledge of IT infrastructure.

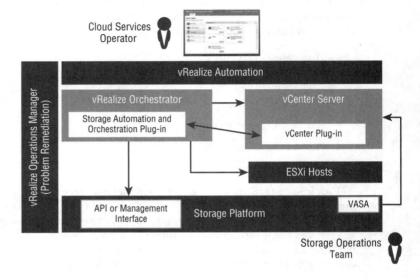

Figure 9.8
STaaS cloud software stack

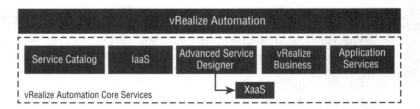

Figure 9.9
vRealize Automation services

vRealize Automation can also help control costs associated with storage and other resources by revealing the monetary value of the resources being requested. This simplifies resource, budget, and capacity management.

vRealize Automation Catalog

The service catalog provides a centralized self-service portal to consumers of the IT services. Authenticated users can view catalog items and request services on demand. They also can track their requests, through vRealize Automation's approval mechanism, and manage their own provisioned services during their life cycle.

It is the role of the cloud services architect to define new services and publish them to the service catalog, with custom forms being used to gather the unique information required to provision those resources. vRealize Automation can also provide other catalog-based capabilities, including these:

- Requesting and managing items in the catalog
- Creating and publishing catalog items

- Providing custom actions and entitlements
- These catalog-based service offerings enable the provisioning of servers, desktops, storage, or *anything-as-a-service* across virtual and physical infrastructures. For server provisioning, the vRealize Automation catalog model works by creating a blueprint of the system, which provides the characteristics of the required virtual machine. Blueprints are then published as catalog items in the service catalog, and when a cloud consumer requests a system, vRealize Automation allows that consumer to manage the complete life cycle of the provisioned entity in line with the original request.

As we mentioned, vRealize Automation also offers XaaS. This approach refers to the ever-increasing diversity of services available via the vRealize Automation cloud management platform and vRealize Orchestrator.

Delivering Anything-as-a-Service

The vRealize Automation platform aims to be able to deliver any IT service on demand, and provides the wizard-driven *Advanced Service Designer* (ASD) as one of its primary mechanisms to achieve this goal.

As illustrated in Figure 9.10, the ASD provides the capability to configure additional custom services, delivering personalized and business-relevant automation.For instance, suppose the cloud services architect wants to deliver additional IT services through vRealize Automation that are not available out of the box. These services might include automating the delivery of all tasks associated with hiring new employees by interacting with Microsoft Active Directory and Exchange to configure new accounts (in addition to other required tasks), or provisioning new storage offerings, such as storage containers or legacy storage LUNs. These can typically be automated through the ASD feature.

By using the ASD feature of vRealize Automation, the cloud services architect can develop custom IT services based on vRealize Orchestrator object types and configure them as items that can be provisioned. From these objects, the cloud services architect can develop blueprints through vRealize Orchestrator workflows. Finally, from these developed workflows, the cloud service architect can develop custom services and provision them as catalog items. By using this workflow mechanism, vRealize Automation, when employed with vRealize Orchestrator, can deliver XaaS through a series of self-service catalog items.

The ASD wizard can be used to design the end-to-end process of delivering a custom service through automation. These custom services, or XaaS, are then published to the vRealize Automation service catalog, along with other application-based, platform-based or infrastructure-based services. The ASD wizard allows cloud service architects to define service-delivery capabilities, user interaction, and entitlements of the service by doing the following:

- Allowing a workflow to be defined that can be used to deliver the service. This may include existing workflows, plug-ins, and custom scripts.
- Using a wizard-based approach to populate a service request. A simple-to-use form is used to gather the required information that the workflow needs in order to deliver the requested service.
- Defining service entitlements, including any necessary workflow approvers, such as managers or IT decision makers.

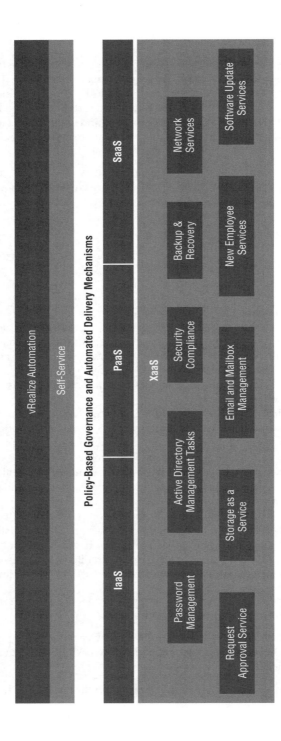

FIGURE 9.10
Advanced Service Design capability examples

CLOUD MANAGEMENT PLATFORM ARCHITECTURAL OVERVIEW

Critical components, addressed later in this chapter, are the workflows and plug-ins supplied by storage vendors and used by ASD to rapidly deploy new storage-related services. However, even without partner-supplied plug-ins for vRealize Orchestrator, the services that can be automated are limited only by the cloud service architect's imagination, or the cloud services operational and development team's skill set (see Figure 9.11).

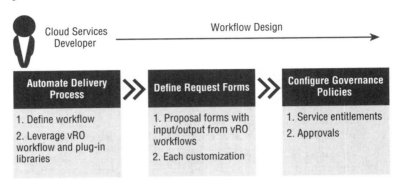

FIGURE 9.11 Advanced Service Designer workflow example

vRealize Orchestrator

VMware vRealize Orchestrator provides the drag-and-drop workflow engine that simplifies the automation and integration of a service offering. This software also allows cloud service developers to create complex automation tasks through the use of its workflow designer. These developed workflows can then be executed either directly from the vSphere Web Client, through vRealize Automation, or a wide range of other integrated applications or cloud management platforms.

In a STaaS design, vRealize Orchestrator facilitates storage-based workflows by integrating the cloud management platform with plug-ins provided by the storage vendor; this allows daily processes to be automated and accelerates and simplifies operations. In addition, by using vRealize Orchestrator to provide cloud-based IT services, you can reduce IT operational expenditures. This reduces the total cost of ownership of the solution and accelerates the transition into offering an ITaaS operational model.

As we have stated, vRealize Orchestrator is designed to allow vSphere system administrators and cloud services operational staff to streamline tasks through automation, and to integrate solutions with third-party providers through various mechanisms. To facilitate this, vRealize Orchestrator comprises the following workflow development and management features:

Workflow Designer The workflow designer facilitates simple or complex workflow creation through a simple drag-and-drop interface.

Workflow Engine The workflow engine creates workflows by using the building blocks created, or by using third-party plug-ins.

Scripting Engine vRealize Orchestrator provides a scripting engine as a mechanism for creating new building-block actions, workflows, and policies across the platform. The scripting engine also provides basic version control, namespace management, variable type checking, and exception handling.

Workflow Library vRealize Orchestrator is deployed with an out-of-the-box library of workflows, providing a wide range of functionality, such as the following:

- Orchestrating group virtual machine snapshots
- Providing email notifications of events, such as powering off a virtual machine

Built-in Version Control vRealize Orchestrator workflows include version history, packaging, and rollback capabilities. These features can facilitate change management during the development life cycle, as well as during the various stages of workflow creation.

Checkpointing Content Database Each section of the workflow is saved in a content database. This functionality enables server restarts, without loss of state and context within the workflow, which is particularly useful for long-running, complex processes.

Central Management vRealize Orchestrator provides a centralized mechanism to manage processes and ensure that operational teams use correctly versioned controlled scripts.

Diagnostics and Troubleshooting The vRealize Orchestrator client provides debugging and error-diagnostic capabilities in order to provide workflow developers with a simple and more efficient user experience.

Software Development Kit The vRealize Orchestrator SDK plug-in not only provides inexperienced workflow developers with a starting point, but also enables advanced developers to integrate SDK features into their workflows. The SDK is composed of the following parts:

- Eclipse add-on for vRealize Orchestrator plug-in development
- Command-line utilities
- Developer guide
- Best-practices guide
- Sample plug-ins

Performance and Scalability vRealize Orchestrator can execute thousands of concurrent workflows on either a minimal or scaled-out distributed architecture. By using the clustering features, cloud service architects can plan vRealize Orchestrator deployments globally, with web-scale scalability. vRealize Orchestrator clusters, when used with external load balancers, also provide increased availability, and allow the dynamic scale-up and scale-down of orchestration capacity. In addition, by using a clustered architecture, if one vRealize Orchestrator server becomes unavailable during the execution of a workflow, a different node can complete the workflow's orchestration tasks without disruption to the service.

Public-Key Infrastructure vRealize Orchestrator provides an internal public-key infrastructure (PKI) to sign and encrypt content that is imported or exported between servers.

Digital Rights Management vRealize Orchestrator provides a digital rights management (DRM) mechanism to control how exported content may be viewed, edited, or distributed.

Secure Sockets Layer vRealize Orchestrator uses encrypted communications between client and server, and also provides HTTPS access to the web-client user interface.

Advanced Access and Rights Management   vRealize Orchestrator's built-in rights management system provides control over access to processes and objects.

Cloud Readiness   vRealize Orchestrator provides deep integration with vSphere, vCloud Director, vRealize Operations Manager, and vRealize Automation. This deep level of product integration enables cloud service architects and developers to automate virtually any task or process within their environment.

Ecosystem Awareness   Published plug-ins and workflows can be obtained from VMware's online Solution Exchange, as well as from multiple third-party independent software vendors (ISVs).

Virtual Appliance Deployment   vRealize Orchestrator provides simple deployment via a prepackaged virtual appliance. This approach significantly reduces the skill level required to deploy the infrastructure, and does not require a Microsoft Windows license, reducing costs.

REST API   vRealize Orchestrator provides a REST API in order to enable other applications to execute workflows. This REST API enables enhanced JavaScript Object Notation (JSON) to support and simplify integration with vCenter Single Sign-On (SSO). This integration also helps facilitate the rapid deployment of vRealize Orchestrator instances for test and development purposes, as well as for scaling up automation capacity as demand increases.

THIRD-PARTY PLUG-IN SUPPORT FROM STORAGE VENDORS

Third-party storage vendors typically provide vRealize Orchestrator plug-ins to support STaaS functionality within their products. In most cases, this greatly simplifies the level of development effort and scripting work required to design a solution. Therefore, even though they are not an essential prerequisite to providing a STaaS design to an enterprise IT organization or service provider, they can, and should, significantly simplify the development and operational overhead associated with building and maintaining catalog-based storage services.

The vendor storage plug-in software is used to connect the key components required by the storage service catalog. VMware vRealize Orchestrator then allows the cloud services developer to discover storage and provisioning capabilities through automation workflows. These predefined workflows, which are typically provided as part of the vendor plug-in, become available in the vRealize Orchestrator workflow library after the installation and configuration of the package. These workflows can then be used by cloud service architects and developers via vRealize Automation to design service blueprints. The design process may include performing actions such as the following:

- Creating a new VMFS datastore
- Extending a VMFS datastore
- Deleting a VMFS datastore
- Creating a new Virtual Volumes datastore
- Adding a Virtual SAN host to a VSAN-enabled cluster

As a result, the workflows become the underlying building blocks called by the orchestration and automation layers (in this design example, vRealize Orchestrator and vRealize Automation).

This means that these workflows are the primary components required to design and build a STaaS catalog.

The storage vendor plug-in packages can typically be downloaded from VMware's Solution Exchange, the hardware vendor's website, or the third-party independent software vendor's website.

By using workflows to carry out daily operational storage tasks, rather than operational teams issuing storage operations manually one at a time, administrators can preplan and automate storage operations within the platform, either directly from vRealize Orchestrator or through the vRealize Automation cloud management platform. By providing this end-to-end efficient provisioning of storage resources, operational overhead can be reduced, and environmental management simplified, as shown in Figure 9.12.

The Combined Solution Stack

The vRealize solution described in this chapter requires tight integration between vRealize Orchestrator, vRealize Automation, vendor plug-ins, and storage hardware to provide this software-defined cloud solution. This integration is essential and must extend across service presentation, delivery, management, orchestration, automation, and hardware, in order to support the STaaS environment described. In addition, for the STaaS platform to provide maximum benefits to operational teams and its end consumers, it must be able to manage all of the associated infrastructure components.

Workflows implemented as part of a STaaS design might also need to interact with other elements, such as a list of virtual distributed switches on which to create port groups, or a configuration management database (CMDB). These settings are typically fixed values stored on the vRealize Orchestrator server as either a configuration or resource element. Other resource elements, such as a list of available VLAN IDs or available IP addresses, could also be retrieved from a CMDB.

Workflow Examples

In the following examples, workflows are used to automate repetitive storage operational tasks. These workflows can be built by automation developers within the IT organization. Alternatively, vendor-provided preexisting workflows can be used or combined with custom automated actions in order to achieve the desired result.

Workflows can be combined or modified to create more-advanced operations, and can also be used as storage service blueprints by the integrated vRealize Automation cloud management platform. The storage cloud services blueprints, once created, become the storage service catalog items available to authenticated users for performing operational tasks.

For instance, the workflow example in Figure 9.13 uses a single form for the basis of provisioning a NAS-based storage device.

In addition, a form might be required to modify access to IP-based storage. This high-level process might require multiple integrated workflows, with multiple custom scripts, as illustrated in Figure 9.14.

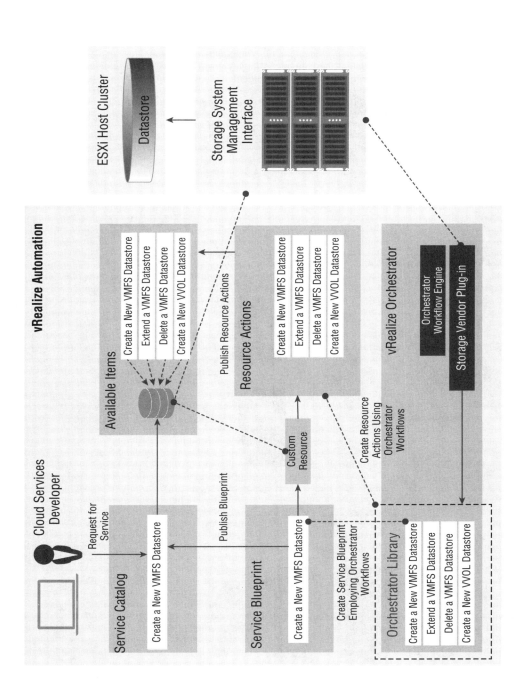

FIGURE 9.12
Example of a workflow's logical configuration

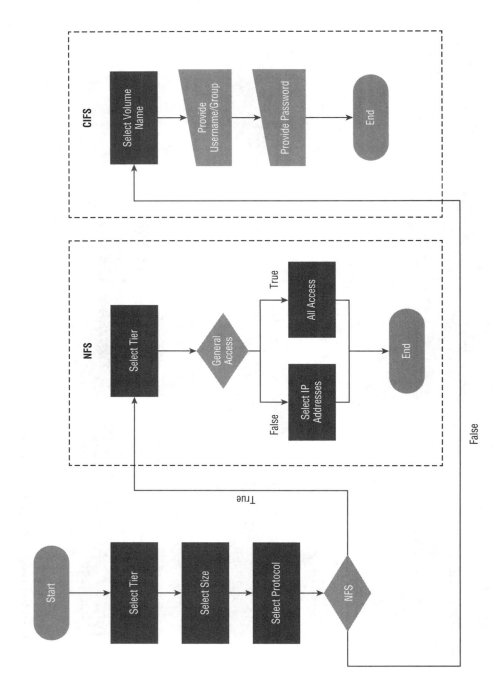

FIGURE 9.13
STaaS NAS form design

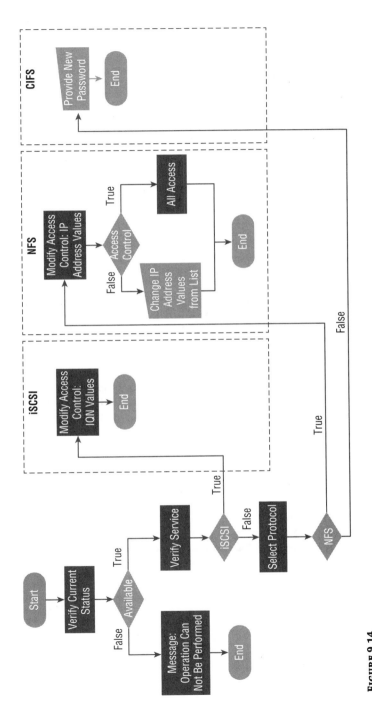

FIGURE 9.14
STaaS access rights modification

This workflow first verifies the catalog item status to ensure its availability to proceed with the action, before determining the service type required:

1. For iSCSI storage, this workflow provides the current storage access rights, based on IQN. The cloud services operator can then modify the listed values.

2. For NFS storage, the cloud services operator can elect either an All Access value, providing access to the storage from any IP address range, or alternatively adopt a more deterministic approach by modifying the current range of IP addresses with existing access to the volume.

3. For CIFS storage, the cloud services operator can modify the user password for the CIFS server or the Microsoft Active Directory group that is currently provisioned with access rights to the volume.

4. Finally, an email is generated to the requestor, notifying them of the completed changes, based on their operation request.

Summary

Software-defined storage, and providing IT organizations with a means by which to deliver storage as a service, is gaining a lot of traction within the industry. vRealize Automation provides an ideal framework for enabling a software-defined cloud storage platform, delivering STaaS to the IT organization to simplify daily operations and reduce costs. The vRealize Orchestrator workflows are key to providing the IT organization or service provider with efficient management and delivery of storage resources using vRealize Automation.

Storage vendor plug-ins for vRealize Orchestrator are a key component to allow VMware platforms to integrate and manage the storage infrastructure, and to allow cloud service architects the control over delivering defined storage resources across back-end disk systems.

In an enterprise IT organization or service provider environment, one of the primary aims of delivering STaaS is to design a solution that can provide support for storage resources offered by multiple storage vendors and that can be managed operationally by a single interface, with simplified operations and quick support for new vSphere deployments. However, collecting hardware system component information from all of the storage systems in a mixed vendor environment, and providing it to the VMware cloud platform for management—and therefore realizing the software-defined storage model of implementing tasks on storage systems across this heterogeneous infrastructure—is a challenge requiring an architect with vision to complete. Nevertheless, such a cloud management platform will deliver the long-term operational benefits described in this chapter, such as operational efficiency, simplification, and a self-service framework.

Chapter 10

Monitoring and Storage Operations Design

As an architect who has designed a resilient data center that removes all single points of failure, the next challenge is to ensure that all the individual components of the storage platform are functioning as expected and are available in accordance with their expectations, in order to meet the service levels required by the business. The primary mechanism for achieving this goal is the monitoring of data-center storage components, both continuously and consistently, to facilitate the resolution of issues and errors in a timely manner.

Storage Monitoring

Monitoring the storage system end to end is essential to all IT organizations, in order to ensure that the underlying platform is available and operational 24/7. In addition, any monitoring solution should ensure that all components are optimized to meet the primary business goal of enabling critical applications to provide users with a consistent and usable experience when performing the activities required by their jobs. To achieve this, the vSphere hosts, network systems, and storage devices should all be monitored to provide metrics on system load, utilization, health, performance, security, and capacity. These metrics then provide correlated data needed to perform not only reactive remediation tasks, but also proactive responses.

With business data trends doubling the level of required capacity every two years or so, automating monitoring and alerting solutions are becoming increasingly important. In addition, closely monitoring the data-center infrastructure can optimize data-center operations and help avoid downtime. When designing a monitoring solution for data-center storage, you should consider the following:

Availability   To provide continuous monitoring to maintain uptime, and provide warnings and error alerts to ensure that issues are corrected proactively.

Scalability   To ensure that the monitoring solution allows for capacity planning and trend analysis, which can provide the necessary metrics and metadata to plan the scaling of the storage solution as the business data requirements grow.

Alerting   A key design factor is to ensure that alerting mechanisms can be used to inform operational teams of failures and potential failures occurring. This approach allows for the appropriate corrective action to be taken to make certain that availability is not impacted.

For end-to-end storage monitoring to be effective, all major components in the data center should be addressed as part of the design, with each component being monitored for health, capacity, performance, and security. The specific list of components will vary depending on the approach being taken to storage in the design, but may include elements such as the following:

- Storage networks, including Fibre Channel and IP SAN or NAS-based IP networks
- Storage arrays and their various hardware components
- Hyper-converged infrastructure (HCI) hosts, their disks, controllers, and other hardware
- Storage application software and firmware

From continuous monitoring of health, capacity, performance, and security of these data-center and application components, it should be possible to ensure the availability and scalability of business-critical data, by notifying appropriate operational personnel with alerts, component failure information, and general trend and utilization data through operational dashboards and reporting. By including such mechanisms as part of a storage design, IT organizations and their operational teams can ensure that corrective actions can be taken quickly without breaching the assigned SLA.

Monitoring Component Health

Health monitoring provides status and availability information on specific storage components or software—for instance, the current status of a SAN device, such as the HBA status or a disk failure.

The health of individual components is important because a failure of any hardware or software element within the storage platform, or its data path, can lead to an outage of numerous components. For example, a failed HBA could in turn cause degraded access to many storage devices, even in a multipathed environment, or would result in a complete loss of access to devices in a single-pathed environment. Therefore, monitoring health is fundamental and can easily be understood and quantified by management teams; a failed component, unless redundancy exists, will lead to loss of data or the loss of access to data.

Monitoring the health of storage components within the data center is critically important, and the monitoring of capacity, performance, and security all depend on the health and availability of the overall system and its components. Even if a business performs no other storage monitoring, health monitoring should be implemented at the very least. Health- related issues should be addressed as a matter of high priority by the operations team.

Monitoring Capacity

The primary aim of *capacity monitoring* is to provide a perspective on the amount of storage resources available at any given time. A design that includes the monitoring of storage capacity as part of its scope should, at the very least, address the following factors:

- The available free and used capacity on a filesystem, datastore, or pool of physical disks
- The amount of disk capacity available on each storage array or an HCI-enabled Virtual SAN cluster
- The number of available ports in a SAN fabric or IP network switch required for expansion
- The level, or speed, of growth of storage capacity, the scalability of the chosen solution, and the requirements for data growth, as defined by the business requirements

A lack of proper monitoring and capacity planning can lead to data becoming unavailable and to an inability to scale the environment. It's critical to provide trend reports from all the storage resources being used, in order to prevent outages before they can occur.

The overall aim is to create a design that is preventative and predictive over health metrics, as opposed to being reactive. For instance, reports should be provided to decision makers on the current trends identified on the storage platform, such as ensuring that the business is aware that 90 percent of a specific datastore is full, and that the datastore's utilized capacity is increasing at a specific rate. Likewise, IT managers should be made aware that 95 percent of the available ports on the SAN fabric have been utilized, so they can budget for new switches to be added if more arrays or host servers are to be joined to the same fabric.

Monitoring Storage Performance

Performance monitoring is required to ensure the efficient operation of components across the storage platform. This type of monitoring includes, for instance, monitoring the amount of I/O that traverses a storage array's front-end ports, the fabric switches' bandwidth utilization, or the amount of write I/O being handled by the flash device in a Virtual SAN disk group, and therefore the response time of an application.

Performance is typically monitored via one of several key measurements. Three distinct metrics are used when it comes to evaluating most aspects of scalability and performance: throughput, latency, and I/O per second (IOPS). As we've noted in previous chapters, IOPS also plays a major part in all aspects of storage design.

The IOPS metric is the most commonly used standard for measuring the performance of the storage system's back-end disks or the performance of the overall array. At its most basic, the term *IOPS* refers to the number of read or write IOPS. The level of IOPS that can be achieved on a storage device will vary, based on factors such as the system's balance of read and write operations; whether the traffic is sequential, random, or mixed; as well as the I/O block size.

It is also common practice to consider the throughput metric alongside IOPS. Both impact performance, but in different ways. For instance, an application with a low-throughput requirement of 100 MB/s may require a much heavier I/O specification of 20,000 IOPS. Even though this might not result in a bandwidth problem, it may put the storage array's controllers under significant pressure. Likewise, a different application might produce a low number of IOPS, but at the same time produce significant throughput, such as sustained periods of data reads, and therefore place the SAN fabric under pressure.

The *throughput* represents the amount of data transferred per data unit, and is typically measured in megabytes per second (MB/s). The level of throughput that can be achieved with a particular storage platform will depend on various factors associated with both the hardware and software. The most important design elements to consider that can directly relate to storage throughput include the following:

- Link speeds, such as Fibre Channel link speed
- The level of outstanding I/O requests
- The number of disks, particularly mechanical disks
- RAID type being used
- SCSI reservations
- Caching or prefetching algorithms and technologies

Latency is the delay in the time taken to complete an I/O request, which is typically measured in milliseconds (ms). As multiple layers of the storage stack traverse data, each layer can impose its own latency delay for each I/O request made. The impact of latency can depend on factors including the following:

- Queue depth
- I/O request size
- Disk hardware properties, such as rotational speed, seek times, and access delays
- Capacity
- SCSI reservations
- Caching or prefetching algorithms and technologies

The primary purpose of monitoring the metric data associated with IOPS , throughput, and latency is to ensure that the storage components work efficiently and optimally for that environment, and also to provide visibility of any element in the storage stack that is pushing performance utilization to an expected limit. In addition, in some cases, the monitoring of these same metrics can identify whether storage components are being underutilized or have sufficient resources available to be provisioned with additional workloads.

Monitoring traditional storage array systems for performance issues and problems can be complicated and is typically seen as the most difficult of the various aspects of data-center operations. This type of monitoring typically requires trained personnel experienced with that vendor's hardware to identify performance bottlenecks and operational issues.

Monitoring Security

Security monitoring prevents and tracks unauthorized access to the storage platform, whether it's accidental or malicious. The level of security monitoring required in a design typically depends on the type of business and any regulatory requirements that must be met. However, for more and more businesses, enforcing security monitoring and recognizing security breaches is becoming a high priority. Storage monitoring will at the very least include identifying the following:

- Repeated login failures
- Unauthorized access to storage arrays
- Unauthorized configuration or reconfiguration of storage hardware
- Physical access to storage systems in the data center, via badge readers, biometric scans, and video surveillance
- Unauthorized access to storage fabric
- Unauthorized zoning and LUN masking in SAN environments, or changes to an existing zone configuration

In addition, login failures and attempts by unauthorized personnel to execute code or launch applications on host servers or storage devices should be closely monitored, to ensure secure operations.

Although login failures are typically accidental mistyping actions, they can also result from a deliberate attempt to execute code or launch applications on host servers or storage devices. Most storage systems allow only two or three successive login failures before not allowing any more attempts. In most environments, this information may be written to a log file but may never be monitored. However, in more and more enterprise IT environments and cloud service provider network operations centers (NOCs), such security events are being monitored; if three or more successive login failures occur, a message is sent to the security administrator, warning of a possible security threat.

Storage Component Monitoring

As we've already discussed, storage component monitoring is a key part of data-center operations and is required in order to ensure uptime of applications. In addition, we've already identified the key areas that any storage operations design should, at the very least, include: health, capacity, performance, and security.

However, as we've also highlighted throughout this book, storage platforms are made up of many components. These components can vary significantly depending on the hardware and the approach taken to data storage within the vSphere environment. In the following section, we address the key hardware components and the various methods for monitoring each element of the storage design.

Monitoring Storage on Host Servers

Enterprise storage arrays and host connectivity are typically designed in a redundant manner. Still, any hardware component failure on a vSphere host server, such as a storage adapter, should be identified immediately by the operational monitoring solution, and replaced or rectified as a top priority. This limits the impact to the environment and prevents a complete outage of the host server.

In addition to hardware components, host servers should be monitored for datastore capacity utilization. By monitoring datastore capacity continuously, the amount of free space and the estimated growth rate of the data can be predicted effectively, identifying when it is expected to become 80 percent and 100 percent utilized. By providing these metrics, corrective action, such as extending the raw LUN capacity or adding datastores, can be taken ahead of time. This action can prevent storage device capacity from becoming exhausted, which is likely to result in an application outage.

Monitoring the Storage Fabric

The SAN fabric design should include the monitoring of components for health, capacity, performance, and security. Each of these four areas should be addressed in order to ensure that the fabric platform can provide the reliable, end-to-end storage connectivity required by shared storage array and HCI designs.

Storage Fabric Health Monitoring

vSphere hosts require continuous and uninterrupted access to data over the storage fabric. This access is completely dependent on the health of the physical and logical components that make up its design. Components such as power supplies, fans in switches, cables and gigabit interface converters (GBICs) make up the physical entities, whereas constructs such as zones and addressing systems make up logical components. The storage monitoring design should include mechanisms to report any failure of these logical or physical components immediately. The key storage fabric components that should be addressed as part of the design include the following:

- Fabric or IP switch errors
- Zoning errors
- Failed switch ports or GBICS
- Port status change or attribute changes
- Fabric and IP switch status changes or attribute changes

For instance, in a Fibre Channel design, errors identified in zoning (such as specifying the incorrect WWN of a port) will cause a failure to access that port. All such errors should be monitored, reported, and rectified as part of the storage operations design.

Storage Fabric Capacity Monitoring

Switch port capacity utilization, which refers to the number of ports available or utilized on different fabric or IP switches, should be monitored. Providing this information in operational reports improves planning and expansion activities when host servers or storage array ports need to be added to the storage fabric.

Capacity monitoring also refers to utilization metrics on the fabric switches themselves, at both an aggregate switch level and port level, in terms of a percentage of utilization. Typically, the most utilized ports that require specific attention as part of the design include Interswitch Links (ISLs), which carry traffic from multiple hosts concurrently. These metrics should also be addressed as part of the storage fabric's performance monitoring design.

Storage Fabric Performance Monitoring

Several SAN fabric performance and statistical metrics can be used to help determine hardware issues, or even predict the failure of a component. For instance, an increasing number of port link failures could indicate a hardware problem and suggest that the port might fail. In addition, loss of signal or loss of synchronization can indicate that imminent port failure is likely.

Performance of the storage fabric's device ports can be measured by the Receive (Rx) or Transmit (Tx) link utilization metrics. These can be addressed either at an aggregate switch level or an individual port utilization level. These values provide a good indication of how busy the switch or switch port is, based on the vendor's published maximum throughput. Switch ports that are heavily utilized can cause delays as queues occur on the host servers, which brings us to the next topic: the HBA queue depth.

The *HBA queue depth* is the number of commands that the HBA can send or receive in a single chunk per storage device. If the queue depth is exceeded on the storage target, performance will be degraded. This is typically caused when too many concurrent I/O operations are sent to a storage device, resulting in the device responding with an I/O failure message, Queue Full (qfull), forcing the vSphere host to retry after a short wait. Some workload use cases may require controlling queue depth on the hosts as part of a design recommendation. VMware provides detailed guidance in its knowledge base.

For instance, when providing storage for large-scale business-critical workloads with intensive I/O patterns that require a queue depth significantly greater than the Paravirtual SCSI adapter's default, it may be necessary to adjust the host's maximum queue depth value as part of a design.

In designs where this is required, two values must be modified on each of the vSphere hosts in the cluster:

- The default Queue Depth values for all of the HBA devices (QLogic, Emulex, or Brocade)
- The Maximum Outstanding Disk Requests parameter (Disk.SchedNumReqOutstanding)

For example, if the design is employing QLogic HBAs, the default queue depth is 64 in ESXi 5.*x* and 6). To modify this configuration to provide a queue depth of 128 on an ESXi 6 host, the following command should be run before rebooting the host:

```
esxcli system module parameters set -p qlfxmaxqdepth=128 -m qlnativefc
```

To configure a new Maximum Outstanding Disk Request value (Disk.SchedNumReqOutstanding), the following command should be run (the value being configured is 256):

```
esxcli storage core device set -d naa.xxx -O 256
```

If a design does require the queue depth to be modified, ensure that this configuration is completed appropriately and in line with the hardware vendor's recommendations. Also ensure to keep the HBA vendor, or at the very least the queue depth value, consistent across all hosts in the cluster.

Modifying these options should be a well-thought-out design decision, and should not be carried out unless there is a clear requirement or design factor being considered. For more information about modifying these values, please refer to VMware's detailed guidance on the subject and their supporting knowledge base articles.

STORAGE FABRIC SECURITY MONITORING

As part of a storage fabric design, you should consider how SAN administration tasks can be limited to a select set of operators, how strict password enforcement can be used, and how physical access to devices can be limited and monitored.

For instance, in a Fibre Channel storage fabric, security monitoring typically includes identifying and notifying compliance personnel when zoning configuration changes are made. Unauthorized zones can compromise data security as well as lead to data inaccessibility. User login attempts and authentication events to storage fabric switches should also be monitored, in order to provide an audit trail of administrative changes.

In addition to those components previously highlighted, when considering the monitoring of IP networks, it is vital that the design should also consider areas more commonly associated with Ethernet-based communication, such as monitoring for collisions, errors, packet loss, network latency, throughput, and bandwidth utilization.

Monitoring a Storage Array System

Whether the design is employing Virtual Volumes or is taking the classic VMFS volumes approach to shared storage systems, enterprise-class devices are typically designed from the ground up to be highly available. End-to-end redundant components provide continuous availability, even when individual components fail, although performance might be degraded during such an outage. However, a storage array operational monitoring design is still required in order to allow the storage team to predict and react to storage events as they occur.

Storage Array Health Monitoring

The primary aim of an operational monitoring design for a storage array is to resume optimal performance and protect the environment from a complete outage caused by single or multiple hardware component failures as quickly as possible. To minimize the impact of any field-replaceable component failure, the failed component should be replaced as quickly as possible. Many storage vendors optimize this process by providing the array with the capability to *call home*; the array's software sends a message to the hardware manufacturer's support team automatically, if a component such as a disk, fan, or power supply module fails.

Storage Array Capacity Monitoring

Examining and scrutinizing capacity across storage systems is a critical aspect of operational monitoring. The capacity on the storage array is typically categorized as either configured or unconfigured. *Configured storage* is the amount of space that has been grouped into either a disk pool (which will be further partitioned into smaller individual SCSI LUNs) or into a storage container (which is used by vSphere to provide a storage entity for Virtual Volumes). *Unconfigured storage* refers to physical disks that have not yet been assigned a role within the array. The next capacity-related terms are allocated and unallocated storage. *Allocated storage* refers to LUNs or a storage container that has been masked for use by hosts, whereas *unallocated storage* refers to storage that has been configured but has not yet been presented to host servers.

Why is this terminology important from an operational perspective? By understanding these terms and how physical disk resources are used by the storage array through the monitoring of the storage array's configured, unconfigured, allocated, and unallocated capacity, storage operational teams are better able to understand, and therefore predict and react to, storage needs before problems occur. This proactive approach allows foresight, time, and planning for the provisioning of new capacity on the arrays, before disk resources are exhausted and reactive decisions need to be made quickly and without due consideration and impact analysis.

Storage Array Performance Monitoring

From a performance perspective, numerous performance and statistical metrics can typically be monitored on the storage array hardware, with the exact requirements depending on the storage systems used. However, some of the key metrics that should be monitored as part of the operational design, regardless of storage vendor, are likely to include the utilization levels of the numerous component parts of the storage platform. Extremely high levels of utilization can lead to performance degradation across the array. In addition to hardware monitoring, it is equally important to monitor storage processes that occur through the storage array's operating environment, software, or firmware. These processes include replication tasks, which could impact disaster-recovery capabilities, and RAID functionality, which could impact both performance and availability of the data.

STORAGE ARRAY SECURITY AND ENVIRONMENTAL MONITORING

Security considerations for storage array monitoring should also be addressed as part of a design. For instance, World Wide Name (WWN) spoofing can be a security concern, allowing unauthorized access to storage. This can occur when a host has been deliberately configured with an HBA that has the identical WWN as an authorized host. If this unauthorized host is connected to the storage array via the same SAN, then zoning and LUN-masking restrictions can be bypassed. However, most storage arrays come with built-in mechanisms that prevent such unauthorized breaches of security.

The storage design should also stipulate which mechanisms are to be employed to ensure that only authorized personnel are able to perform administrative tasks, such as device configuration, storage device presentation, replication operations, and port configurations. These tasks should be performed only by an authorized administrator, and should be audited for appropriate change request approval.

Physical data-center and hardware environmental monitoring is a key consideration for any infrastructure design. Monitoring the internal environment of the physical data center is just as critical as monitoring the individual storage components. As with all electrical equipment, storage arrays and HCI hardware are extremely sensitive to temperature, humidity, airflow, voltage fluctuations, and hazards such as water, dust, and smoke. Therefore, a data-center design must always account for the correct levels of ventilation, providing accurate control over the temperature and humidity, as well as ensuring that power supplies are uninterruptable and can provide corrections to any voltage fluctuation that might occur. The overall aim of physical data-center monitoring is to enable immediate reporting of changes to environmental conditions to operational teams, in order to ensure the stability of the platform.

The final consideration as it relates to the environment is physical access and security. Typically, such environments are monitored 24/7 by cameras and security staff, with authorized employees requiring an access card, biometric scan, or both, to allow entry to the data center. The aim is to limit physical access to the hardware within the data-center facility to those personnel with a specific requirement, as dictated by their job function.

Storage Monitoring Challenges

As outlined previously, the core data-center elements that require end-to-end monitoring are the storage arrays or converged storage platforms, the networks, the host servers, and the applications and databases themselves. In addition, as you have seen, in a vSphere environment, storage arrays might be based on Fibre Channel, IP SAN, IP NAS, or even DAS hardware. Some infrastructures may even have attached additional hardware, such as tape units, disk libraries, or bare-metal servers, installed with Open Systems or Microsoft Windows.

Therefore, one of the key challenges to storage platform monitoring is the heterogeneous nature of most SAN and vSphere estates maintained by enterprise IT organizations and service providers, and the means by which you can integrate multivendor hardware and software under a single management and monitoring solution, as shown in Figure 10.1. It is relatively simple for operational teams to monitor an individual array. However, monitoring many SAN switches, a multitude of storage platforms, and other equipment together, and then correlating that data into meaningful metrics and alerts is a far more complex challenge.

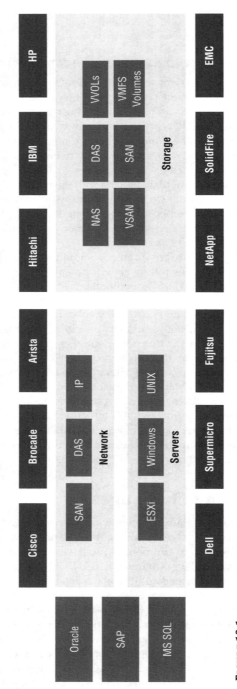

FIGURE 10.1
Storage monitoring challenges

This situation has largely come about by IT organizations and service providers being given the flexibility to select the most suitable products for a particular application or project, or alternatively shopping around for the best deal on hardware and support. However, without defined standards, policy-based management can be extremely complicated across this type of mixed-vendor environment, which poses a big problem for these types of heterogeneous storage infrastructures.

Even though each vendor typically provides monitoring and management tools for its own components, the primary challenge for any operational monitoring solution is to report on each of the multitude of components that make up a mixed-vendor storage platform.

The monitoring and management of these environments requires storage administrators to learn multiple data-center operational tools, and even different terminology adopted by various vendors. This lack of cross-platform integration into a single solution can prove to be challenging. It can be next to impossible to get a complete picture of what is occurring in the environment on a single pane of glass. Furthermore, this lack of cross-platform monitoring can make it difficult to correlate the information from all the storage platforms in an environment into a single place. This therefore limits the ability of operational teams to make informed decisions on capacity, performance, and availability.

Common Storage Management and Monitoring Standards

Operational architects face serious challenges when attempting to integrate multivendor hardware and software under a single management framework. Although it is relatively simple for SAN administrators to report on a single vendor's fabric switches or storage array type, monitoring hardware from different vendors and correlating the performance and event data onto a *single pane of glass* is a far more complex challenge.

Several attempts have been made to formulate common standards across manufacturers of storage hardware and software, with mixed levels of success. Without a common management platform, policy-based management and providing storage as a service can be extremely difficult to achieve. This creates a serious challenge in heterogeneous storage environments made up of hardware from multiple vendors.

Without common standards, no common access layer can exist between vendor-specific management and monitoring applications. In addition, there can be no multivendor automated discovery of hardware, and policy-based management across an entire class of storage devices is unlikely to be possible, resulting in legacy storage systems providing high operational overhead to administrative teams. The following section covers, at a high level, common standards that may alleviate some, if not all, of these pain points when designing an operational solution for a multivendor, heterogeneous storage platform.

Until relatively recently, the Simple Network Management Protocol (SNMP) was the standard, and possibly the only choice to effectively manage and maintain a multivendor storage and SAN environment. However, SNMP does not typically provide adequate detail, because of its well-known protocol limitations. These limitations include providing no common object model, generally employing a reactive mechanism, poor security (unless version 3 is supported), and failing to provide autodiscovery functions. Although it is true that SNMP still retains some level of adoption for SAN and storage management and monitoring, particularly in heterogeneous environments, newer and emerging standards for deep storage and SAN visibility are likely to change this.

The Storage Networking Industry Association (SNIA) has been working on an initiative to develop common, open storage and SAN management interface standards, based on the Distributed Management Task Force (DMTF) Common Information Model (CIM). This initiative is referred to as the *Storage Management Initiative* (*SMI*), and the specification itself is known as the *Storage Management Initiative Specification* (*SMI-S*).

The primary aim of this initiative is to create common standards that can be implemented by all storage and SAN hardware and software manufacturers, in order to improve interoperability between vendor storage systems, and therefore vastly improve the management and monitoring capabilities of multivendor and heterogeneous environments. By providing this interoperability, operational teams can be furnished with a *single* user interface for executing most, if not all, common operations. The result is substantial benefits to both storage administrators and vendors.

In addition to interoperability and the benefits that this brings to operational management and monitoring, SMI-S allows independent software vendors (ISVs) and their development teams to work from a single unified object model, with a single source of documentation, providing all of the required knowledge for them to interact with and manage diverse SAN and storage components. As a result, SMI-S-enabled products can lead to easier and faster deployment times, and accelerate the adoption of policy-based storage management in heterogeneous environments.

As shown in Figure 10.2, SMI-S forms a layer that sits between the managed objects and the management application, and comprises two different technologies: the Common Information Model (CIM) and the Web-Based Enterprise Management (WBEM) initiative. These combined technologies allow the SMI-S agents to interrogate a storage or switch device, extract the required management data from the CIM-enabled hardware, and provide the collected information upstream to the requesting management application.

The CIM provides the language and methodology for describing management data, allowing this information to be used to perform tasks. The CIM schema includes models for host systems, software applications, network hardware, and other devices. This schema can enable software from different vendors, on different platforms, to describe management data for objects in a standard format, thus allowing it to be used across a diverse set of management applications.

The WBEM provides a common set of management and Internet standard architectures that were developed by the DMTF to help bring together the management of enterprise data-center environments. The DMFT core set of standards that make up WBEM include the following:

- A data model
- The CIM standard
- An encoding specification
- An XML CIM encoding specification
- A transport mechanism
- CIM operation over HTTP

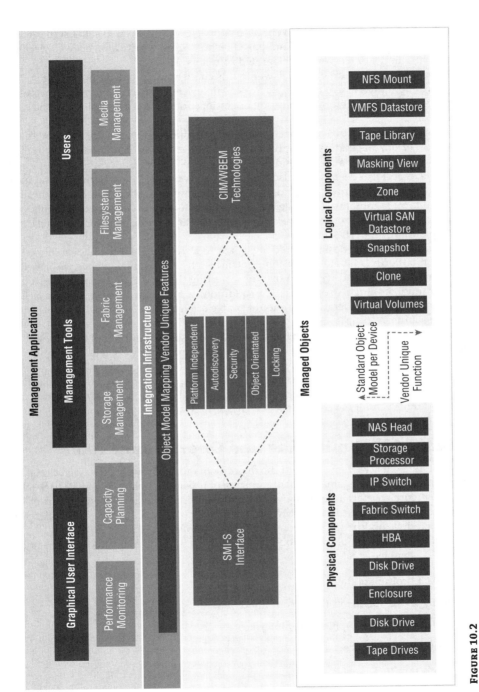

FIGURE 10.2
SMI-S design and specification

Together, these standards enable WBEM to deliver a well-integrated set of standards-based management tools using web and cloud-based technologies, which, alongside CIM, can be used to develop ISV and storage-vendor-integrated solutions.

This results in a software-based answer to this challenging problem, providing monitoring and management coverage across complex enterprise IT organizations and service provider estates, and therefore simplifying operations for those who employ multivendor storage platforms.

Any solution adopted in a heterogeneous storage design should ideally monitor all data-center objects from all vendors, and provide a correlation engine that will feed the metrics collected to operational teams via a single user interface. The ideal monitoring solution should also be able to perform deep root-cause analysis of all connected environments, and indicate how individual component failures might affect applications. It should also have the ability to identify multiple symptoms, and trigger alerts to inform operational teams of events via various mechanisms, such as email, and generate reports that can provide consolidated metric data and capacity planning analysis.

The overall aim of the design should always be to have a single integrated solution that is capable of monitoring all data-center components (such as network switches, SAN switches, storage hardware, HCI and vSphere hosts) and alerting the user of any problem with those components through its single user interface. In addition, as noted previously, in order to facilitate on-call, or out-of-hours operations, it should enable other types of alerting, such as email or cell phone messaging.

Finally, any enterprise monitoring and management solution should also include the functionality to schedule daily tasks and other operational procedures, such as reporting on capacity and performance. However, as you can see, this type of management platform is getting to be complex, and may even require a suite of integrated applications to perform tasks that run across multiple data centers, in order to simplify the job of managing and monitoring multicomponent and multivendor environments (see Figure 10.3).

Virtual SAN Monitoring and Operational Tools

In this chapter so far, we have focused the majority of our attention on the operational aspects associated with the management and monitoring of third-party storage array and SAN-based hardware. Before we start addressing vRealize Operations Manager, we will first provide an overview of the tools provided by VMware to carry out monitoring, management, and troubleshooting activities on Virtual SAN–enabled storage resources.

VMware provides the following tools for monitoring, management, and troubleshooting of Virtual SAN–enabled clusters:

- ESXCLI command line
- Ruby vSphere Console (RVC) command line
- VSAN Observer graphical user interface
- Performance service
- Health service
- vRealize Operations Manager with Management Pack for Storage Devices
- vRealize Log Insight with Virtual SAN Content Pack

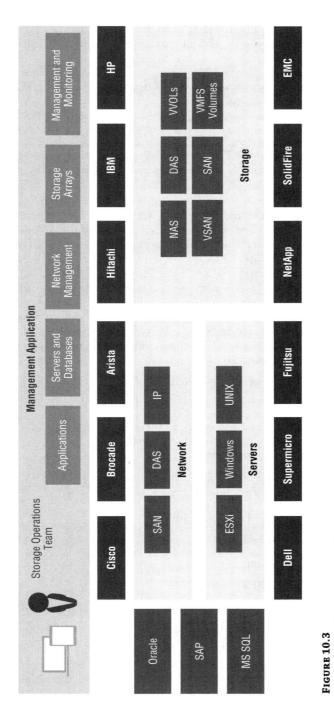

FIGURE 10.3
Target solution for storage and platform monitoring

As illustrated in Figure 10.4, the ESXCLI command-line interface is accessed per host and therefore can provide a host-only view of the state of a Virtual SAN cluster.

FIGURE 10.4
Virtual SAN ESXCLI namespace options

```
[root@hp-001:~] esxcli vsan
Usage: esxcli vsan {cmd} [cmd options]

Available Namespaces:
  cluster              Commands for VSAN host cluster configuration
  datastore            Commands for VSAN datastore configuration
  network              Commands for VSAN host network configuration
  storage              Commands for VSAN physical storage configuration
  faultdomain          Commands for VSAN fault domain configuration
  maintenancemode      Commands for VSAN maintenance mode operation
  policy               Commands for VSAN storage policy configuration
  trace                Commands for VSAN trace configuration
```

The RVC provides a command-line-based user interface for ESXi and vCenter Server that can use Virtual SAN commands for a cluster-centric view of the environment. The RVC can be used to monitor, manage, and troubleshoot the platform through a set of commands, a subset of which are highlighted in Figure 10.5.

The RVC is bundled with both the vCenter Server Appliance (VCSA) and the Windows version of vCenter Server, and is one of the most comprehensive tools available for the management and troubleshooting of a Virtual SAN–enabled cluster.

The vsan.observer is executed from the RVC, and provides a comprehensive list of counters at an engineering level, delivering deep visibility into the mechanics of Virtual SAN. To initiate a vsan.observer session, from an RVC command window, navigate to your Virtual SAN cluster and run the following command:

```
vsan.observer ~/computers/VSAN <options>
```

Available Options

```
Live Monitoring   : --run-webserver --force
Offline Monitoring : --generate-html-bundle
RAW Stats Bundle  : --filename
```

In addition to providing performance information from the whole Virtual SAN cluster, the vsan.observer can display granular data such as the following:

- SSD IOPS, latency, and read cache hit ratio
- Magnetic disk performance
- VSAN CPU utilization
- Virtual machine VMDK performance
- Data refreshed every 60 seconds to give the "average" for that 60-second period

As a result, vsan.observer, among other things, provides the ability to identify bottlenecks anywhere within the Virtual SAN environment.

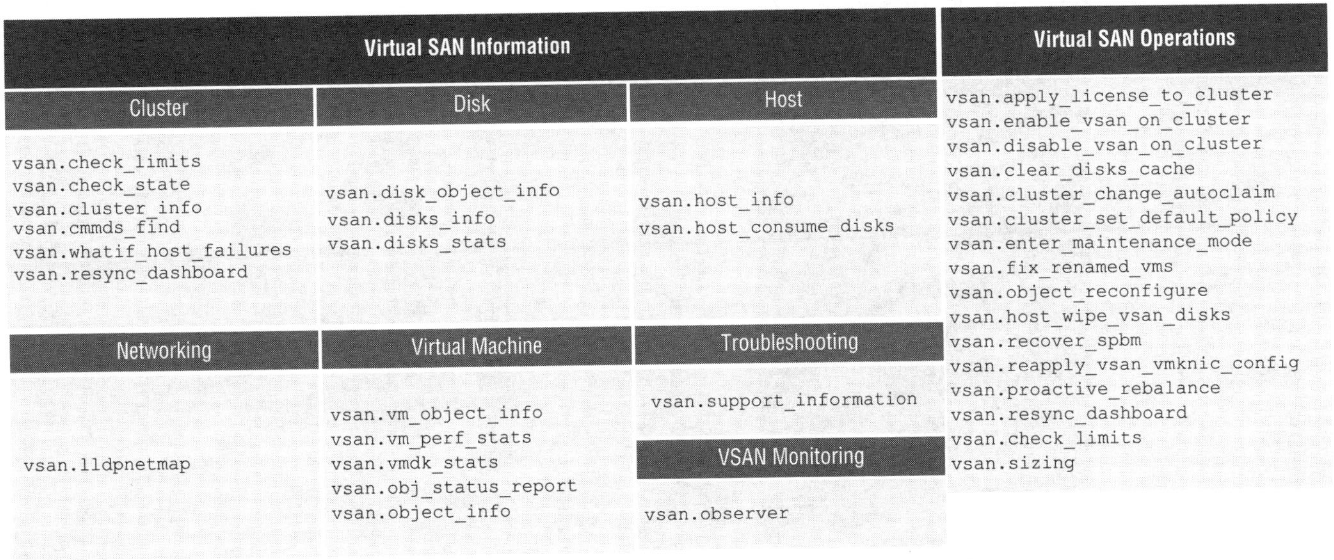

Virtual SAN Information

Cluster	Disk	Host
vsan.check_limits vsan.check_state vsan.cluster_info vsan.cmmds_find vsan.whatif_host_failures vsan.resync_dashboard	vsan.disk_object_info vsan.disks_info vsan.disks_stats	vsan.host_info vsan.host_consume_disks

Networking	Virtual Machine
vsan.lldpnetmap	vsan.vm_object_info vsan.vm_perf_stats vsan.vmdk_stats vsan.obj_status_report vsan.object_info

Virtual SAN Operations

vsan.apply_license_to_cluster
vsan.enable_vsan_on_cluster
vsan.disable_vsan_on_cluster
vsan.clear_disks_cache
vsan.cluster_change_autoclaim
vsan.cluster_set_default_policy
vsan.enter_maintenance_mode
vsan.fix_renamed_vms
vsan.object_reconfigure
vsan.host_wipe_vsan_disks
vsan.recover_spbm
vsan.reapply_vsan_vmknic_config
vsan.proactive_rebalance
vsan.resync_dashboard
vsan.check_limits
vsan.sizing

Troubleshooting

vsan.support_information

VSAN Monitoring

vsan.observer

FIGURE 10.5
Virtual SAN RVC namespaces options

However, as illustrated in Figure 10.6, although this tool provides extensive and detailed data through its GUI, most administrators don't consider it a tool that is intuitive enough to be used for day-to-day operations. In addition, it is unable to provide any historical data, as it displays only a real-time view. As a result, vsan.observer has proven itself to have limited value to storage teams for daily monitoring tasks, and has been predominantly used for troubleshooting activities only. For this reason, among others, with the release of Virtual SAN 6.2, VMware introduced the Performance Service.

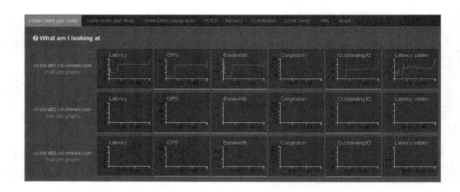

FIGURE 10.6
VSAN Observer user interface

The Performance Service provides out-of-the-box, in-depth Virtual SAN metric information, which unlike VSAN Observer, can easily be viewed from within the vSphere Web Client, illustrated in Figure 10.7. However, unlike other vSphere performance metrics, the Performance Service runs on a distributed database that is stored on the Virtual SAN datastore itself, and not in the vCenter Server database, resulting in no additional overhead on the vCenter management components.

FIGURE 10.7
Performance service status and policy configuration

The Performance Service provides various views, such as cluster latency, throughput, and IOPS, as well as illustrating more-granular views of metrics such as per disk performance, read cache hit ratios, and per disk group statistics. In addition, the Performance Service delivers to administrators an aggregated view of the current status across the entire Virtual SAN cluster, onto a *single pane of glass*, illustrating what the current load and latency look like (see Figure 10.8).

Although all of these charts can be viewed without moving outside the vCenter Server Web Client, both the real-time data and historical statistics can also be accessed via the Virtual SAN API, allowing third-party or custom-written solutions to leverage these metrics.

FIGURE 10.8
Performance Service monitoring and reporting

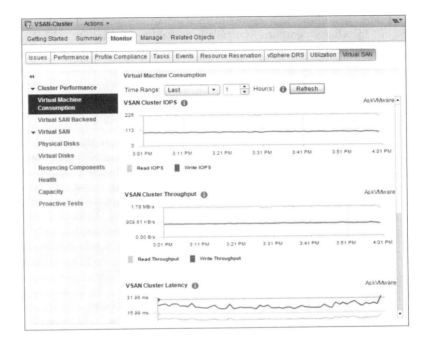

As highlighted in Figure 10.7, the Performance Service database is assigned a storage policy, just like any other object on a Virtual SAN datastore, to control its consumption of disk capacity, availability, and performance. The historical metric database, associated with the Performance Service, can consume up to approximately 255 GB of capacity on the Virtual SAN datastore. If a non-fault-tolerant capability, such as a Number of Failures to Tolerate = 0, is assigned to this object, and the object becomes unavailable as the result of a host failure, the performance history for the cluster will also become unavailable until the object is restored.

Finally, the Health Service feature illustrated in Figure 10.9, which is closely linked to the Performance Service and is also fully integrated into the vSphere Web Client, proactively monitors the health of the Virtual SAN cluster every 60 minutes. This monitoring ensures that hardware compatibility, network connectivity, and storage utilization are maintained. This feature generates an alarm if a threshold or support status is breached.

FIGURE 10.9
Virtual SAN Health Service feature

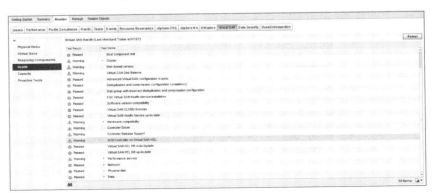

In addition to the monitoring mechanisms and tools we have outlined, VMware provides vRealize Operations Manager and vRealize Log Insight, which can alleviate many storage operational pain points. These tools deliver a complete data-center operational solution by providing end-to-end monitoring for homogeneous and heterogeneous storage, network, and compute environments.

vRealize Operations Manager

VMware vRealize Operations Manager provides a highly scalable, fully integrated operations platform that can unify performance, configuration, and capacity management across an entire IT organization or service provider's diverse multivendor storage estate, as shown in Figure 10.10.

VMware vRealize Operations Manager comes in multiple versions with different feature sets. However, the goal of most storage designs is to provide performance, availability, and capacity management information for the vSphere and storage infrastructure. vRealize Operations achieves this by using management packs and adapters, allowing the visualization of storage and fabric metrics through the VMware Management Pack for Storage Devices and third-party management packs such as EMC Storage Analytics (ESA) or the Blue Medora vRealize Operations Management Pack for NetApp systems.

Management Pack for Storage Devices

The vRealize Operations Management Pack for Storage Devices (MPSD), shown in Figure 10.11, provides deep end-to-end visibility into the storage devices and the storage fabric. The MPSD also provides a correlation between compute, storage, and networking issues to help reduce delays in troubleshooting by taking a holistic approach to metric data.

When monitoring storage end to end, various components are in play, from the hosts to the storage system. The MPSD aims to provide visibility to simplify troubleshooting, by determining the layer in which the problem is located. The MPSD provides this visibility into the storage environment by using many of the common protocols outlined earlier in this chapter. With the MPSD, you can collect performance and health data from various storage devices. The predefined dashboards allow operational teams to follow the path end to end, from the virtual machine to the storage volume, helping to simplify the identification of any issue that may exist along that path. Note that in order to get this end-to-end visibility through the MPSD, you must register the storage provider in vCenter.

The MPSD provides the following key features:

- Support for Virtual SAN metrics
- An end-to-end view of the data path through the SAN or NAS, providing visibility from virtual machine to the storage volume
- Support for NFS, iSCSI, Fibre Channel, and FCoE protocols
- Access to storage devices via standardized protocols, including CIM, SMI-S, and VASA.
- Out-of-the-box dashboards for health and performance monitoring.
- Analytics for commonly identified storage events, such as All Paths Down (APD) and Permanent Device Loss (PDL).

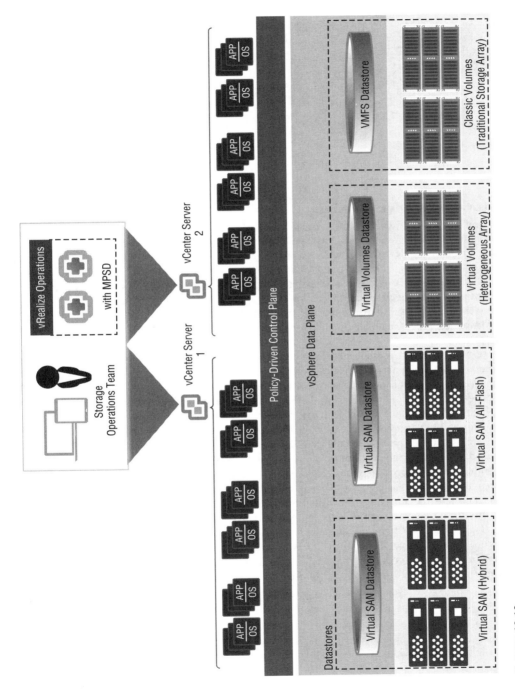

FIGURE 10.10
vRealize Operations Manager logical design

FIGURE 10.11
Management Pack for Storage Devices dashboard view

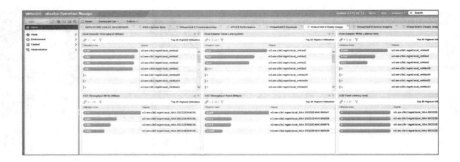

Storage Partner Solutions

Many storage hardware manufacturers and independent software vendors produce vRealize Operations management packs for specific storage and fabric hardware, in order to provide a *single pane of glass* view that can be easily integrated into an existing monitoring and operational solution. Typically, the third-party management packs provide end-to-end visibility, analytics, and capacity planning for workloads running on those storage devices. A third-party management pack uses the vRealize Operations Manager analytics engine to provide deep insight into the storage hardware, delivering automated analysis to proactively expose any indication of impending performance or capacity problems before they can negatively impact business applications.

However, where is the line drawn between what the MPSD can offer operational teams, and what a third-party vendor's specific solution is able to offer, in providing a deeper level of scrutiny into their own storage hardware and software. This question typically needs to be addressed as part of an operational design that plans to use these storage monitoring features, which are available within vRealize Operations Manager.

Figure 10.12 and Figure 10.13 provide some general guidance on this question. Both illustrate where the demarcation line falls by showing what the MPSD can provide visibility into, as compared to a third-party, vendor-specific solution. However, these solutions vary from vendor to vendor, and new product features are added regularly to these offerings, which typically come at an additional cost. Therefore, it is important to provide customers with the latest comparison for their specific hardware, and illustrate to the customer what features are provided by which solution, so that an informed design decision can be made.

Figure 10.13 further compares the features offered by the MPSD, and solutions provided by either the storage vendor or third-party independent software developers. However, once again, it is important to recognize that each vendor's solution will be specific to their hardware and its capabilities, so features and functionality are likely to vary significantly.

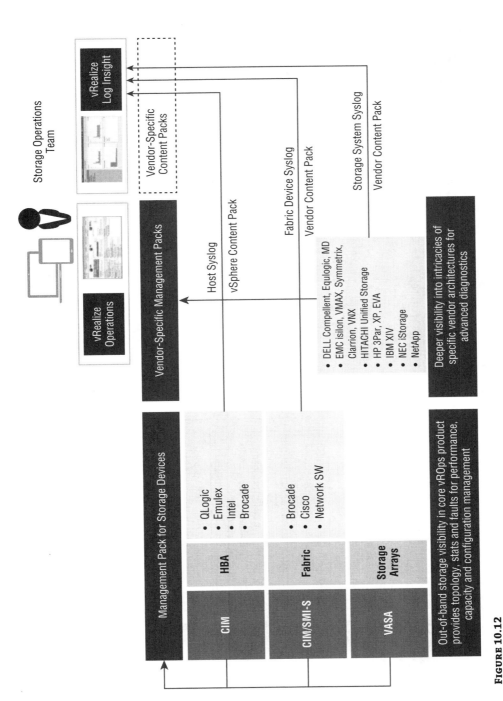

FIGURE 10.12
Overview of vRealize Operations Manager integrated solution

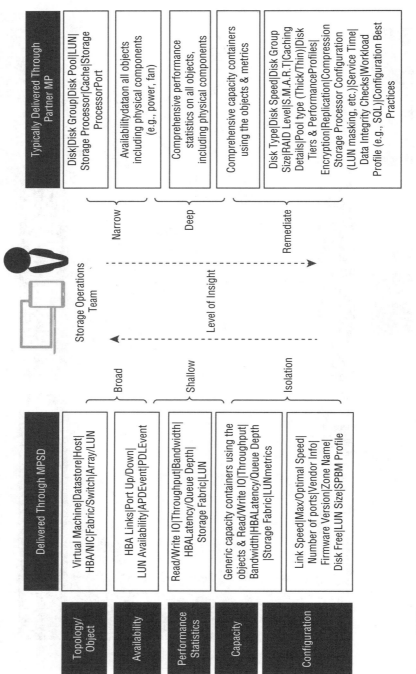

FIGURE 10.13
Feature comparison—MPSD and storage vendor management packs

vRealize Log Insight

VMware vRealize Log Insight is a syslog-based solution for providing real-time log administration, not only for the VMware platform, but also any-third party network, storage, application, or operating system that supports the forwarding of log files through the syslog protocol.

The syslog protocol is a standard framework, defined in RFC 5424, by which a wide range of devices can send event messages to a centralized logging server for analysis, troubleshooting, and security auditing. However, unlike SNMP, syslog cannot be used to poll devices to gather information. Instead, it simply forwards messages to the central location, where special event handling can be triggered by receipt of specific log messages or a defined pattern of messages.

As the syslog protocol has a defined, standard framework, every syslog message should contain five distinct fields, as illustrated in Figure 10.14.

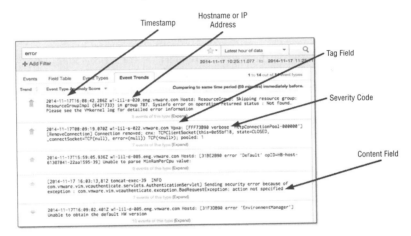

FIGURE 10.14
Syslog message structure

As part of an end-to-end operational monitoring design, hardware and software systems should typically be configured to forward log messages through the syslog protocol, to an external and centralized system message-logging destination. This improves system administration and provides the security and investigative capabilities that most organizations now require. By configuring logging to external syslog servers from all hardware in the data center, including vSphere hosts, storage, network, and application components, vRealize Log Insight provides the framework for the aggregate analysis of log messages on a centralized management appliance, and provides granular visibility into events that are affecting multiple hosts, storage components, or other data-center hardware.

Although vRealize Log Insight is only one log aggregation and analysis tool available on the market, when employed alongside vRealize Operations Manager, it gives operational teams the ability to consolidate logs, monitor and troubleshoot vSphere storage and third-party infrastructure, and perform security auditing, compliance testing, log querying, aggregation, correlation, and retention.

In addition, through the creation of custom dashboards based on saved queries, administrative views can be exported, shared, and integrated with vCenter Server and vRealize Operations Manager. This provides a single unified approach to dashboard creation, monitoring, and operational management. For instance, by adopting this integrated product approach, operational teams can, all from a single user interface, interrogate events and warnings that require administrative attention (such as datastores becoming full or soft media errors). The teams also can be alerted to failures or errors that require immediate attention, such as the following:

- Power-supply unit failures
- Memory-module failures
- SAN or network switch failures
- Disk failures

In addition, this integrated approach can provide continuous monitoring, in combination with automated and smart alerting, to enable operational teams and administrators to proactively avert failures by looking at trends in utilization and performance.

Monitoring data-center systems in this way also allows operational teams to assign different severity levels to different types of alerts or conditions identified across the platform. For example, health-related alerts, such as disk failures, are typically classified as critical, identifying this alert as having immediate adverse consequences. Other events can be assigned an appropriate level of priority:

Informational Useful information currently requiring no IT operational intervention, such as an event notifying operational teams that an authorized user has logged in to a specific audited system.

Warning IT operations attention is required, although the status of the situation is not currently deemed critical. For instance, a virtual disk file has reached the 80 percent capacity full threshold. For this alert, the operations team has time to decide what action should be taken to resolve the issue.

Critical Immediate attention is required by the operational team. If occurring outside business hours, this alert requires on-call intervention, as the current condition will affect system performance or availability. An example is a storage device failure alerting that the assigned operational administrator must ensure that it is brought back online as quickly as possible.

Log Insight Syslog Design

In the following example design, the NOC is located in London, alongside one of three data centers. The two other primary and secondary data centers are located at remote sites. In this sample design, syslog data must be secured end to end, between source and destination. Sensitive syslog messages are secured on the local network and MPLS network by using SSL, and are secured across the public Internet by using a private tunnel connection.

The three data centers send log messages from vSphere hosts, network, and storage devices to a three-node vRealize Log Insight cluster used as part of a full VMware management infrastructure that includes vRealize Operations Manager.

In this example design, the customer has more than 1,500 devices sending log data, and requires a two-node vRealize Log Insight cluster to meet the ingestion requirements of the three data centers. To allow for host failure, without causing impact to log message ingestion, a three-node vRealize Log Insight cluster is configured. In addition, as illustrated in Figure 10.15, each data center uses a pair of syslog aggregators to forward logs to the central NOC in London. The NOC ingests syslog data via an external load balancer provided through a highly available pair of NSX Edge devices, allowing syslog messages to be distributed evenly across the vRealize Log Insight nodes.

It is noteworthy enough to mention again that despite there being a vRealize Log Insight content pack for Virtual SAN, *VSAN trace files* cannot be forwarded by the ESXi syslog daemon to the syslog server, because extensive overhead on the host system would be incurred. VSAN trace files are available locally on the vSphere host only.

End-to-End Monitoring Solution Summary

As outlined throughout this chapter, end-to-end monitoring is the key to providing a robust operational support mechanism for any storage platform. The ability to quickly analyze the impact of a single or multiple failure event can often require an integrated toolset to correlate information from different sources. By using a holistic and integrated approach to tooling, as illustrated in Figure 10.16, the operational team should be able to deduce when a set of seemingly unrelated issues, symptoms, or failings have a single common root cause. An enterprise monitoring solution must be able to not only alert on individual issues, but also correlate and analyze multiple failure events in order to proactively limit the impact to the business from any pending component failure.

Storage Capacity Management and Planning

Achieving true elasticity is a challenge that every enterprise IT organization and cloud service provider faces. Allowing applications and databases to scale up and scale down their infrastructure based on demand requires carefully considered capacity planning, hardware contingency, and trend metrics.

IT organizations and service providers of all sizes must establish a strategy for the storage capacity, management, and availability of their storage infrastructure, as well as other resources such as the compute and network platforms. This strategy should ensure that sufficient storage capacity exists within the infrastructure to meet the current and future needs of the business. It is critical for businesses to reserve sufficient storage, which must then be maintained to prevent applications having to contend with one another for either performance or capacity resources. However, even though it is important that new storage resources be provisioned in a timely manner, in order to avoid contention, it is also equally essential that resource is not overprovisioned or deployed too early, resulting in large amounts of disk remaining idle for a considerable period of time.

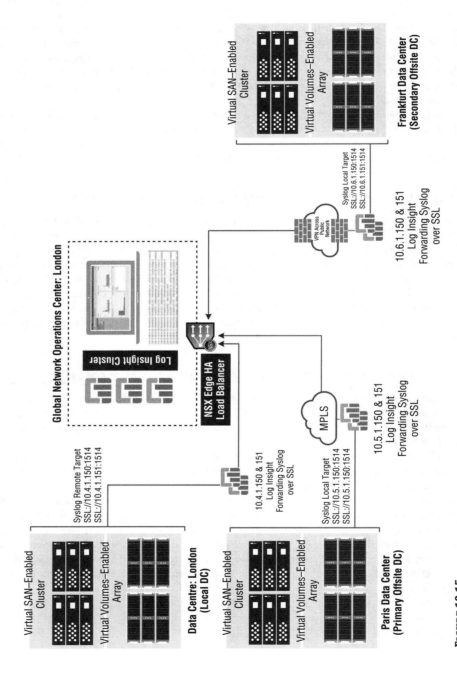

FIGURE 10.15
Design scenario

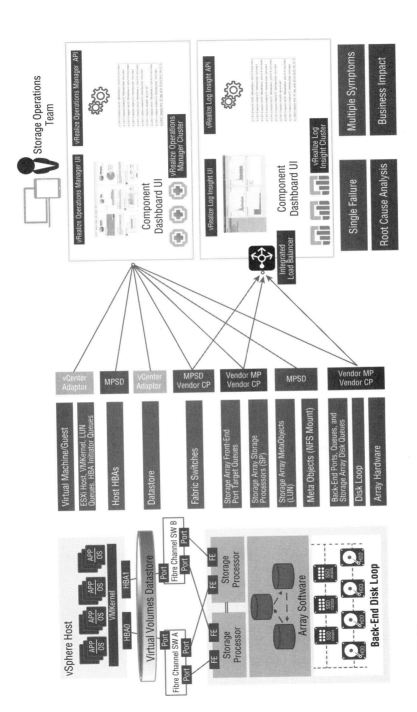

Figure 10.16
End-to-end monitoring

Management Strategy Design

To meet the following objectives, the process and business strategy to support a centralized approach to capacity management needs to be developed and adopted with defined activities, such as timeframes and the roles and responsibilities of those involved in the process. The primary objectives of capacity management may include the following:

- Supporting the organization's existing budgetary planning process and timeframe for approval and procurement
- Optimizing the capacity and performance utilization of the existing storage platform, in order to maximize the return on investment (ROI)
- Delivering reliable performance and maintaining SLAs for existing applications and workloads.
- Ensuring prompt provisioning of new storage without stressing the provisioning process or making panic purchases of additional resources, such as hosts, disks, or storage network equipment.

In doing so, the storage capacity and management strategy should also aim to do the following:

- Define the total current capacity of the platform.
- Define a storage capacity planning process.
- Define a storage capacity model.
- Identify requirements for storage capacity model implementation.
- Identify and set storage provisioning trigger points.
- Ensure that a storage capacity and a budget forecast exists.
- Establish capacity thresholds and capacity monitoring procedures.
- Ensure ongoing financing for the existing and new storage requirements.

Defining and implementing a storage capacity strategy, such as that outlined here, also plays into virtually all aspects of a business's organization and operating model, which is likely to include these elements:

- Organizational structure
- Cross-organizational integration and alignment
- Business support systems (BSS)
- Operations support systems (OSS)
- IT service management (ITSM)
- Process retrenching and improvement
- Reviewing the applicability of current IT services

Process and Approach

The capacity management process and its related activities aim to support the preceding objectives through simple predefined and planned actions. These actions must typically be supported by accurate storage capacity and utilization data, the verification of metadata, and regular capacity growth, through daily data-center operational tasks.

The approach required to manage storage capacity involves various steps (shown in Figure 10.17) to determine both the current unused or reserved capacity, and the forecasting of new requirements, as well as the planning for the deployment of new or additional storage resources. In addition, the continuous evaluation and improvement of the workflow defined as part of the capacity management strategy is key to minimizing the impact of the process to the business applications and workloads.

Figure 10.17 illustrates an example of a defined capacity and performance management strategy for storage resources, and the steps required by the process.

By using regularly scheduled capacity reports and monitoring storage resources, operational teams can plan and proactively evaluate the storage infrastructure and its resource consumption. However, providing and guaranteeing available usable capacity levels is also tied into daily IT operational activities, such as these:

- Monitoring storage infrastructure for alerts and events
- Collecting performance and utilization data
- Verifying available usable capacity, and validating change requests when required
- Investigating storage performance events as they occur, to determine whether overutilization of storage capacity is the root cause of the issue
- Initiating and managing the procurement and provisioning of additional storage resources in the data center

In addition, from a continuous improvement perspective, the reporting and planning activities and their supporting tools, must be tuned and kept aligned with business and operational requirements. For instance, the process being used must be capable of comparing the capacity utilization levels on storage devices to the observed levels of I/O, and fine-tune the process to drive greater use of resources, without sacrificing availability or performance. Also, these processes need to provide the ability to measure the delivery timeframe for hardware components (such as hosts, storage network devices, and disks) and to take into consideration the time required to gain access to data-center facilities. All of these considerations must be built into the process in order to understand the dependencies, and to optimize the actual provisioning timeframe and procurement life cycle, resulting in shorter deployment times and therefore making the outcomes more predictable.

Finally, the capacity management strategy and its processes clearly cannot stand alone, as they are closely tied to both performance and availability management. The personnel responsible for capacity management should also be intrinsically involved with operational issues or incidents, which are linked to either the performance or availability of storage, in order to ensure that the defined process operates as part of an integrated approach. By adopting this methodology, the IT organization can ensure that capacity utilization and provisioning is not addressed as a stand-alone process, without regard for other aspects of storage design and implementation.

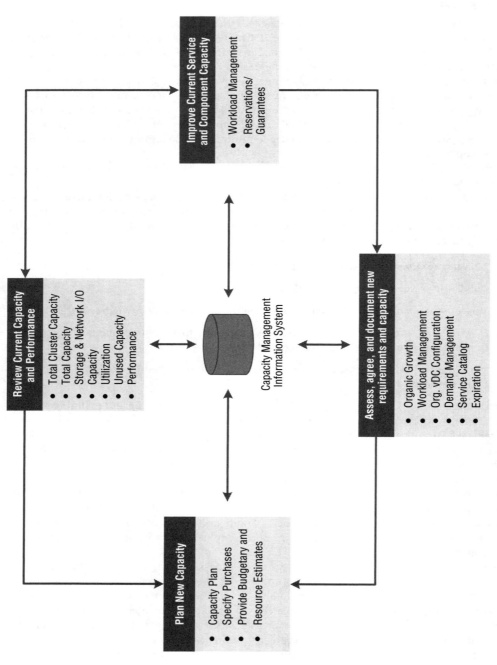

Figure 10.17
Capacity and performance management process

Capacity Management for Virtual SAN

The approach to capacity management for Virtual SAN–enabled storage will differ from traditional array-based systems in a various ways. Traditionally, storage capacity management and scalability was, to some extent, governed by the storage vendor's hardware architecture, with clear and specific guidance provided by the vendor on expanding the storage array into additional disk bays. This often had to be considered as part of the physical design, in terms of the layout of storage cabinets in the data center, as illustrated in Figure 10.18.

However, Virtual SAN does not typically require complex forecasting such as this. It also provides an elastic model, which can grow or shrink on demand, by granularly providing the ability to add single nodes or disks without disruption or application downtime.

The Virtual SAN HCI building-block model aims to provide the ability to procure what you need when you need it, in order to scale to add capacity or to increase performance, without disrupting operations (see Figure 10.19).

As outlined earlier in this book, Virtual SAN allows you to either scale up or scale out. Scaling the infrastructure out by adding nodes increases capacity and performance, in terms of both compute and storage. This is realized through the capacity disks increasing storage resources, new caching flash devices contributing to I/O performance enhancement, and new CPUs and memory adding compute resources.

Likewise, depending on the level of compute utilization in the host, available disk slots, network utilization, and the building-block design approach being adopted in the design, it may be more appropriate to scale up storage resources through the addition of new disks or disk groups in existing Virtual SAN hosts. New capacity disks being added to an existing disk group increases storage capacity only, with the provisioning of new disk groups allowing the host to be scaled up in terms of both capacity and performance, through the addition of new caching devices. However, Virtual SAN does not automatically distribute data to the newly added storage devices. Proactive rebalance operations must be initiated manually by using the RVC command `vsan.proactive_rebalance` when adding or removing storage.

Either way, the end result provides IT organizations and service providers the ability to elastically and granularly scale storage resources in a nondisruptive way, effectively creating building blocks out of what you need to modularize the data center. This approach typically reduces the need for large up-front CapEx investment, and provides a more linear level of predictability for capacity forecasting. It also aligns nicely with the current trend of applications moving toward a scale-out or distributed architecture model.

Summary

This chapter has identified and addressed all of the primary concerns related to storage platform monitoring and management, in order to maintain an environment that supports optimum health, capacity, performance, and security of business-critical data.

Finally, let me leave you with these last few words. The software-defined data center is not just a part of VMware's marketing department's vision; it is an approach, which hundreds of IT teams around the world are embracing. The software-defined enterprise model decouples the software from the hardware, after many years of trying to bring them closer together, to provide improved agility, flexibility, and efficiency.

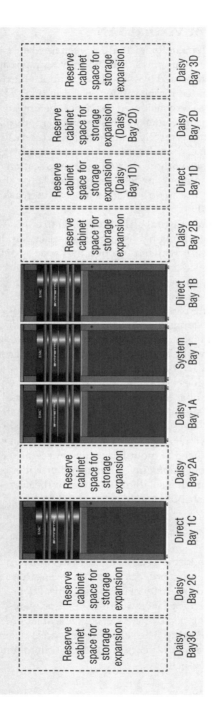

FIGURE 10.18
EMC Symmetrix VMAX layout and expansion

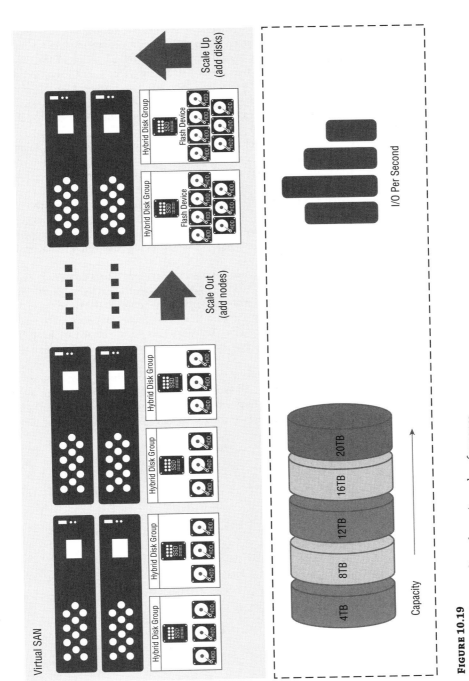

FIGURE 10.19
Virtual SAN elastic scaling of capacity and performance

For most IT organizations and service providers, the shift toward the software-defined data-center model is a journey made up of numerous steps. For most, software-defined compute starts the transformation. *Software-defined compute* refers to the virtualization of the $x86$ platform, as pioneered by VMware during the late 1990s. The virtualization of the compute platform has been well established in data centers over the last 15 years, and provides a wide range of data-center availability and performance-based optimizations, including the expansion of the software-defined compute platform to all application types in the data center, and enabling cross-data center mobility.

The other steps are typically more interchangeable when it comes to the order of adoption. This includes the move toward transforming IT operations through the introduction of automation, in order to replace the manual approach to IT tasks through orchestrated workflows, and also the introduction of software-defined networking into the data center.

Software-defined networking refers to the virtualization of the network layer and the transport technologies found in the data center, facilitating highly scalable logical networks and software-based network devices, such as firewalls, routers, and load balancers. The primary aim of network virtualization is to increase speed, provide efficiency of operations, and enable simplified disaster-recovery models.

Then, as per the focus of this book, software-defined storage introduces a new operational model, which is driven by applications and not hardware capability. With the traditional classic storage arrays no longer meeting many IT organizations' requirements, VMware's software-defined storage model is leading the way in the IT industry to drive efficiency, automation, and simplification, while maintaining enterprise-class features and performance, making Virtual SAN and Virtual Volumes the ideal storage platforms for the next-generation of virtualized applications.

I think we have seen enough in the IT industry already to know that these software-defined concepts are going to completely transform all aspects of data-center infrastructure over the next few years, and finally allow these hypervisor-driven concepts to bind together the compute, network, and storage layers for the optimization of the next-generation of virtual platforms.

Index

Note to the reader: Throughout this index **boldfaced** page numbers indicate primary discussions of a topic. *Italicized* page numbers indicate illustrations.

Numerals

3D NAND TLC, 228
*x*86 architecture
    compute virtualization of, **2**
    server platform, 18
6296UP Fabric Interconnect devices, 358

A

absent failure event, **289**
access control lists, with NFS, **156**
access control, zoning and, 125
active/active disk arrays, vs. active/passive, 172
active/active NAS environment, 62
active data, deduplication of, 60
active/passive NAS system, 62
admission control policy, configuration, *336*
Advanced Service Designer (ASD), 463, *464*
    workflow example, *465*
affinity group configuration, 334–335
aggregated switch IP SAN, design example, *148*
AHCI SATA adapter, 72
airflow, monitoring storage, 481
alerts
    assigning different severity levels, 498
    in monitoring solution, 473
all-flash disk arrays, **65**
all-flash erasure coding, **256–260**, *257*
all-flash solution, nonvolatile cache for, 227
all-flash Virtual SAN model, 245
All Paths Down (APD), 78
allocated storage, **480**
ALUA (Asymmetric Logical Unit Access), 173, *174*
Amazon
    cloud storage platforms, 3–4
    Simple Storage Service (S3), 457
anti-affinity rule guidelines, for cloud management cluster applications, 408
anything as a service (XaaS), 461, **463**, **465**
application-based pools, 25
applications
    assessment, and storage-policy design, **268–271**, *272*
    control with Virtual Volumes, **450**
    dependencies, 270
Arbitrated Loop (FC-AL), 115–116
array-based snapshots, **448**
array-based thin provisioning, **449**
as a service, 458, *459*
ASD (Advanced Service Designer), 463, *464*, *465*
Asymmetric Logical Unit Access (ALUA), 173, *174*
asynchronous replication, bandwidth requirements, 63
atomic test-and-set (ATS), 90
authentication, Kerberos for nonroot user, 156
AUTH_SYS mechanism, 156
automated tiered storage, **52–54**, *53*
automated virtual machine placement, 335
autotiering solutions, 50
availability. *See also* High Availability (HA)
    designing for, **287–306**, *288*
        hardware component failure, **289–292**
    fault domains, **302–306**, *304*
    of monitoring solution, 473
    requirements, 269, *270*
    and storage design, 7

B

backup server, VADP to back up virtual machines from, 92
backups, storage array-based, **62–63**
bandwidth, 315
    calculating for stretched clusters, **325–326**
    as data-center interconnect design factor, 354
    and replication, 62–63
banner messages, 427
baseline hardening configuration, 424
Best Effort system class, 163
binding operations, for Virtual Volumes, **442–444**, *444*
blade servers, 216
Blade systems, 106
block zeroing, for Virtual Volumes, 436
block zeroing primitive, 89–90
boot devices

architecture, 404
local disks and, 398
in Virtual SAN, **221–222**
Boot from SAN (BfSAN), **92–94**, **132**, *133*, *134*
iSCSI, **148**
bootstrap, 419
bottlenecks, vsan.observer to identify, 488
budget, and storage design, 8
business community, quantifying objectives for, 6
business data, and business success, 13
business goals, quantifying, 6
BusLogic adapter type, 72

C

cables, 101
SAN, 107
cache-tiering technology, 52
caching
algorithms in Virtual SAN, **205–206**
client-side read, 205
capacity
business data trends on, 473
calculating, 417
elastic scaling of, *507*
expansion in switched fabric, 117
management and planning, **499–505**, *504*
management strategy design, **502**
process and approach, **503**
for Virtual SAN, **505**
monitoring, **474**, 478
for storage array, **480**
reduced waste for Virtual Volumes, **450**
and storage design, 10
capacity disk failure, 368
capacity disk sizing, **282–287**
capital expenditures (CapEx), 11, 15
CBT (Changed Block Tracking), 92
centralized logging, 426
centralized server management, Boot from SAN and, 93
Challenge-Handshake Authentication Protocol (CHAP), security, **139–140**
change and configuration management, **427**
Changed Block Tracking (CBT), 92
channels, in CWDM, 342
chassis, 10

checkpointing content database, in vRealize Orchestrator, 466
Cisco C-Series hardening baseline, 426
Cisco C-Series management hardware, 401
Cisco C-Series servers, 400
Cisco FlexFlash SD cards, 404
Cisco Integrated Management Controller (IMC), 426
Cisco, LISP configuration, **363**
Cisco Nexus 5548UP hardening baseline, 427
Cisco Overlay Transport Virtualization (OTV), 322, 356
Cisco UCS C240 rack-mount servers, 401, 415
Cisco UCS environment physical connectivity, Virtual SAN with, *403*
Cisco UCS Fabric Interconnects, 401
Cisco UCS Virtual SAN Ready Nodes, 401
Cisco Unified Computing System C240 M4 rack-mount servers, 358
Cisco vPC
over DWDM and dark fiber, **358**, **360**, *361*
Overlay Transport Virtualization (OTV), **360**, *362*, **363**
classes of service (CoS), 163
classic storage model, 19, **21–66**, 22
challenges, 189, 429
vs. Virtual Volumes, *433*
vSphere APIs for Storage Awareness (VASA), **90–91**, *91*
vSphere storage policies, **94–95**, *95*
vStorage APIs for Array Integration (VAAI), 89–90
client-side read cache, 205
CLOM (Cluster Level Object Manager), **310**
daemon, 302, *304*
clomd.log file, 218
clones, in classic storage, **63–64**, *64*
cloud computing, common services, *460*
cloud management platform (CMP), 58, 453
architectural overview, **461–468**
vRealize Automation, **461–465**
clusters, 395, 397
anti-affinity rule guidelines for applications, 408
enterprise, *399*, 401
overview, **458–461**, *459*
security, **423–426**
for virtual machines, **424–425**

storage policy-driven, *449*
cloud service providers, 3
  disk pools from, 59
Cluster Across Boxes (CAB) virtual machine clustering, 79, *80*, 81
Cluster Level Object Manager (CLOM), **310**
  daemon, 302, *304*
Cluster Monitoring Membership and Directory Services (CMMDS), 287, **308**, **310**
clusters. *See also* stretched clusters
  architecture, **415**
  capacity calculation, 283
  configuration, **404–408**
    vSphere High-Availability (HA), **404**, **406**
  datastore, **82–83**, *83*
  host design and planning, for host failure, **292–299**
  rebalance operations, **290**, *291*, **292**
  tiered workload virtual SAN, *235*
CMP. *See* cloud management platform (CMP)
coarse wavelength-division multiplexing (CWDM), 340
  in data center interconnect design, 355–356
cold data, deduplication of, 60
Common Information Model (CIM), 484, 486
communication protocol, 107
communications, between cloud management virtual machines, 408
compatibility, Boot from SAN and, 93
complexity, 8
compliance, and storage design, 7
component-level scrubbing, 216
compression, for all-flash solution, **202–205**, *203*
compute and storage-layer security hardening, **426**
compute nodes, determining number in cluster, 292
configured storage, **480**
congested datastore, 86
Congestion Notification, 165
connectivity, fabric, 101
  and storage I/O architecture, summary, **184–186**
constraints, 398
control plane, in SDS, 189
controllers, failure, multipathing software for access after, 61

converged enhanced Ethernet (CEE). *See* data center bridging (DCB)
converged network adapters (CNAs), 143, **165–166**, *166*
core, aggregation, and access model, 242, *243*
core dump, 219
core-edge fabric topology
  core-edge, **117–118**, *118*, **121**, 134
  dual-core, *119*
  edge-core-edge, *121*
  single-core, *119*
core storage, establishing requirements, 15–16
core tier, 118
costs, 8
  as data-center interconnect design factor, 355
  of dedicated IP SAN fabric, 137
  hardware for Virtual SAN cluster, 372
  of SAN storage, 94
  and storage protocol decision, 186
CPU
  off-loading, 105
  overhead, **223**
Critical event, 498
custom IT services, 463
customer design requirements and constraints, **398**, **400–404**
  host compute design, **400–402**
customer profiling, 13
CWDM. *See* coarse wavelength-division multiplexing (CWDM)
cyclic redundancy check (CRC32) algorithm, 215

D

dark fiber, **340**, *341*
  Cisco vPC over, **358**, **360**, *361*
  in data center interconnect design, 355–356
  OTV over, **360**, *362*, **363**
  stretched VLANs over, *349*
DAS (direct-attached storage), **180–182**, *181*
das.isolationAddress option, 298, 336
das.useDefaultIsolationAddress option, 298, 336
data at rest, 61
Data Center Bridging Capabilities Exchange (DCBX), 165
data center bridging (DCB), 162
  attributes, 165
data centers, 3

interconnect design considerations, **354–356**
web-scale pod scale-out strategy, *375*
data compression, **60–61**
data in flight, 61
data inflight encryption, 353
data locality, in Virtual SAN, **205–206**
data loss, from storage array-based snapshots, 64
data plane, **437–444**
    protocol endpoint (PE), **440–442**, *442*
        provisioning process, *443*
    in SDS, 189
    storage containers (SCs), **437–440**
Data Protection, 313
data routers, 107
data, value changes over time, 14
datastore, *73*, **73–75**
    capacity disk sizing, **281–282**
    creating, 311–312
    design and sizing, **271–287**
        capacity disk sizing, **282–287**
        disk stripes per object, **276**
        hosts per cluster, **273**, **275**
        Object Space Reservation (OSR), **276**
        usable capacity calculation, **285**
    monitoring capacity utilization, 477
    number of virtual machines sharing, 75
    signatures, **75–77**
    types, *76*
datastore clusters, **82–83**, *83*
datastore heartbeat system, 294
dedicated linked clones, storage policy for, 394
dedicated VMware, 98
deduplication, **60–61**, *204*, **449**
    for all-flash solution, **202–205**, *203*
    in backup, 63
default controller, for virtual machines, 71
default gateway routes, **327**
default storage policy, **267–268**
degraded failures, **289**
Delete Status (Dead Space Reclamation), 90
denial-of-service (DoS) attacks, 60
dense wavelength-division multiplexing (DWDM), 340, 342, *343*
    Cisco vPC over, **358**, **360**, *361*
    in data center interconnect design, 355–356
    OTV over, **360**, *362*, **363**
dependent hardware iSCSI adapters, 140, **141**, *142*
deploying

stretched VLANs, **347–354**, *348*
    over dark fiber, *349*
    in vRealize Orchestrator, 467
design sequence methodology, *5*
destaging mechanism
    performance, 260
    in Virtual SAN, **206**
digital rights management (DRM), in vRealize Orchestrator, 466
direct-attached storage (DAS), **180–182**, *181*
Director class Fibre Channel switches, 107
dirty reads, 231
disaster avoidance, stretched cluster environment for, 318
disaster recovery
    architecture example, *385*
    datastore UUID significance for, 77
    quantifying objectives for, 6
    from replicated data, 383
    with SAN, 93
    storage array-based, **62–63**
    storage platform for, 386
disk arrays, active/active vs. active/passive, 172
disk drives, risk of contention, 25
disk failure
    hot spares for, 42
    risk of multiple, 43
disk groups
    configuring multiple, **276**, *277*, **278**
    in Virtual SAN, **194–198**, *195*
        configuration example, *199*
disk pools
    homogeneous or heterogeneous, 52
    provisioning, 24–25
    shared or separate, 59
    vs. traditional model of LUN, 23–24, *24*
disk scrubbing, **215–216**
disk stripes per object, sizing impact of number, **276**
distance, as data-center interconnect design factor, 354
Distributed Management Task Force (DMTF), 484
Distributed Object Manager (DOM), **310**
    object mirror I/O path, *311*
distributed parity blocks, data disks with, 38–39, *40*
Distributed Power Management (DPM), 298, 313
distributed RAID, **214–215**

Distributed Resource Scheduler (DRS), 77, 312, **332–335**, *333*, 398
    for clusters, **406–408**
    in Virtual SAN
        parameters, 299
        resource balancing, 300
        transparent maintenance, 300
distributed storage system, 205
distributed virtual switches, vs. vSphere standard, **238**
DOM (Distributed Object Manager), **310**
    object mirror I/O path, *311*
DPM (Distributed Power Management), 298, 313
DRAM, cache-tiering between SSD and, 52
drive writes per day (DWPD), 229
DRS. *See* Distributed Resource Scheduler (DRS)
DRS (Distributed Resource Scheduler), 312
dual-core, core-edge fabric topology, *119*
DWDM. *See* dense wavelength-division multiplexing (DWDM)

E

eager zeroed thick (EZT) format, **69–70**, 261
ECMP (equal cost multipathing), 377
edge-core-edge fabric topology, 134
    dual-core, *121*
edge Fibre over Ethernet model, **167**, *169*
edge switches, 106
edge tier, 117
EMC CloudArray, 395
EMC Symmetrix VMAX, layout and expansion, *506*
EMC, ViPR, 58
eMLC (enterprise multilevel cell) device, 228
encapsulation, 350
encryption, 61
    data inflight, 353
end of frame (EOF), 163
end-to-end Fibre Channel over Ethernet, 170, *171*
end-to-end monitoring solution, **499**, *501*
endpoint (N_Port), 110
endurance-class devices, 227, 228
endurance flash device, 201
endurance metrics, 229
Enhanced Transmission Selection (ETS), 165
Enhanced vMotion, 82, 312
Ensure Accessibility option, **301**
Enterprise-class switches, form factors, 182

enterprise IT organizations, VMware change to, 3
enterprise multilevel cell (eMLC) device, 228
enterprise storage
    heterogeneity of infrastructure, 12
    software-defined, *188*
    system development, 187
enterprises, 3
environment
    current state analysis of, 6
    sizing for customer needs, 210
environmental monitoring, **481**
EoMPLS (Ethernet over MPLS), 350
E_Port (expansion port), 110
equal cost multipathing (ECMP), 377
erasure coding, 214
    all-flash, **256–260**, *257*
    capacity and configuration requirements, 260
    web client configuration, *259*
ESXCLI command-line interface, *488*, 488
    for configuring Virtual SAN datastore, 419
    vsan policy setdefault command, 267
    vsan trace set command, 221
ESXi
    host communication with storage devices, 75
    host hardware, specifications, 401–402
    memory dump configuration, 219
ESXi hypervisor, 217
    iSCSI software adapter in, **141**
    thin provisioning and, 46
/etc/vmware/esx.conf file, *175*
Ethernet frames, tunneling, 350
Ethernet over MPLS (EoMPLS), 350
Ethernet, storage over, 136
expansion port (E_Port), 110
Extended Unique Identifier (EUI), 138
extensibility, 373
EZT (eager zeroed thick) format, **69–70**, 261

F

Fabric Binding, 131
fabric connectivity, 101, 116
    and storage I/O architecture, summary, **184–186**
fabric controller, 108
Fabric Login (FLOGI), 110
fabric loop port (FL_Port), 110
fabric port (F_Port), 110
fabric services, **108–109**

components, **106–107**
core-edge topology, **117–118**, *118*, **121**, 134
    dual-core, *119*
    edge-core-edge, *121*
    single-core, *119*
fabric switch zoning, **125–126**, *126*
failover, in iSCSI storage multipathing, *179*
failure
    number to tolerate, **254–260**, *255*
        all-flash erasure coding, **256–260**, *257*
        sizing impact of, **275**
    tolerance for, 198
failure domain, 368
Fault Domain Manager (FDM), 294
fault domains, **302–306**, *304*
    architecture, **318**, *319*
    design, *305*
    preferred sites and, 329
    sample architecture, *307*
fault tolerance, 23
    absence in RAID 0, 33
    RAID 1 and, 34
    RAID and, 26
    in Virtual SAN, 193
FCoE. *See also* Fibre Channel over Ethernet (FCoE)
federated overlay software, 58
federated storage solutions, 4
Fibre Channel, 101
    frame, 163, *164*
    hubs, 106
    storage protocol design factors, *185*
    summary, **132**, **134–135**
Fibre Channel over Ethernet (FCoE), 101, 161–170, *162*
    design options, **167–170**
        edge Fibre over Ethernet model, **167**, *169*
        end-to-end Fibre Channel over Ethernet, *170*, *171*
    distance limitations, 163
    infrastructure, 167, *168*
    physical components, **165–167**
        converged network adapters (CNAs), **165–166**, *166*
        switches, *166*, **166–167**
    storage protocol design factors, *185*
    summary, **170**

fibre channel SAN, **102–135**
    architecture, **104**
    components, **104–105**, *105*
        fabric components, **106–107**
        fabric services, **108–109**
        Fibre Channel address mechanism, *109*, **109**
        Fibre Channel port naming, **110**, *111*
        host components, **105–106**
        node ports, 109
        storage components, *107*, **107–108**
    Fibre Channel protocol, **102–103**
        vs. iSCSI, or NAS, 150
        layers, **103–104**, *104*
    login types in switched fabric, **110**
    SAN management software, **112**, **114**
    switch-based fabric architecture, **117–125**
    topologies, **115–117**
    World Wide Name device addressing, **110**, **112**, *112*, *113*
Fibre Channel Security Protocol (FC-SP), 131
Fibre Channel switch, 106
file services, Virtual SAN, **395–404**
file sharing, DAS model for localized, 181
firewall, **249**
Fixed pathing policy, 175
fixed switches, 182
flapping, avoiding unnecessary, 158
flash devices, 225
    for boot media, 222
    life expectancy, 202
    pooling, 194
    for read-caching and write-buffering, 193
Flash Read Cache Reservation, *263*, **263–264**
flattened storage, 12
-flat.vmdk file, 68, 69
FL_Port (fabric loop port), 110
folded clos design, 377
Force Provisioning, *264*, **264–265**
F_Port (fabric port), 110
front-end port, failure, multipathing software for access after, 61
full clone desktops, storage policy for, 394
full copy primitive, 89
full mesh topology, *123*
full workload data-center failure, in Virtual SAN stretched cluster, 364
fully autotiered environment, *97*, **97–98**

G

gateway NAS, **152**, **154**, *154*
gateway routes, default, **327**
generic port (G_Port), 110
GIF/GRE tunnels, 322
Global Traffic Manager (GTM) devices, 358
goals, and storage design, 8–9
Google, cloud storage plaforms, 3–4
governance, and storage design, 9
G_Port (generic port), 110
groups, node ports segmented to, 125
GTM (Global Traffic Manager) devices, 358
guest operating system, on virtual machine, 2

H

hard disk drives
    capacity improvements, 1981 to 2014, *11*
    cost per gigabyte, 1981 to 2014, *10*
hardware-assisted locking, 90
hardware initiators, 135
hardware iSCSI adapter, independent, **143**
hardware resources
    designing for component failure, **289–292**
    high availability, **61–62**
    scaling up vs. out, 57
    for virtual machine, 2
HBA (host bus adapter), 101
HBA queue depth, **478–479**
Health Insurance Portability and Accountability Act of 1996 (HIPAA), 7
health monitoring, **474**
    for storage array, **480**
    of storage fabric, **478**
Health Service, in Virtual SAN cluster, 491
heterogeneous disk pools, 52
High Availability (HA)
    of hardware resources, **61–62**
    iSCSI adapters, **143**
    with SAN, 93
    for stretched clusters, **335–338**
    in Virtual SAN, parameters, 299
    in vSphere
        heartbeat datastores, **297**
        host isolation addresses, **297–298**
        network communication, *295*
        operational comparison, 294
        quorum logic design and, **302**, *303*
        and Virtual SAN, **292**, *293*, **294–298**
        WAN interconnect design, **353**
historical metric database, and Performance Service, 491
homogeneous disk pools, 52
Horizon portfolio (VMware), **392–395**, *393*
host adapter (host interface device), 101
host boot architecture, **217–221**
host bus adapter (HBA), 101
host cluster, design and planning, for host failure, **292–299**
host components, **105–106**
    maximum component count per, 210
host compute design, **400–402**
host CPU overhead, for Virtual SAN, **223**
host form factor, design requirements, **216–217**
host interface device, 101
Host Profiles, 313
host resources, 402
host servers, monitoring storage on, **477**
hosts per cluster, **273**, **275**
hot-added virtual disks, 78
hot-removed virtual disks, 78
hubs, in JBOD environments, 116
humidity, monitoring storage, 481
hybrid cloud model, *460*, 460
hybrid disk group, and performance, 273
hybrid switches, 183
hyper-converged infrastructure (HCI) hardware architecture model, 12, 18
hypervisor layer, 2
hypervisor platform, 1

I

IaaS (infrastructure as a service), 458
IEEE 802.1Qbb standard, 137
IGMP (Internet Group Management Protocol), 248
ILM. *See* Information Lifecycle Management (ILM)
IMC (Integrated Management Controller), 426
in-memory cache, 201
INCITS (International Committee for Information Technology Standards), 102
independent data disks, with independent parity schemes, 40–41
independent hardware iSCSI adapter, **143**
independent software vendors (ISVs), 17
Information Lifecycle Management (ILM), 13–14
    key challenges, *14*

Informational event, 498
infrastructure as a service (IaaS), 458
initiator, for iSCSI, 135
installing, vCenter Server, **418–419, 422–423**
Integrated Managetment Controller (IMC), 426
integration of components within design, 5
intelligent compression, 60. *See also* deduplication
inter-array tiered storage, 52, 54
Inter-Switch Links (ISLs)
    encryption, 131
    and switch design evaluation, 183–184
    in switched fabric, 117
International Committee for Information Technology Standards (INCITS), 102
International Organization for Standardization (ISO), Open Systems Interconnect (OSI) model, 103
Internet Group Management Protocol (IGMP), 248
Internet (IP), in data center interconnect design, 355–356
Internet Protocol (IP), 101
    blocking, 426
Internet Protocol Security (IPSec), 353
Internet Small Computer System Interface (iSCSI), 101
    Boot Firmware Table (iBFT), 148
    Boot from SAN, **148**
    device-naming standards, **138–139**
    multipathing, **177–178**
        failover and load balancing, *179*
    vs. NAS or Fibre Channel, 150
    network adapters, **140–143**
    protocol summary, **148**
    storage protocol design factors, *185*
    traffic isolation, **137–138**
intra-array tiered storage, 52
intra-cluster communication, multicasting for, 248
I/O blender effect, 273, *274*
I/O controller, and performance, 224
I/O failure message, Queue Full (qfull), 479
I/O per second (IOPS), 30, 475
    Performance Service for, 490
    translating MB/s into, 86
iormstats.sf file, 86
IP (Internet), in data center interconnect design, 355–356
IP SAN, 102
IP Source Guard, 427

IPSec (Internet Protocol Security), 353
iSCSI. *See* Internet Small Computer System Interface (iSCSI)
iSCSI adapters, high availability, **143**
iSCSI HBAs, 141
iSCSI off-load adapter, comparison, *142*
iSCSI Qualified Name (IQN) format, *139*, 139
iSCSI software adapter, in ESXi hypervisor kernel, **141**
iSCSI storage transport protocol, **135–148**
    components, **135–137**, *136*
island clusters, 383, *384*, 386
isolated edge cluster design, in NSX (networking platform), *387*
isolation address, 297–298, 336
IT as a service, 458, 461

J

JBOD (just a bunch of disks) set, 10, 26
jumbo frames, **138**, **245**, **247**, 412
    data path configuration, *138*

K

Kerberos
    NFS protocol support by, 151
    for nonroot user authentication, 156
KISS standard, 8

L

LACP (Link Aggregation Control Protocol), 158, 239
latency, 315
    as data-center interconnect design factor, 354
    for mechanical disk, 231
    from network, 206
    over Ethernet, 136
    in performance monitoring, 476
    Performance Service for, 490
    in shared storage, 85
    threshold for Storage I/O Control, 87
latent sector errors, 215
layer 2 extension, *323*
    for interconnect between 2 data centers, 325
Layer 2 Tunneling Protocol version 3 (L2TPv3), 350
    stretched VLANS over, *352*
layer 3-enabled Nexus 5548UP switches, 401

Layer 3 networks, Virtual SAN over, **248**
lazy zero thick (LZT) format, 47, **69**, 261
leaf-spine model, 242, *244*
    oversubscription factor, 245, *247*
    for web-scale architecture, **377–379**, *378*
leased WAN, 339
legacy storage, problems, **187–190**
Lenovo, Flex SEN with x240 Blade Series, *182*
licensing, Virtual SAN, **233–234**
Link Aggregation Control Protocol (LACP), 158, 239
LISP (Locator/ID Separation Protocol), configuration, **363**
live migrations, 315
load balancing, 159, 335
    in iSCSI storage multipathing, *179*
Local Log-Structured Object Manager (LSOM), 287, **310–311**
Locator/ID Separation Protocol (LISP), configuration, **363**
.locker directory, 415
log files
    for virtual machines, 68
    for Virtual SAN, **217–219**
Log Insight Syslog, design, **498–499**, *500*
logging, centralized, 426
logical failure zones, 302
logical unit numbers (LUNs), 21, 23, 73
    Boot from SAN and, 94
    categories, 23
    maintenance, 222
    masking, **128**
    sizing, 74, 81
    storage based on, 190
    storage provisioning mechanisms, 24
    thick, 46
    thin, 44, 46
    use by multiple virtual machines, 64
    Virtual SAN and, 18
login server, as fabric service, 108
login types in switched fabric, **110**
LSI Logic Parallel controller, 72
LSI Logic SAS adapter type, 72
LSOM (Local Log-Structured Object Manager), 287, **310–311**
LUN ID, 73
LUNs. *See* logical unit numbers (LUNs)
LZ4 lossless data-compression algorithm, 203

M

magnetic disks, identifiers for, 422
maintenance LUN (VMFS-formatted LUN), 222
MAN (metropolitan area network), 340
manageability, **184–186**
    and storage design, 8
managed snapshot, in Virtual Volumes, 448
management
    of change and configuration, **427**
    patch and update practices, **428**
    quantifying objectives for, 6
    storage policy-based framework, **250–251**, *251*, *252*
management ports, on switches, 107
management server, 109
manual tiering of storage, 50
master node, 294
Maximum Outstanding Disk Request value, configuring, **479**
maximum transmission unit (MTU), default on network devices, 138
MaxVolumes parameter, 155
mechanical disks, 225, *231*
    Virtual SAN requirements, **229–233**
media quality, for ESXi hypervisor installation, 217
memory, 68, **223**
memory cells, 65
metadata tag, 251
metropolitan area network (MAN), 340
Microsoft
    cloud storage plaforms, 3–4
    clustering and Boot from SAN, 94
migration, of workload, 386
mirroring, 215
mirroring with RAID, 26, *29*, **29**, 34, *35*
    vs. parity-based RAID sets, 43
mission-critical storage, RAID 1 for, 34
MLC (multilevel cell) device, 228
modular switches, 182
monitoring storage, **473–477**
    capacity, **474**
    challenges, **481**, *482*, **483**
    components, **477–479**
    health monitoring, **474**
    performance, **475–476**
    of storage fabric, **478–479**
    security monitoring, **476–477**

standards, **483–486**
storage array system, **480–481**
storage fabric security, **479**
Most Recently Used (MRU) policy, 175–176
mount points, in NFS, 155
MPLS. *See* Multiprotocol Label Switching (MPLS)
MPP (Multipathing Plug-in), 177
MTU (maximum transmission unit), default on network devices, 138
multi-level cell (MLC) technology, 65
multicasting, **248**
    Virtual SAN, 412
multilevel cell (MLC) device, 228
multipathing module, **170**, *172*, **172–180**, *176*
    iSCSI, **177–178**
    for NAS-based storage, **178**, **180**
    NFS version 3 and, 157
    Pluggable Storage Architecture (PSA), **174–177**
Multipathing Plug-in (MPP), 177
Multiprotocol Label Switching (MPLS), **344**, *346*, **347**
    in data center interconnect design, 355–356
    stretched VLANS over, *351*
multitasking application environments, RAID 5 for, 39
multitenanted environments
    disk pool in, 25
    storage design, **58–59**
mutable Virtual SAN objects, 279–280

N

name server, as fabric service, 108
NAS. *See* network-attached storage (NAS)
Native Multipathing Plug-in (NMP), 174–175
Near Line Serial Attached SCSI (NL-SAS), 232
nested RAID, 35
    vs. parity-based RAID sets, 43
NetApp, RAID-DP, 42
network adapters
    iSCSI, **140–143**
    for NFS storage transport protocol, **157**
network-attached storage (NAS), 75
    implementation, **152–157**
        gateway NAS, **152**, *154*, **154**
        unified NAS system, **152**, *153*
    vs. iSCSI, and Fibre Channel, 150
    multipathing, **178**, **180**

network clients, *151*
vs. SAN, **149**
Network File System (NFS) storage transport protocol, 75, 101, **149–161**
    access control lists, **156**
    advanced host configuration, **155–156**
    components, **149–152**
    configuration example, *180*
    exports, **154**, *155*
    mount points, 155
    NAS vs. SAN, **149**
    network adapters for, **157**
    protocol endpoint for, 440
    single virtual switch/multiple network design, **159–160**, *160*
    single virtual switch/single network design, **157–159**, *158*
    storage protocol design factors, *185*
    summary, **161**
    version 4.1
        configuration example, *181*
        limitations with vSphere 6, **161**
    virtual switch design, **157**
Network I/O Control (NIOC), **409–412**, *410*
    virtual switches, **143–144**
        design example, *145*
    vSphere Distributed Switch, **143–144**, **239–240**, *241*
        configuration, **409–412**
        design example, *145*
network interface card (NIC), 101
network layer
    design, use case library, **408–412**
    hardening, **426–427**
    virtualization of, 508
network partition scenario, failure in Virtual SAN stretched cluster, 364
network port binding, 141, 143
network traffic, in Virtual SAN, 236
networking, software-defined, **2–3**
NexentaConnect for Virtual SAN, 395
Nexus 7710 devices, 358
NFS. *See* Network File System (NFS) storage transport protocol
NFS volumes, **78–79**
NL-SAS (Near Line Serial Attached SCSI), 232
NL_Port (node loop port), 110
NMP (Native Multipathing Plug-in), 174–175

node loop port (NL_Port), 110
nodes, 109
noise, 107
noisy neighbor scenario, 85, 265
nonpersistent linked clone, storage policy for, 394
nonrecoverable error, 215
nonroot user authentication, Kerberos for, 156
nonvolatile memory, 65
no_root_squash parameter, 156
N_Port (endpoint), 110
N_Port ID Virtualization (NPIV), *131*, **131–132**, *133*
N_Port Virtualization (NPV), *131*, **131–132**, *133*, 135
NSX (networking platform), 3
    isolated edge cluster design in, 387
NTP, 426, 427
Number of Disk Stripes per Object, **260–261**, *262*
Number of Failures to Tolerate (FTT), **254–260**, *255*
    all-flash erasure coding, **256–260**, *257*
    for fault domains, 306
    for ROBO architecture, 392
NumberofDiskStripesperObject option, for controlling striping, 215
NumberofFailurestoTolerate option, 215
.nvram file, 68

O

object-level scrubbing, 216
object policy, defaults, 271
Object Space Reservation (OSR), **261**, 276
Object Storage File System (OSFS), 207, **311–312**
    and Virtual SAN datastore, **279–281**
object storage, VMware overview, **191–192**
objectives, quantifying, 6
objects
    components, **280–281**
    count calculation, 283–284
    in Virtual SAN, 279, **280**
on-disk formats, **212**, **214**
Open Space Reservation, 205
operating systems, for virtual machines, 69
operational and management costs (OpEx), 11, 15
operational simplification, SPBM framework and, **450**
optical cabling, 107
Oracle RAC (Real Application Clusters), 313
OSFS (Object Storage File System), **311–312**

osfsd.log file, 218
OSR (Object Space Reservation), **261**
OTV (Overlay Transport Virtualization), **360**, *362*, **363**
out-of-band cloud management infrastructure, cluster design to support, 397–398
outage, performance levels during, 61
Overlay Transport Virtualization (OTV), **360**, *362*, **363**
overprovisioning, acceptable level, 46
oversubscription, and switch design evaluation, 183

P

Parallel NFS (pNFS), 157
parallel processing, RAID 5 for, 39
parity, 27, *28*
    in RAID 3, 37
    in RAID 6, 40–41
parity-based RAID sets, processing overhead, 31
partial mesh topology, *124*
pass-through mode, for storage controller, 417
password, 426
    for CHAP, 139
    strong, 427
patch and update practices, management of, **428**
Path Selection Plug-ins (PSPs), 174–175, 177
path-thrashing, 172–173
Payment Card Industry Data Security Standard (PCI DSS), 7
PBW (petabytes written), 229
PCIe drives, 226
    Virtual SAN requirements, **229**
PEC (program-erase cycle), 225, 227
percentage-based admission control policy, 404, *405*
performance
    arbitrated loop topology limitations, 116
    bandwidth for replication, 63
    Boot from SAN and, 93
    calculating I/O per second RAID penalty, **30–32**
    compression and uncompression of data, 203
    of destaging mechanism, 260
    elastic scaling of, *507*
    fabric connectivity and storage I/O, 184
    I/O controller and, 224

latency in shared storage and, 85
monitoring
    for storage, **475–476**
    storage array, **480–481**
    storage fabric, **478–479**
multiple vs. single disk pools, 25
quantifying objectives for, 6
SAN hybrid disk groups and, 273
and storage design, 9
striping and, 27, *28*
tiered storage and, 54
for Virtual SAN cluster, 372
Virtual Volumes and, **450**
in vRealize Orchestrator, 466
performance-class devices, 201
    solid-state drives, 227, 228
Performance Service
    and historical metric database, 491
    monitoring and reporting, *491*
Permanent Device Loss (PDL), 78
permissions, configuration on storage device, 156
petabytes written (PBW), 229
physical access, monitoring, 481
Physical and Virtual Machine clustering technologies, 79, *80*, 81
physical IP SAN, creating for storage I/O traffic, 137–138
physical network
    bandwidth requirements, **245**
    building-block rack architecture, **245**
    core, aggregation, and access model, 242, *243*
    firewall, **249**
    high-level design, *413*
    Internet Protocol (IP) version 6, **249**
    jumbo frames, **245**, **247**
    leaf-spine model, 242, *244*
    multicasting, **248**
    quality of service (QoS), **248–249**
    requirements, **240–242**
physical ports, for isolating tenant, 59
physical server, virtual machines on, 2
physical storage, provisioning resources, 453, *454*
physical switch, configuration overview, 412, *413*
platform as a service (PaaS), 458
Pluggable Storage Architecture (PSA), 92, **174–177**, *175*
Point-to-Point (FC-P2P), **115**

policy-driven storage
    common management platform for, *445*
    overview, **190–191**
policy engine, for inter-array storage-tiering, 52
pools
    configuring resources into, 17
    of raw disk capacity, 19
port count, on switches, 106
port density, and switch design evaluation, 183
Port Login (PLOGI), 110
Port Security, 131
ports, 101
    naming, **110**, *111*
positional latency, for mechanical disk, 231
PowerCLI, 446
Priority Flow Control (PFC), 137, 165
private cloud platform, 460
private WAN, 339
Process Login (PRLI), 110
program-erase cycle (PEC), 225, 227
protocol, 101
protocol endpoint (PE), 430, 438, **440–442**, *442*
    provisioning process, *443*
provisioning storage, 24, *24*
    thick, example, 45
    traditional, vs. virtual, 45
    virtual, *44*, **44–49**
proximity extension, in switched fabric, 117
PSA (Pluggable Storage Architecture), 92, **174–177**, *175*
PSPs (Path Selection Plug-ins), 174–175, 177
public cloud infrastructure, 460
public-key infrastructure, in vRealize Orchestrator, 466
Public StaaS platform, 457
PVSCSI (VMware Paravirtual) controller, 72

Q

quality of service (QoS), **59–60**, 85
    for physical network, **248–249**
    in storage policy design, **265**, *266*, *267*, **267**
queue, and drive performance, 225–226
Queue Full (qfull) I/O failure message, 479
quorum logic design
    failure scenario, *303*
    and vSphere High-Availability, **302**, *303*

R

rackmount servers, 216–217
    Cisco Unified Computing System C240 M4, 358
    memory, 223
RAID-DP (NetApp), 42, 43
RAID (redundant array of independent disks), 25–43
    calculating I/O per second penalty, **30–32**
    distributed, **214–215**
    hot spares, 42
    levels explained, **32–42**
        RAID 0 (striped disk), **32–33**, *33*, 210, 224–225, 417
        RAID 1 (mirroring and duplexing), 34, *35*, 259
        RAID 1+0 (mirroring and striping), 35–36, *36*
        RAID 3, 37, *38*
        RAID 03 (0+3), 43
        RAID 5, 38–39, *40*
        RAID 6, 40–41, *41*
        RAID 50 (5+0), 43
    nested, 30
    summary, 42–43
RAID sets
    advantages, 26
    mirroring with, *29*, **29**
    striping in, *27*, **27**
    Virtual SAN and, 18
RAIN (redundant array of interdependent nodes), 287
RAM disk, log messages on, 218
raw device mapping (RDM), **79–81**, 313
    connection topology, *79*
    vs. VMFS or NFS, **81**
RDT (Reliable Datagram Transport), 248, 308
read-ahead cache, for storage controller, 224
read cache, 201, 278
    for flash device, 226
read misses, 231
read operations, I/O penalty, 31
read requests, in stretched cluster, 326
read/write locality, for stretched clusters, **329**, **332**
read-write permissions, 156
rebalance operations, **290**, *291*, **292**
rebind mechanism, for Virtual Volumes, 442

Receive (Rx) link utilization metric, 478
recoverability, and storage design, 9
recoverable data, 215
recovery point objective (RPO), 9, 62
    SPBM framework and, 191
    stretched clusters and, 318
recovery time objective (RTO), 9, 60, 62
    SPBM framework and, 191
    stretched clusters and, 318
redundancy
    as data-center interconnect design factor, 355
    in disk mirroring, *29*
    from RAID, 26
redundant array of interdependent nodes (RAIN), 287
redundant components, 353
redundant links, 353
registered state change notifications (RSCNs), 108, 125
regulatory compliance, 7
Reliable Datagram Transport (RDT), 248, 308
remote office/branch office (ROBO) site
    two-node design, **386–392**, *388*, *390*
    vSphere host at, 182
remote syslog, 427
replicated data, disaster recovery from, 383
replication of data, 62
Representational State Transfer (REST), 15
requirements, 398
resource balancing, 300
REST API, from vRealize Orchestrator, 467
resynchronization, **289–290**
rotational latency, for mechanical disk, 231
Round Robin (RR) policy, 176
route based on physical NIC load algorithm, 240
route based on port ID algorithm, 239
RPO. *See* recovery point objective (RPO)
RTO. *See* recovery time objective (RTO)
Ruby vSphere Console (RVC), 487
    namespace options, *489*
    vsan.proactive_rebalance command, 292

S

SAN. *See* Boot from SAN (BfSAN); fibre channel SAN; storage area network (SAN)
SAN-based block protocols, protocol endpoint for, 440

SAN-based datastore, 74–75
SAN-based storage systems, 75
SAN cables, 107
SAN fabric design, **477–479**
SAN islands, 57
SAP, 313
Sarbanes-Oxley Act of 2002, 7
SAS drives, 226
SAS expanders, 368
SAS (Serial Attached SCSI), 232
SATA drives, 226
SATA (Serial Advanced Technology Attachment), 232
SATADOM (Serial ATA Disk on Module), 221
    boot devices, **222**
SATP (Storage Array Type Plug-in), 174–175, 177
scalability
    arbitrated loop topology limitations, 116
    of building-block web-scale architecture, *375*
    of monitoring solution, 473
    and physical resources design for web-scale, **373–377**
    quantifying objectives for, 6
    and storage design, 9–10, **54–57**, *56*
    in vRealize Orchestrator, 466
scale-out approach, 55, 505
    building-block storage infrastructure from, 392
    for web-scale Virtual SAN platform, 367, **370**, *371*
scale-up approach, 55, 505
    in Virtual SAN, **368**, *369*, *370*
    for web-scale Virtual SAN platform, 367
scratch directory, VMFS volume for, 218
scratch partitions, 415
    Boot from SAN and, 94
    ESXi and, 217
scripting engine, in vRealize Orchestrator, 465
SCSI LUN, in classic storage, 23
SCSI UNMAP command, 90
SDDC (software-defined data center), 1, **187–190**, *188*
    primary aim of, 1
SDH (Synchronous Digital Hierarchy), 342
SDRS (Storage Distributed Resource Scheduler), **83–85**, 313
    affinity rules, *84*
secret, for CHAP, 139

secret source software, 15
sectors, on mechanical disks, 231
Secure Sockets Layer (SSL), in vRealize Orchestrator, 466
security
    for CHAP, **139–140**
    for cloud management platform (CMP), **423–426**
    for communications, **353–354**
    compute and storage-layer hardening, **426**
    Kerberos for nonroot user authentication, 156
    monitoring for storage array, **481**
    monitoring storage fabric, **479**
    SAN, **125–131**
        LUN masking, **128**
        options, 130–131
        Virtual Fabric design, **128–130**, *129*
    and storage design, 9
    for storage device, **61**
    for Virtual SAN cluster, 372
seek time, for mechanical disk, 231
self-provisioning, STaaS to enable, 457
sequential data transfer, RAID 3 for, 37
Serial Advanced Technology Attachment (SATA), 232
Serial Attached SCSI (SAS), 232
server-side locking, 151
servers. *See also* vCenter Server
    x86 architecture platform, 18
    Cisco C-Series, 400
    host, monitoring storage on, **477**
    login, 108
    management, 109
    name, 108
    rackmount, 216–217
        Cisco Unified Computing System C240 M4, 358
        memory, 223
    virtual machines on physical, 2
service-level agreements, 9
session trunking, 151, 157
share values, Storage I/O control and, 85
shared VMware, 98
shared WAN, 339
shares, 86
silent corruption, 215
Simple Network Management Protocol (SNMP), 483–484

single-core, core-edge fabric topology, *119*
single-host failure, in Virtual SAN stretched cluster, 363
single initiator/single target zoning strategy, 125
single-instance storage, 60. *See also* deduplication
single-level cell (SLC) device, 228
single pane of glass, 4
single point of failure, Boot from SAN and, 93
single virtual switch/single network design, **157–159**, *158*
SIOC. *See* Storage I/O Control (SIOC)
Site Recovery Manager, 161
sizing
    capacity disk, **282–287**
    datastore, 74, **271–287**
    datastore capacity disk, **281–282**
    endurance flash device, **278–279**
    environment for customer's needs, 210
    and swap object, **284–285**
slave node, datastore heartbeat system verifying operational, 294
SLC (single-level cell) device, 228
SMP-FT (Symmetric Multiprocessing Fault Tolerance), 312
snapshot delta VMDK objects, 207
snapshot state file, 68
snapshots, 214, 280, 313
    array-based, **448**
    recovering individual files from, 64
    storage array-based, in classic storage, **63–64**, *64*
SNMP (Simple Network Management Protocol), 483–484
software as a service (SaaS), 458
    workflow example, **468**, **472**
        logical configuration, *469*
software checksum, **215–216**
    reporting actions of, 216
software-defined compute, **2**, 508
software-defined data center (SDDC), 1, **187–190**, *188*
    primary aim of, 1
software-defined enterprise (SDE), **187–190**, *188*, 505
software-defined networking, 2–3, 508
software-defined storage, **3–4**
    goal of, 4
    strategy implementation, 15–16

summary, **16–19**
software development kit, in vRealize Orchestrator, 466
software initiators, 135
solid-state drives (SSDs), **65**, **225–229**
    cache-tiering between DRAM and, 52
    identifiers for, 422
    performance-class, 228
    for RAID set, 26
SONET (Synchronous Optical Networking), **342**, **344**, *345*
    in data center interconnect design, 355–356
Spanning Tree Protocol (STP), 240, 350
sparse swap (swap efficiency), 214
SPBM (Storage Policy-Based Management Framework), 189, 251, **312**, 430
    with Virtual Volumes (VVOLs), **444–449**
speed, as data-center interconnect design factor, 354
sps.log file, 312
squash root access, 156
SSH, 427
stakeholders, gathering information from, 5
standardization, of configuration for web-scale infrastructure, 373
standards, and storage design, 9
start of frame (SOF), 163
static routes, **327**
static storage tiering, **96**, *96*
storage
    abstraction of advanced functions, 17–18
    capacity for witnesses, 389
    economics of, **10–14**
    multitenanted design, **58–59**
    next generation model, *431*
    out-of-band, for management-only clusters, 414
    problems of legacy, **187–190**
    security for, **61**
    software-defined, **3–4**. *See also* software-defined storage
storage area network (SAN), 101
    multivendor operational challenges, 58
    vs. NAS, **149**
    security, **125–131**
        fabric switch zoning, **125–126**, *126*
        options, 130–131
        Virtual Fabric design, **128–130**, *129*

storage array-based snapshots, in classic storage, **63–64**, *64*
storage array system, monitoring, **480–481**
Storage Array Type Plug-in (SATP), 174–175, 177
storage-as-a service (STaaS), 58, **453–472**
    cloud software stack, *462*
    definition, **457**
    workflow
        access rights modification, *471*
        example, *456*
        NAS form design, *470*
storage components, **225–233**
    disk pools
        homogeneous or heterogeneous, 52
        provisioning, 24–25
        shared or separate, 59
        vs. traditional model of LUN, 23–24, *24*
    in fibre channel SAN, *107*, **107–108**
    flash devices, 225
        for boot media, 222
        life expectancy, 202
        pooling, 194
        for read-caching and write-buffering, 193
    solid-state drives (SSDs), **65**, **225–229**
        identifiers for, 422
        performance-class, 228
        for RAID set, 26
storage containers (SCs)
    architecture, *439*
    provisioning process, *441*
    in Virtual Volumes, **437–440**
    vs. Volumes/LUNs, 438
storage controllers, *71*, **71–72**, *73*
    caches for, 224
storage design factors, **6–10**. *See also* tiered storage
    availability, 7
    budget, 8
    business drivers and, 7
    capacity, 10
    compliance, 7
    goals, 8–9
    manageability, 8
    performance, 9
    recoverability, 9
    scalability, 9–10, **54–57**, *56*
    security and governance, 9
    standards, 9
    usability, 8
Storage Distributed Resource Scheduler (SDRS), **83–85**, 161, 313
    affinity rules, *84*
storage fabric, monitoring, **477–479**
Storage I/O Control (SIOC), **85–88**, *87*, 161, 313
    architecture summary, **184–186**
    disabling and enabling, 87
storage I/O controllers per host, as design factor, 368
storage layer design, use case, **412–417**
storage management
    standards, **483–486**
    tools, **57–58**
Storage Management Initiative (SMI), 484
Storage Management Initiative Specification (SMI-S), 15, 484, *485*
Storage Networking Industry Association (SNIA), 15, 484
Storage Policy-Based Management Framework (SPBM), 189, 251, **312**, 430
    with Virtual Volumes (VVOLs), **444–449**
storage policy, configuration, **418–423**
storage policy design
    for stretched clusters, **327**
    for Virtual SAN, **250–271**
storage primitives, defining, 89
storage provider, 436
storage provisioning mechanisms, complex, *455*
storage resources, 3
    efficient use, 12
    life expectancy, 11
    total cost of ownership calculation, *11*, **11–13**
storage tier-based pools, 25
Storage vMotion (SvMotion), **81–82**
STP (Spanning Tree Protocol), 240, 350
stretched clusters, 65, 315, *316*, 347
    deployment scenarios, **327**, *328*
    Distributed Resource Scheduler (DRS), **332–335**
        affinity rule configuration, *333*, *334*
    failure scenarios, **363–365**
    fault domain architecture, *319*
    High Availability (HA), **335–338**
    host isolation advanced settings, *337*
    interoperability, **365**
    local read operation, anatomy, *330*

network design requirements, **320–326**
  bandwidth requirements calculations, **325–326**
  distance and latency, **322**, **325**
  overview, *324*
  preferred and nonpreferred site concepts, **329**
  read/write locality, **329**, **332**
  solution architecture example, **356–363**, *357*
  storage policy design, **327**
  use cases for, **317–318**
  WAN interconnect design, **339–347**
    dark fiber, **340**, *341*
  write operation, anatomy, *331*
stretched layer 2 extension, 322, *323*
stretched VLANs, deploying, **347–354**, *348*
  over dark fiber, *349*
  over L2TP version 3, *352*
  over MPLS, *351*
stripe width, 260
striping, 26, *27*, **27**, 215
stub file, 52, 60
sub-LUN tiering, 88
  access, **98–100**
    VMware dedicated disk subsystem, 98, *99*
    VMware shared disk subsystem, *99*, **99–100**
suspend state file, 68
SvMotion (Storage vMotion), **81–82**
swap efficiency (sparse swap), 214, 261
swap file, 68
swap object, 280
  and sizing, **284–285**
  for virtual machines, 207
SwapThickProvisionedDisabled option, 214, 261, 284
sweet spot, 223, 271, 275
switch-based fabric architecture, **117–125**
  core-edge fabric topology, **117–118**, *118*, **121**
switch port capacity utilization, 478
Switched Fabric (FC-SW), 115, **116–117**
switches. *See also* virtual switches
  aggregation, 184
  evaluating design characteristics, **182–184**
  Fibre Channel over Ethernet (FCoE), *166*, **166–167**
symlink, 218
Symmetric Multiprocessing Fault Tolerance (SMP-FT), 312
synchronous data, in stretched cluster, 315

Synchronous Digital Hierarchy (SDH), 342
Synchronous Optical Networking (SONET), **342**, **344**, *345*
  in data center interconnect design, 355–356
synchronous replication, 62–63
syslog collector, central, 219
syslog protocol, 497
  message structure, *497*
system logging, 426

T

T11 Network Address Authority (NAA), 138
target, iSCSI, 136
TCO. *See* total cost of ownership (TCO)
TCP off-load engine (TOE), 140
  network adapter with, 157
TcpipHeapMax parameter, 156
TcpipHeapSize parameter, 155
Technical Committee T11, 102
temperature, monitoring storage, 481
thick disk, 261
thick LUNs, 46
thick-on-thick storage allocation, 47–49
thick-on-thin storage allocation, 47–49
thick provisioning, example, 45
thin disks format, **69–70**
thin LUN, 44, 46
thin-on-thin storage allocation, 47–49
thin provisioning, 44. *See also* virtual provisioning
  array-based, **449**
thin-provisioning stun primitive, 90
throughput, 475
  Performance Service for, 490
  and switch design evaluation, 183
tier-1 storage, 12
tiered storage, 12–13, *13*, **49–54**, *50*, *51*
  automated, **52–54**, *53*
  design considerations, *54*, 98
  design models in vSphere, **95–98**
  I/O Control latency values, 88
  mechanisms, *53*
  mixed storage tiering, **96**, *97*
  static storage tiering, *96*, **96**
  sub-LUN, 88
TLC (triple-level cell), 228
TOE adapters. *See* TCP off-load engine (TOE)

tolerance for failure, 198
total cost of ownership (TCO), 11, 281, *282*
    calculating for storage resources, *11*, **11–13**
    for network service providers, 353
    per I/O operation, 202
    quantifying objectives for, 6
    simplified annual, *12*
TPI (tracks per inch), 229
trace files, in Virtual SAN, **219–221**
track density, 229
tracks, 229
traditional model of LUN, vs. disk pools, 23–24, *24*
transfer rate, in RAID 5, 39
transfer speed, for mechanical disk, 231
Transmission Control Protocol (TCP), 136
Transmit (Tx) link utilitization metric, 478
transport mode in IPSec, 353
triple-level cell (TLC), 228
troubleshooting, in vRealize Orchestrator, 466
trunked ISLs, and Fibre Channel switches, 130
tunnel mode in IPSec, 353

U

unallocated storage, **480**
unbind mechanism, for Virtual Volumes, 442
unconfigured storage, **480**
unified NAS system, **152**, *153*
unmanaged snapshot, in Virtual Volumes, 448
UNMAP primitive, 90
update practices, management of, **428**
usability, and storage design, 8
utilization efficiency, 12
UUID (universal unique ID)
    for datastore, 75–76
    for Virtual SAN cluster, 422

V

VAAI (vStorage API for Array Integration), **89–90**, 436
    and Virtual Volumes, **451**
VADP (VMware vSphere Storage APIs for Data Protection), **91–92**
VAMP (vStorage API for Multipathing), **91–92**
vCenter Server, 193
    bootstrap option, *421*
    installing, **418–419, 422–423**
    migration option, *420*

vendor prices, 10
version control, in vRealize Orchestrator, 466
VirstoFS format, 212
virtual datastore, 19, 438
virtual desktop infrastructure (VDI), Virtual SAN and, **392–395**, *393*
virtual disks (VMDKs), **68–70**, 280
    creating in EZT format, 89
    eager zeroed thick (EZT) format, **69–70**
    formats, 47–49
    lazy zero thick (LZT) format, **69**
    thin disks format, **69–70**
virtual environments, networking to, 2
Virtual Extensible LAN (VXLAN), 322
Virtual Fabric design, **128–130**, *129*
    sample use case, *130*
virtual LAN. *See* VLANs
Virtual Machine File System (VMFS), **77–78**, *78*
virtual machines, 2
    automated placement, 335
    calculating capacity requirements, **285–287**
    defining service level assigned to, 432
    inadvertent rollback, 64
    live-migrating, 81–82
    memory, 68
    namespace object, 207, 280
    object storage, 386
    objects associated with, 320
    security, **424–425**
    storage controllers (vSCSI adapters), *71*, **71–72**, *73*
    swap file creation, 214
    thin provisioning, 313
Virtual Private LAN Services (VPLS), 350
virtual provisioning, *44*, **44–49**
    design considerations, **46–47**
    layering, **47–49**, *48*
Virtual SAN, 4, 17, 18, 19, 187
    architecture, **194–216**
        all-flash disk group, *197*, *201*, 202
        disk groups, **194–198**, *195*
        hybrid, *196*, *200*, 200
        hybrid, vs. all-flash models, **200–202**
    basics, **192–193**
    boot devices, **221–222**
    capacity management for, **505**
    with Cisco UCS environment physical connectivity, *403*

data locality and caching algorithms, **205–206**
datastore
    creating, 311–312
    design for multiple, *209*
    installing vCenter Server on, **418–419**, **422–423**
    usable capacity calculation, **285**
datastore design and sizing, **271–287**
    capabilities, **275–276**
    capacity disk sizing, **282–287**
    hosts per cluster, **273**, **275**
    multiple disk group configuration, **276**, *277*, **278**
    Object Space Reservation (OSR), **276**
    Object Storage File System and, **279–281**
    sizing impact of disk stripes per object, **276**
deduplication and compression, **202–205**, *203*
design requirements, **216–234**
    host boot architecture, **217–221**
    host form factor, **216–217**
    log files, **217–219**
destaging mechanism, **206**
disk components, *211*
distributed datastore, **206–207**, *208*
distributed RAID, **214–215**
elastic scaling of capacity and performance, *507*
endurance flash device, sizing, **278–279**
as generic object storage platform, *396*
HA and DRS parameters, 299
hardware requirements, **222–233**
    host CPU overhead, **223**
    memory, **223**
    storage controllers, **223–225**
    storage devices, **225–233**
HCI building-block model, 505, *507*
hybrid disk group, configuration, **417**
hyper-converged infrastructure (HCI) architecture, 193
integration and interoperability, **312–313**
internal component technologies, **308–312**, *309*
    Cluster Level Object Manager (CLOM), **310**
    Cluster Monitoring Membership and Directory Services (CMMDS), **308**, **310**
    Distributed Object Manager (DOM), **310**, *311*
    Local Log-Structured Object Manager (LSOM), **310–311**
    Object Storage File System (OSFS), **311–312**
    Reliable Datagram Transport (RDT), 308
    Storage Policy-Based Management Framework (SPBM), **312**
licensing, **233–234**
maintenance mode operations, *300*, **300–302**
    Ensure Accessibility option, **301**
    Full Data Migration option, **301**
    No Data Migration option, **301–302**
major releases, 192–193
maximums, 379
monitoring and operational tools, **486–492**, *487*
multicasting, 412
network design summary, **249–250**
network fabric design, **236–250**
    logical network design, *237*
    network I/O control, **239–240**, *241*
    network teaming design, **238–239**
    physical network requirements, **240–242**
    VMkernel network configuration, 236, 238
    vSphere network requirements, **236–239**
network I/O control, sample cluster policy, 240
network partition scenario, *296*
object types, 207
on-disk formats, **212**, **214**
optimum rack design, *246*
over Layer 3 networks, **248**
PCIe-based flash devices configuration, *230*
policy-based storage-provisioning mechanism from, 190
Ready Nodes, **233**, 401
scale-up approach in, **368**, *369*, *370*
storage controllers, **417**
storage policy design, **250–271**
    application assessment and, **268–271**, *272*
    default, **267–268**
    Flash Read Cache Reservation, *263*, **263–264**
    Force Provisioning, *264*, **264–265**
    management framework, **250–251**, *251*, *252*
    Number of Disk Stripes per Object, **260–261**, *262*
    Number of Failures to Tolerate (FTT), **254–260**, *255*
    Object Space Reservation (OSR), **261**

Quality of Service (QoS), **265**, *265, 266, 267*, **267**
rule, **251**
rule sets, **253**
tiered workload clusters, *235*
trace files, **219–221**
traffic isolation, **247**
use case library, **381–428**
    cluster architecture, **415**
    cluster configuration, **404–408**
    customer design requirements and constraints, **398**, **400**
    disaster recovery architecture example, *385*
    Horizon and virtual desktop infrastructure, **392–395**, *393*
    network-layer design, **408–412**
    overview, *382*
    remote office/branch office (ROBO) two-node design, **386–392**, *388*
    storage layer design, **412–417**
    Virtual SAN file services, **395–404**
and VDI architecture, *393*
version 6
    6.2 hybrid configuration, *414*
    object placement, *304*
and VMkernel adapters, *238*
and vSphere High-Availability interoperability, **292**, *293*, **294–298**
Virtual SAN cluster, UUID for, *422*
Virtual SAN-enabled clusters, *381*
virtual switches, **143–148**
    distributed, vs. vSphere standard, **238**
    management vSphere, *409*
    multiple, *147*, **147–148**
    Network I/O Control (NIOC), **143–144**
        design example, *145*
    NFS design, **157**
    single, *146*, **146–147**
        multiple network design, **159–160**, *160*
Virtual Volumes (VVOLs), 4, **18–19**, 161, 187, 381, **429–430**
    benefits of designing for, **449–450**
    binding operations, **442–444**, *444*
    component technology architecture, **434**, *435*
        object architecture, **434**, **436**
    data plane, **437–444**
        protocol endpoint (PE), **440–442**, *442*
        storage containers (SCs), **437–440**
    key design requirements, **450–451**

management plane, **436–437**
    VASA provider, **436–437**
policy-based storage-provisioning mechanism from, 190
space reclamation, **449**
storage capabilities, **448–449**
    array-based snapshots, **448**
    array-based thin provisioning, **449**
storage policy-based management (SPBM), **444–449**
    published capabilities, **446**, **448**
technology basics, **430–436**
and VAAI, **451**
VLANs
    deploying stretched, **347–354**, *348*
    over dark fiber, *349*
    for isolated broadcast domain, 137
VMcore partition, for vSphere releases, 221
.vmdk file, 68–69
    maximum size, 69
VMDK object, 207
.vmem file, 68
VMFS filesystem, 75
VMFS-formatted LUN (maintenance LUN), 222
vmfsSparse, for snapshots, 214
VMKdiagnostic partition, 221
VMkernel adapters, and Virtual SAN, 238
VMkernel port, 412
    configuration, 238
VMkernel software Data Mover driver, 89
VMkernel, Virtual SAN embedded into, 18
vMSC. *See* vSphere Metro Storage Cluster (vMSC)
.vmsd file, 68
.vmsn file, 68
.vmss file, 68
VMware
    Distributed Power Management (DPM), 298
    Horizon portfolio, **392–395**, *393*
    mechanical disk support, 232
    monitoring and operational tools, 487
    object storage overview, **191–192**
    vRealize Automation, 446
VMware Compatibility Guide for Virtual SAN, 222
VMware Dump Collector, 219
VMware ESXi software, 1
VMware Global Support Services, Virtual SAN trace files from, 219

VMware Integrated Open Stack (VIO), 446
VMware NSX layer 2 gateway services, 322
VMware Paravirtual (PVSCSI) controller, 72
VMware Site Recovery Manager, 77
VMware storage environments
    design, **4–10**
        establishing storage factors, **6–10**
        technical assessment and requirements gathering, **5–6**
VMware vRealize Log Insight, **497–498**
VMware vSphere Storage APIs for Data Protection (VADP), **91–92**
.vmx file, 68
voltage fluctuations, monitoring storage, 481
volumes, 74, 279–280. *See also* Virtual Volumes (VVOLs)
    NFS, **78–79**
VPLS (Virtual Private LAN Services), 350
vRealize Automation, 446, **461–465**
    catalogs, 453, **462–463**
    storage service catalog example, *458*
vRealize Log Insight, 426, 492
vRealize Operations Manager, **492–494**
    logical design, *493*
    Management Pack for Storage Devices (MPSD), **492**
        dashboard view, *494*
    storage partner solutions, **494**, *495*, *496*
vRealize Orchestrator, 453, **465–468**
    integration, **468**
    support from storage vendors, **467–468**
VSAN Observer, user interface, *490*
VSAN trace files, 499
VSAN.ClomMaxComponentSizeGB parameter, 281
VSAN.DOMOwnerForceWarmCache parameter, 329
vsan.observer command, 488, 490
vsan.proactive_rebalance command, 505
vsantraces, 219
    location of, 220
vsanvpd.log file, 218
vSCSI adapters (storage controllers), 71, **71–72**, *73*
vSphere
    comparing standard with distributed virtual switches, **238**
    Distributed Resource Scheduler (DRS), 77, 398, **406–408**

Fault Tolerance feature, 392
High Availability (HA), 77, 312, 398
    for clusters, 65, **404**, **406**
    example design values, 406
    heartbeat datastores, **297**
    host isolation addresses, **297–298**
    network communication, *295*
    operational comparison, 294
    quorum logic design and, **302**, *303*
    restart priorities, 338
    and Virtual SAN, **292**, *293*, **294–298**
host cluster design for web-scale, **372**
operational priorities, 432
technologies available to, 3
vSphere 6, NFS protocol support by, 150
vSphere API, 446
vSphere APIs for Storage Awareness (VASA), **90–91**, *91*, 250–251
vSphere APIs for Storage Awareness 2.0 (VASA 2.0), 430, **436**
    control path, *437*
vSphere Distributed Switch
    Network I/O Control (NIOC), **143–144**, **239–240**, *241*
        configuration, **409–412**, *410*
        design example, *145*
vSphere Dump Collector, 221
vSphere host
    connection with Virtual Volumes, 430
    data placement management, 432
vSphere layer, demarcation between storage array layer and, 67
vSphere Metro Storage Cluster (vMSC), **65**, *66*
    design, *66*
vSphere network requirements, **236–239**
vSphere Replication, 313
vSphere stack, Virtual SAN and, 193
vSphere Storage APIs for Array Integration (VAAI), 161
vSphere storage technologies, **67–100**
    datastore, *73*, **73–75**
    NFS volumes, **78–79**
    raw device mapping (RDM), **79–81**
        connection topology, *79*
        vs. VMFS or NFS, **81**
    storage I/O control, **85–88**
    tiered storage model, **95–98**
        fully autotiered environment, *97*, **97–98**

mixed storage tiering, **96**, *97*
static storage tiering, *96*, **96**
virtual disks, **68–70**
  eager zeroed thick (EZT) format, **69–70**
  lazy zero thick (LZT) format, **69**
  thin disks format, **70**
vSphere svMotion, 312
vSphere vMotion, 312, 398
vStorage API for Array Integration (VAAI), 89–90, 436
  and Virtual Volumes, **451**
vStorage API for Multipathing (VAMP), **91–92**
.vswp file, 68, 284
VXLAN (Virtual Extensible LAN), 322

W

WAN accelerator, 63
WAN interconnect design
  high availability, **353**
  for stretched clusters, **339–347**
    dark fiber, **340**, *341*
Warning event, 498
wavelength division multiplexing, **340**, **342**
Web-Based Enterprise Management (WBEM) initiative, 484
web client, erasure coding configuration, *259*
web-scale pod, 373
  logical architecture, *374*
web-scale Virtual SAN platform, 4, 367–368
  building-block clusters and scale-out architecture, **372–373**
  leaf-spine model, **377–379**, *378*
  physical resources design, scalability and, **373–377**
  scale-out approach for, **370**, *371*
  scale-up approach for, **368**, *369*, *370*
  vSphere host clusters design for, **372**
Windows Server Failover Clusters (WSFC), 313
witness appliance, **318**, **320**
  sizing configuration options, 320
witness-site failure, in Virtual SAN stretched cluster, 364
witness virtual appliance fault domain, 318

witnesses, 210, 212
  metadata, **281**
  metadata failure scenario, *213*
  object location, 389
  object metadata architecture, *391*
  storage capacity for, 389
workflow designer, in vRealize Orchestrator, 465
workflow engine, 465
workflow library, in vRealize Orchestrator, 466
workload
  balancing, 298
  initial placement in datastore cluster, 84
  migration of, 386
workload-data-center to witness-network failure, in Virtual SAN stretched cluster, 365
World Wide Name (WWN)
  device addressing, **110**, *112*, **112**, *113*
  spoofing, 481
World Wide Node Name (WWNN), 112
World Wide Port Name (WWPN), 112
World Wide Web Consortium, Representational State Transfer (REST), 15
write buffering, 201, 278
  for flash device, 226
write cache, for storage controller, 224
write data transactions, in RAID 6, 40
write hot data, 203
write I/O, bandwidth and, 326
write-intensive systems, RAID 5 and, 39
WSFC (Windows Server Failover Clusters), 313
WWN. *See* World Wide Name (WWN)
WWNN (World Wide Node Name), 112
WWPN (World Wide Port Name), 112

X

X-vMotion, 82
XCOPY, for Virtual Volumes, 436

Z

zone configurations, 125
zone set, 125, *127*
zoning fabric switch, **125–126**, *126*